Solid-Phase Microextraction and Related Techniques in Bioanalysis

Solid-Phase Microextraction and Related Techniques in Bioanalysis

Editor

Hiroyuki Kataoka

MDPI • Basel • Beijing • Wuhan • Barcelona • Belgrade • Manchester • Tokyo • Cluj • Tianjin

Editor
Hiroyuki Kataoka
School of Pharmacy
Shujitsu University
Okayama
Japan

Editorial Office
MDPI
St. Alban-Anlage 66
4052 Basel, Switzerland

This is a reprint of articles from the Special Issue published online in the open access journal *Molecules* (ISSN 1420-3049) (available at: www.mdpi.com/journal/molecules/special_issues/SPME_Bioanalysis).

For citation purposes, cite each article independently as indicated on the article page online and as indicated below:

LastName, A.A.; LastName, B.B.; LastName, C.C. Article Title. *Journal Name* **Year**, *Volume Number*, Page Range.

ISBN 978-3-0365-7047-1 (Hbk)
ISBN 978-3-0365-7046-4 (PDF)

© 2023 by the authors. Articles in this book are Open Access and distributed under the Creative Commons Attribution (CC BY) license, which allows users to download, copy and build upon published articles, as long as the author and publisher are properly credited, which ensures maximum dissemination and a wider impact of our publications.

The book as a whole is distributed by MDPI under the terms and conditions of the Creative Commons license CC BY-NC-ND.

Contents

About the Editor vii

Preface to "Solid-Phase Microextraction and Related Techniques in Bioanalysis" ix

Hiroyuki Kataoka
Solid-Phase Microextraction and Related Techniques in Bioanalysis
Reprinted from: *Molecules* **2023**, *28*, 2467, doi:10.3390/molecules28062467 1

Yesenia Mendoza García, Ana Luiza Coeli Cruz Ramos, Ana Cardoso Clemente Filha Ferreira de Paula, Maicon Heitor do Nascimento, Rodinei Augusti and Raquel Linhares Bello de Araújo et al.
Chemical Physical Characterization and Profile of Fruit Volatile Compounds from Different Accesses of *Myrciaria floribunda* (H. West Ex Wild.) O. Berg through Polyacrylate Fiber
Reprinted from: *Molecules* **2021**, *26*, 5281, doi:10.3390/molecules26175281 5

Ana P. X. Mariano, Ana L. C. C. Ramos, Afonso H. de Oliveira Júnior, Yesenia M. García, Ana C. C. F. F. de Paula and Mauro R. Silva et al.
Optimization of Extraction Conditions and Characterization of Volatile Organic Compounds of *Eugenia klotzschiana* O. Berg Fruit Pulp
Reprinted from: *Molecules* **2022**, *27*, 935, doi:10.3390/molecules27030935 19

Yuri G. Figueiredo, Eduardo A. Corrêa, Afonso H. de Oliveira Junior, Ana C. d. C. Mazzinghy, Henrique d. O. P. Mendonça and Yan J. G. Lobo et al.
Profile of *Myracrodruon urundeuva* Volatile Compounds Ease of Extraction and Biodegradability and In Silico Evaluation of Their Interactions with COX-1 and iNOS
Reprinted from: *Molecules* **2022**, *27*, 1633, doi:10.3390/molecules27051633 33

Zhenying Liu, Ye Fang, Cui Wu, Xian Hai, Bo Xu and Zhuojun Li et al.
The Difference of Volatile Compounds in Female and Male Buds of *Herpetospermum pedunculosum* Based on HS-SPME-GC-MS and Multivariate Statistical Analysis
Reprinted from: *Molecules* **2022**, *27*, 1288, doi:10.3390/molecules27041288 53

Lijun Cai, Sarina Macfadyen, Baozhen Hua, Haochuan Zhang, Wei Xu and Yonglin Ren
Identification of Biomarker Volatile Organic Compounds Released by Three Stored-Grain Insect Pests in Wheat
Reprinted from: *Molecules* **2022**, *27*, 1963, doi:10.3390/molecules27061963 65

Qasim Ahmed, Manjree Agarwal, Ruaa Alobaidi, Haochuan Zhang and Yonglin Ren
Response of Aphid Parasitoids to Volatile Organic Compounds from Undamaged and Infested *Brassica oleracea* with *Myzus persicae*
Reprinted from: *Molecules* **2022**, *27*, 1522, doi:10.3390/molecules27051522 75

Keita Saito, Yoshiyuki Tokorodani, Chihiro Sakamoto and Hiroyuki Kataoka
Headspace Solid-Phase Microextraction/Gas Chromatography–Mass Spectrometry for the Determination of 2-Nonenal and Its Application to Body Odor Analysis
Reprinted from: *Molecules* **2021**, *26*, 5739, doi:10.3390/molecules26195739 89

Sehyun Kim and Sunyoung Bae
In Vitro and In Vivo Human Body Odor Analysis Method Using GO:PANI/ZNRs/ZIF–8 Adsorbent Followed by GC/MS
Reprinted from: *Molecules* **2022**, *27*, 4795, doi:10.3390/molecules27154795 99

Kamil Łuczykowski, Natalia Warmuzińska, Sylwia Operacz, Iga Stryjak, Joanna Bogusiewicz and Julia Jacyna et al.
Metabolic Evaluation of Urine from Patients Diagnosed with High Grade (HG) Bladder Cancer by SPME-LC-MS Method
Reprinted from: *Molecules* **2021**, *26*, 2194, doi:10.3390/molecules26082194 . . . . **113**

Andrea Speltini, Francesca Merlo, Federica Maraschi, Giorgio Marrubini, Anna Faravelli and Antonella Profumo
Magnetic Micro-Solid-Phase Extraction Using a Novel Carbon-Based Composite Coupled with HPLC–MS/MS for Steroid Multiclass Determination in Human Plasma
Reprinted from: *Molecules* **2021**, *26*, 2061, doi:10.3390/molecules26072061 . . . . **125**

Hiroyuki Kataoka and Daiki Nakayama
Online In-Tube Solid-Phase Microextraction Coupled with Liquid Chromatography–Tandem Mass Spectrometry for Automated Analysis of Four Sulfated Steroid Metabolites in Saliva Samples
Reprinted from: *Molecules* **2022**, *27*, 3225, doi:10.3390/molecules27103225 . . . . **141**

Hiroyuki Kataoka, Sanae Kaji and Maki Moai
Risk Assessment of Passive Smoking Based on Analysis of Hair Nicotine and Cotinine as Exposure Biomarkers by In-Tube Solid-Phase Microextraction Coupled On-Line to LC-MS/MS
Reprinted from: *Molecules* **2021**, *26*, 7356, doi:10.3390/molecules26237356 . . . . **153**

Atsushi Ishizaki and Hiroyuki Kataoka
Online In-Tube Solid Phase Microextraction Coupled to Liquid Chromatography–Tandem Mass Spectrometry for the Determination of Tobacco-Specific Nitrosamines in Hair Samples
Reprinted from: *Molecules* **2021**, *26*, 2056, doi:10.3390/molecules26072056 . . . . **163**

Yuping Zhang, Ning Wang, Zhenyu Lu, Na Chen, Chengxing Cui and Xinxin Chen
Smart Titanium Wire Used for the Evaluation of Hydrophobic/Hydrophilic Interaction by In-Tube Solid Phase Microextraction
Reprinted from: *Molecules* **2022**, *27*, 2353, doi:10.3390/molecules27072353 . . . . **173**

About the Editor

Hiroyuki Kataoka

Hiroyuki Kataoka has received a BSc degree (Pharmaceutical Science, Nagasaki University, Japan, 1977), MSc degree (Osaka University, Japan, 1979), and PhD degree (Tohoku University, Japan, 1986). Since working as a Research Associate and Associate Professor in Okayama University 1979–2003, he has been serving as a Professor in Shujitsu University. In 1998–1999, he worked with Professor Janusz Pawliszyn as a postdoctoral research fellow in the Solid Phase Microextraction Group at University of Waterloo (Waterloo, Ontario, Canada) developing in-tube solid-phase microextraction. His research interests include the development of selective and sensitive methods for the analysis of bioactive and potentially harmful compounds in living systems, foods and environments. His present research projects cover the development of automated sample preparation methods in combination with LC-MS/MS or GC-MS, and their applications in environmental and biomedical fields. Using these techniques, he is developing methods to assess biological exposure to toxic chemicals, and non-invasive analysis of stress and fatigue-related biomarkers to protect health by detecting early signs of disease. The goal of the research is to innovate analytical technology and contribute to more comfortable life and health promotion. He has authored over 170 scientific publications in peer reviewed journals and his Hirsch Index is 43 (January, 2023). He was named one of the top 2% of scientists in the world by a 2022 Stanford University research team survey for the fourth consecutive year. He is a member of the Editorial Board of Analytica Chimica Acta, Journal of Chromatography A, Analytica Chemistry Letters, Journal of Environmental & Analytical Toxicology, Journal of Analytical Methods in Chemistry, Journal of Translational Biomarkers & Diagnosis, Journal of Bioanalytical Techniques, Chromatography and Molecules.

Preface to "Solid-Phase Microextraction and Related Techniques in Bioanalysis"

Bioanalysis of endogenous substances, metabolites, and contaminants poisons is important in analyses of biological functions, metabolomics, forensic toxicology and patient diagnosis. In these analyses, methods of sample preparation are essential for the isolation and concentration of target analytes from complex biological matrices, including blood, urine, saliva, hair and tissue. However, preparation processes are time-consuming, labor-intensive and error-prone, and markedly influence the reliability and accuracy of determining target molecules. Thus, efficient sample preparation techniques and their integration with analytical methods have become significant.

Solid-phase microextraction (SPME) is a simple and convenient sample preparation technique that enables automation, miniaturization, high-throughput performance and online coupling with analytical instruments. Moreover, SPME reduces analysis times, as well as solvent and disposal costs. Since SPME was first introduced in the early 1990s by Arthur and Pawliszyn, more robust fiber assemblies and coatings with higher extraction efficiencies, selectivity and stability have become commercially available. Furthermore, new geometries have been designed for extraction as alternatives to fibers, such as capillary tubes, magnetic stir bars or thin films; moreover, novel intelligent polymer coatings with great sorption capacity or good selectivity have been developed for use as extraction phases.

This Special Issue, entitled "Solid-Phase Microextraction and Related Techniques in Bioanalysis", consists of 14 original, peer-reviewed papers written by research groups worldwide. The topics covered include headspace fiber SPME (HS-SPME) gas chromatography-mass spectrometry (GC-MS), HS-in-needle microextraction GC-MS, thin film SPME liquid chromatography-tandem mass spectrometry (LC-MS/MS), magnetic solid phase extraction LC-MS/MS, in-tube SPME LC-MS/MS and in-tube SPME LC-UV. The samples analyzed include a wide range of plant-derived volatile organic compounds; body odor from the skin; metabolites in urine, plasma and saliva sample; biomarkers of exposure to tobacco smoke in hair and environmental estrogens in water.

I thank all contributors, authors, and colleagues who contributed to this Special Issue, as well as the reviewers who devoted their time, effort and expertise to evaluating the submissions and ensuring the quality of the published articles. I would also like to thank the publisher MDPI and the editorial staff of the journal for their constant professional support and for their invitation to edit this Special Issue. Furthermore, I would like to thank all readers, and I hope that the content of this book will provide new perspectives and ideas for starting and continuing further research.

Hiroyuki Kataoka
Editor

Editorial

Solid-Phase Microextraction and Related Techniques in Bioanalysis

Hiroyuki Kataoka

Laboratory of Applied Analytical Chemistry, School of Pharmacy, Shujitsu University, Nishigawara, Okayama 703-8516, Japan; hkataoka@shujitsu.ac.jp; Tel.: +81-86-271-8342

Living organisms, such as microorganisms, plants and animals, are composed of complex constituents, which may include bioactive components that maintain their functions. In addition, these organisms may contain harmful external contaminants. Bioanalysis of these endogenous substances, metabolites and contaminant poisons is important in analyses of biological functions, metabolomics, forensic toxicology, patient diagnosis and the biomonitoring of human exposure to hazardous chemicals. In these analyses, sample preparation is essential for the isolation and concentration of target analytes from complex biological matrices, including serum/plasma, urine, saliva, hair, breath and tissue. However, preparation processes are time-consuming, labor-intensive and error-prone, and they markedly influence the reliability and accuracy of determining molecules of interest. Thus, effective sample preparation techniques and their integration with analytical methods have become a prominent research topic.

Solid-phase microextraction (SPME) is a simple and convenient sample preparation technique that enables automation, miniaturization, high-throughput performance and online coupling with analytical instruments. Moreover, SPME reduces analysis times, as well as solvent and disposal costs. Since SPME was first introduced in the early 1990s [1], more robust fiber assemblies and coatings with higher extraction efficiencies, selectivity and, stability have become commercially available. Furthermore, new geometries have been designed for extraction as alternatives to fibers, such as capillary tubes, magnetic stir bars or thin films; moreover, novel intelligent polymer coatings with great sorption capacity or good selectivity have been developed for use as extraction phases.

This Special Issue, entitled "Solid-Phase Microextraction and Related Techniques in Bioanalysis", consists of 14 original, peer-reviewed papers written by research groups worldwide. The topics covered include headspace fiber SPME (HS-SPME) gas chromatography–mass spectrometry (GC-MS) [2–8], HS-in-needle ME (HS-INME)-GC-MS [9], thin film SPME (TF-SPME) liquid chromatography–tandem mass spectrometry (LC-MS/MS) [10], magnetic solid-phase extraction (MSPE) LC-MS/MS [11], in-tube SPME (IT-SPME) LC-MS/MS [12–14] and IT-SPME LC-UV [15]. The samples analyzed include a wide range of plant-derived volatile organic compounds [2–7]; body odor from the skin [8,9]; metabolites from urine [10], plasma [11] and saliva [12] samples; biomarkers of exposure to tobacco smoke in hair [13,14]; and environmental estrogens in water [15]. An overview of these papers is provided below.

Profiles of volatile organic compounds (VOCs) emitted by plants were analyzed using HS-SPME GC-MS [2–4]. The tastes and aromas released during fruit ripening play an important role in the identification of cultivars and the quality of fruits and their products, including their characteristic flavors. Fragrance components include a variety of low-molecular-weight VOCs, such as alcohols, esters, acids, aldehydes, ketones, aliphatic and aromatic esters, terpenes, hydrocarbons, phenols and sulfur compounds. An HS-SPME GC-MS analysis of the volatility profile in nine species of rumberry (*Myrciaria floribunda*) fruits identified 36 VOCs, among which the sesquiterpenes caryophyllene and γ-selinene were found to be flavor-determining components [2]. HS-SPME GC-MS conditions were optimized to extract, detect and quantify volatile components from the pulp of *Eugenia klotzschiana* O. Berg, a landrace of the Cerrado biome with important nutritional value, and its aroma was found to be composed of 38 sesquiterpene and monoterpene compounds [3]. HS-SPME GC-MS also identified 22 VOCs in the essential oil of dioecious aroeira seeds of *Myracrodruon urundeuva* using HS-SPME GC-MS, and their anti-inflammatory properties were analyzed using a chemoinformatics approach [4]. A combination of HS-SPME GC-MS and multivariate statistical analysis was used to isolate and identify VOCs in *Herpetospermum pedunculosum*, a dioecious plant that has been used as a traditional Tibetan medicine for the treatment of hepatobiliary diseases [5]. These analyses showed that the levels of nine VOCs, including isoamyl alcohol, (Z)-3-methylbutanal oxime and 1-nitropentane, differed significantly in female and male flower buds.

The HS-SPME GC-MS method was also used to identify damage to crops caused by pests [6,7]. Because stored crops can be affected by pests and parasites, the prevention of damage and maintenance of quality parameters over time are important for ensuring global food security. Rapid and appropriate methods of sampling, pest detection and data analysis are therefore required to control crop quality in real time. VOCs from crops and pests can serve as biomarkers for monitoring pest damage. For example, HS-SPME GC-MS identified wheat- and insect-specific VOCs, including benzoquinone homologues, released by three stored grain pests, suggesting that these compounds can act as biomarkers of crop damage [6]. Similarly, HS-SPME GC-MS was used to compare VOCs from cabbage plants that were and were not infected with the green peach aphid *Myzus persicae* [7]. Among the 28 VOCs detected in these plants, several, including propane; 2-methoxy; alpha- and beta pinene; myrcene; 1-hexanone; 5-methyl-1-phenyl-; limonene; decane; γ-terpinen and heptane; and 2,4,4-trimethyl propane, were more abundant in infected plants; this indicated that these compounds were responsible for aphid attraction, and therefore, useful in screening for *M. persicae* infection [7].

The odors and emanations released from the human body can provide important information about an individual's health status and the presence or absence of disease. Among the various VOCs emitted from human skin, trans-2-nonenal, benzothiazole, hexyl salicylate, α-hexyl cinnamaldehyde, and isopropyl palmitate are important indicators of the degree of aging [9]. Because these compounds often emanate from the body's surface in very small quantities, simple sampling and sensitive analytical methods are required. Two methods are available for sampling body odor using SPME: an in vivo method, in which SPME fibers are placed on exposed skin, and an in vitro method, in which SPME fibers indirectly extract body odor from a cotton swab or T-shirt. For example, an in vitro method for the HS-SPME GC-MS analysis of trans-2-nonenal consisted of wiping body odor from the skin's surface with gauze, followed by an analysis of changes in body odor in response to lifestyle changes [8]. In contrast, in vitro and in vivo HS-INME GC-MS analyses of body odor VOCs involved solid dynamic microextraction using the adsorbent-coated inner wall of the needle of a gas-tight syringe, or an adsorbent-coated stainless steel wire inserted into the needle [9].

Cystoscopy is an invasive and uncomfortable procedure for patients with bladder cancer, and clinical tests, such as cytology of urine sediment, have low sensitivity and specificity in the monitoring of early-stage bladder cancer. The untargeted metabolomic/metabolomic profiling of biological fluids may be a more effective and less invasive approach to identifying biomarkers of bladder cancer, along with the development of new biomarker-based diagnostic tools. For example, the metabolomic profiling of urine samples from bladder cancer patients and healthy controls has been performed using high throughput TF-SPME LC-MS [10].

Starting with cholesterol, steroid hormones are biosynthesized by various enzymes in the adrenal cortex, gonads and brain, and are subsequently metabolized via phase I oxidation and reduction reactions and phase II conjugation reactions. However, these biosynthetic and metabolic pathways are complex, and their molecular roles are not fully understood. The analysis of steroid hormones and their metabolites is important for elucidating biological regulatory mechanisms and diagnosing diseases related to them. One method of achieving this is dispersive MSPE, in which a magnetic sorbent is dispersed in a sample solution, such as plasma, the solution stirred to extract the compounds of interest, and the extracted compounds eluted from the magnetic sorbent are used in MSPE LC-MS/MS analysis of glucocorticoids, estrogens, progestogens and androgens [11]. Furthermore, an automated analysis system that couples on-line IT-SPME and LC-MS/MS was constructed using an open-tube fused silica capillary with a coated inner surface as the extraction device, enabling simultaneous selective and sensitive analysis of the metabolism of sulfated steroids in saliva samples [12].

Tobacco smoking and exposure to environmental tobacco smoke are considered risk factors for cancers, cardiovascular diseases and respiratory disease, and these health effects have become a serious problem. Tobacco-related compounds, such as nicotine, its metabolite cotinine and tobacco-specific nitrosamines, have been used as biomarkers of exposure to tobacco smoke. Levels of nicotine and cotinine in the hair of non-smokers were analyzed as biomarkers of exposure to tobacco smoke using IT-SPME LC-MS/MS, in order to determine the risks of passive smoking in different lifestyle environments [13]. In addition, online IT-SPME LC–MS/MS was used to simultaneously measure the content of five tobacco-specific nitrosamines in hair as biomarkers of exposure to tobacco smoke [14].

Finally, an IT-SPME method using extraction tubes filled with hydrophilic, superhydrophilic, superhydrophobic and UV-irradiated superhydrophilic Ti wires as sorbents was used for on-line IT-SPME LC-UV analysis of six estrogen-like hormones in water samples [15].

Funding: This research was funded by a Grant-in-Aid for Basic Scientific Research (C, No. 20K07007) and the Smoking Research Foundation (2022).

Institutional Review Board Statement: Not applicable.

Informed Consent Statement: Not applicable.

Data Availability Statement: Not applicable.

Acknowledgments: As Guest Editor of this Special Issue, I thank all of the authors for their contributions, and hope the contents of this publication will help readers to further develop their research.

Conflicts of Interest: The author declares no conflict of interest.

References

1. Arthur, C.L.; Pawliszyn, J. Solid phase microextraction with thermal desorption using fused silica optical fibers. *Anal. Chem.* **1990**, *62*, 2145. [CrossRef]
2. García, Y.M.; Ramos, A.L.C.C.; de Paula, A.C.C.F.F.; do Nascimento, M.H.; Augusti, R.; de Araújo, R.L.B.; de Lemos, E.E.P.; Melo, J.O.F. Chemical Physical Characterization and Profile of Fruit Volatile Compounds from Different Accesses of *Myrciaria floribunda* (H. West Ex Wild.) O. Berg through Polyacrylate Fiber. *Molecules* **2021**, *26*, 5281. [CrossRef] [PubMed]
3. Mariano, A.P.X.; Ramos, A.L.C.C.; de Oliveira, A.H.; García, Y.M.; de Paula, A.C.C.F.F.; Silva, M.R.; Augusti, R.; de Araújo, R.L.B.; Melo, J.O.F. Optimization of Extraction Conditions and Characterization of Volatile Organic Compounds of *Eugenia klotzschiana* O. Berg Fruit Pulp. *Molecules* **2022**, *27*, 935. [CrossRef] [PubMed]
4. Figueiredo, Y.G.; Corrêa, E.A.; de Oliveira, A.H.; Mazzinghy, A.C.d.C.; Mendonça, H.d.O.P.; Lobo, Y.J.G.; García, Y.M.; Gouvêia, M.A.d.S.; de Paula, A.C.C.F.F.; Augusti, R.; et al. Profile of *Myracrodruon urundeuva* Volatile Compounds Ease of Extraction and Biodegradability and In Silico Evaluation of Their Interactions with COX-1 and iNOS. *Molecules* **2022**, *27*, 1633. [CrossRef] [PubMed]
5. Liu, Z.; Fang, Y.; Wu, C.; Hai, X.; Xu, B.; Li, Z.; Song, P.; Wang, H.; Chao, Z. The Difference of Volatile Compounds in Female and Male Buds of *Herpetospermum pedunculosum* Based on HS-SPME-GC-MS and Multivariate Statistical Analysis. *Molecules* **2022**, *27*, 1288. [CrossRef] [PubMed]
6. Cai, L.; Macfadyen, S.; Hua, B.; Zhang, H.; Wei Xu, W.; Ren, Y. Identification of Biomarker Volatile Organic Compounds Released by Three Stored-Grain Insect Pests in Wheat. *Molecules* **2022**, *27*, 1963. [CrossRef] [PubMed]
7. Ahmed, Q.; Agarwal, M.; Alobaidi, R.; Haochuan Zhang, H.; Ren, Y. Response of Aphid Parasitoids to Volatile Organic Compounds from Undamaged and Infested *Brassica oleracea* with *Myzus persicae*. *Molecules* **2022**, *27*, 1522. [CrossRef] [PubMed]
8. Saito, K.; Tokorodani, Y.; Sakamoto, C.; Kataoka, H. Headspace Solid-Phase Microextraction/Gas Chromatography–Mass Spectrometry for the Determination of 2-Nonenal and Its Application to Body Odor Analysis. *Molecules* **2021**, *26*, 5739. [CrossRef] [PubMed]
9. Kim, S.; Bae, S. In Vitro and In Vivo Human Body Odor Analysis Method Using GO:PANI/ZNRs/ZIF−8 Adsorbent Followed by GC/MS. *Molecules* **2022**, *27*, 4795. [CrossRef] [PubMed]

10. Łuczykowski, K.; Warmuzińska, N.; Operacz, S.; Stryjak, I.; Bogusiewicz, J.; Jacyna, J.; Wawrzyniak, R.; Struck-Lewicka, W.; Markuszewski, M.J.; Bojko, B. Metabolic Evaluation of Urine from Patients Diagnosed with High Grade (HG) Bladder Cancer by SPME-LC-MS Method. *Molecules* **2021**, *26*, 2194. [CrossRef] [PubMed]
11. Speltini, A.; Merlo, F.; Maraschi, F.; Marrubini, G.; Anna Faravelli, A.; Profumo, A. Magnetic Micro-Solid-Phase Extraction Using a Novel Carbon-Based Composite Coupled with HPLC–MS/MS for Steroid Multiclass Determination in Human Plasma. *Molecules* **2021**, *26*, 2061. [CrossRef] [PubMed]
12. Kataoka, H.; Nakayama, D. Online In-Tube Solid-Phase Microextraction Coupled with Liquid Chromatography–Tandem Mass Spectrometry for Automated Analysis of Four Sulfated Steroid Metabolites in Saliva Samples. *Molecules* **2022**, *27*, 3225. [CrossRef] [PubMed]
13. Kataoka, H.; Kaji, S.; Moai, M. Risk Assessment of Passive Smoking Based on Analysis of Hair Nicotine and Cotinine as Exposure Biomarkers by In-Tube Solid-Phase Microextraction Coupled On-Line to LC-MS/MS. *Molecules* **2021**, *26*, 7356. [CrossRef] [PubMed]
14. Ishizaki, A.; Kataoka, H. Online In-Tube Solid-Phase Microextraction Coupled to Liquid Chromatography–Tandem Mass Spectrometry for the Determination of Tobacco-Specific Nitrosamines in Hair Samples. *Molecules* **2021**, *26*, 2056. [CrossRef] [PubMed]
15. Zhang, Y.; Wang, N.; Lu, Z.; Chen, N.; Chengxing Cui, C.; Chen, X. Smart Titanium Wire Used for the Evaluation of Hydrophobic/Hydrophilic Interaction by In-Tube Solid Phase Microextraction. *Molecules* **2022**, *27*, 2353. [CrossRef] [PubMed]

Disclaimer/Publisher's Note: The statements, opinions and data contained in all publications are solely those of the individual author(s) and contributor(s) and not of MDPI and/or the editor(s). MDPI and/or the editor(s) disclaim responsibility for any injury to people or property resulting from any ideas, methods, instructions or products referred to in the content.

Article

Chemical Physical Characterization and Profile of Fruit Volatile Compounds from Different Accesses of *Myrciaria floribunda* (H. West Ex Wild.) O. Berg through Polyacrylate Fiber

Yesenia Mendoza García [1], Ana Luiza Coeli Cruz Ramos [2], Ana Cardoso Clemente Filha Ferreira de Paula [3], Maicon Heitor do Nascimento [3], Rodinei Augusti [4], Raquel Linhares Bello de Araújo [2], Eurico Eduardo Pinto de Lemos [1] and Júlio Onésio Ferreira Melo [5,*]

1. Centro de Ciências Agrárias, Campus A. C. Simões, Universidade Federal de Alagoas, Rio Largo 57072-970, Brazil; jenny_thesiba@hotmail.com (Y.M.G.); eurico@ceca.ufal.br (E.E.P.d.L.)
2. Departamento de Alimentos, Faculdade de Farmácia, Campus Belo Horizonte, Universidade Federal de Minas Gerais, Belo Horizonte 31270-901, Brazil; analuizacoeli@gmail.com (A.L.C.C.R.); raquel@bromatologiaufmg.com.br (R.L.B.d.A.)
3. Departamento de Ciências Agrárias, Instituto Federal de Educação, Ciência e Tecnologia de Minas Gerais, Campus Bambuí, Bambuí 38900-000, Brazil; ana.paula@ifmg.edu.br (A.C.C.F.F.d.P.); maicon-naza@hotmail.com (M.H.d.N.)
4. Departamento de Química, Campus Belo Horizonte, Universidade Federal de Minas Gerais, Belo Horizonte 35702-031, Brazil; augusti.rodinei@gmail.com
5. Departamento de Ciências Exatas e Biológicas, Campus Sete Lagoas, Universidade Federal de São João Del-Rei, Sete Lagoas 36307-352, Brazil
* Correspondence: onesiomelo@gmail.com

Abstract: Among the many species of native fruit of Brazil that have been little explored, there is *Myrciaria floribunda* (also known as rumberry, cambuizeiro, or guavaberry), a species with significant variability, which has fruits of different colors (orange, red, and purple) when ripe. The physical-chemical characteristics evaluated were fruit weight (FW), seed weight (SW), pulp weight (PW), number of seeds (NS), longitudinal diameter (LD), transverse diameter (TD), format (LD/TD), hydrogen potential (pH), soluble solids (SS), titratable acidity (TA), and ratio (SS/TA); further, the volatile organic compounds (VOCs) of nine accesses of rumberry orchards were identified. The averages of the variables FW, SW, PW, NS, LD, TD, shape, and firmness were 0.76 g, 0.22 g, 0.54 g, 1.45, 10.06 mm, 9.90 mm, 1.02, 2.96 N, respectively. LD/TD data showed that the fruits have a slightly rounded shape (LD/TD = 1). The averages for pH, SS, TA, and SS/TA were 3.74, 17.58 Brix, 4.31% citric acid, and 4.31, respectively. The evaluated parameters indicated that the fruits can be consumed both in natura and industrialized, with the red-colored fruits presenting a good balance of SS/TA, standards demanded by the processing industries. Thirty-six VOCs were identified, with emphasis on the sesquiterpenes. Caryophyllene (21.6% to 49.3%) and γ-selinene (11.3% to 16.3%) were the most predominant compounds in rumberry fruits.

Keywords: Myrtaceae; native fruit; volatile compounds; sesquiterpenes

1. Introduction

Myrciaria floribunda (H. West ex Willd.) O. Berg is a species belonging to the Myrtaceae family, from which the fruits are popularly known as 'camboim,' 'jabuticabinha,' 'myrtle,' 'duke,' 'goiabarana,' 'araçazeiro,' 'cambuí,' and 'rumberry.' In Brazil, this species can be found mainly in the Atlantic Forest biome, easily observed in the Northeast of the country, specifically in the states of Sergipe and Alagoas [1].

The rumberry has excellent potential for commercial export, widely cultivated due to its edible fruits, consumed both in natura and industrialized. These are globose fruits, with a slightly acidic flavor, of varied color (orange, red, and purple) [2], with high sugar content and are rich in bioactive compounds (carotenoids, flavonoids, and phenolic acids) [3]. When ripe, the fruits become attractive because of the intense citrusy and slightly sweet aroma of the fleshy and succulent pulp [4].

Chemical compounds are responsible for the characteristic flavor of tropical or subtropical fruits, playing an important role in the quality of the fruits and their products, determined by several parameters, such as appearance, flavor, nutritional value, and food security [5,6]. However, the chemical composition of rumberry fruits is still unknown, which prevents further tapping into market potential.

The acceptance of fruits is directly related to taste and aroma, released mainly during the ripening phase of the fruit. These parameters are among the most important quality criteria of fresh and processed foods, since the aroma and taste of the fruit provoke sensory sensations that stimulate the desire to consume it [7]. These sensations are attributed to hundreds or thousands of volatile and non-volatile substances that are present in food and which are measured by sensory organs. [8]. In general, the aromatic compounds of fruits are low-molecular-weight substances, partially soluble in water and volatile at room temperature, belonging to a wide variety of chemical families such as alcohols, esters, acids, aldehydes, ketones, aliphatic and aromatic esters, terpenes, hydrocarbons, and phenolic and sulfur compounds, which are found in different concentrations [9].

Recent studies have shown that the leaves of *M. floribunda* produce essential oils rich in monoterpenes (53.9%) and, among them, 1,8-cineole is the main constituent (38.4%) [10]. The essential oils of *M. floribunda* showed pharmacological potential in terms of insecticidal activity [10], inhibition of acetylcholinesterase [11], and antimicrobial and antitumor activity [12].

The determination of volatile compounds contributed to the discovery of a new aroma. For this, there are techniques such as solid-phase microextraction (SPME), which is based on the extraction and rapid concentration of volatile and semi-volatile organic compounds without the use of organic solvents [7]. In this type of extraction, the analyte is adsorbed on a silica fiber coated with a polymer, which is inserted in the headspace bottle for subsequent thermal desorption, injecting the analytes in a gas chromatograph (GC) that, coupled with mass spectrometry (MS), provides speed and practicality in the analysis of the volatile profile of the fruits.

In the absence of studies evaluating the volatile profile of *Myrciaria floribunda* fruits using the solid-phase microextraction technique, the objective of the present work was to characterize the volatile profile of nine rumberry accesses through the polyacrylate fiber, using SPME-HS/GC–MS.

2. Results and Discussion

2.1. Physicochemical Characterization of Rumberry Fruits

Significant differences ($p \leq 0.05$) were observed between the different accesses evaluated, for the physicochemical characteristics, except for pH (Table 1). The fruits had an average weight of 0.76 g, with the highest values found in AC132. These results agree with those reported by Silva et al. [13] in the evaluation of the biometry of fruits of the same species. In guabiju fruits (*Myrcianthes pungens*), the average weight was 2.87 g [14], and for *Campomanesia rufa*, 10.88 g [15], species belonging to the Myrtaceae family.

Table 1. Physical and physicochemical parameters of the nine accesses of rumberry.

Access	FW (g)	SW (g)	PW (g)	NS	LD (mm)	TD (mm)	Format	Firmness (N)	pH	SS	TA	SS/TA
AC67	0.58 a	0.17 a	0.41 a	1.18 a	8.91 b	9.95 b	0.90 b	1.46 a	3.53 a	21.30 d	5.15 b	4.15 a
AC92	0.43 a	0.14 a	0.29 a	1.19 a	7.10 a	8.94 a	0.86 a	2.59 a	3.83 a	22.78 e	5.05 b	4.51 a
AC112	0.79 b	0.25 b	0.54 b	1.71 c	10.81 d	9.81 b	1.10 d	3.26 a	3.68 a	16.88 b	4.93 b	3.60 a
AC132	1.14 c	0.49 c	0.64 b	1.66 c	11.64 e	10.43 c	1.12 d	1.44 a	3.99 a	17.65 c	4.93 b	3.58 a
AC136	0.81 b	0.21 b	0.61 b	1.48 b	10.69 d	9.65 b	1.11 d	3.57 a	3.75 a	16.53 b	4.90 b	3.55 a
AC137	0.77 b	0.16 a	0.61 b	1.50 b	10.76 d	9.54 b	1.13 d	2.61 a	3.99 a	18.30 c	2.70 a	6.87 b
AC153	0.82 b	0.24 b	0.58 b	1.42 b	9.50 b	10.47 c	0.91 b	1.98 a	3.90 a	16.10 b	4.35 b	3.79 a
AC156	0.91 b	0.22 b	0.69 b	1.91 c	10.66 d	11.50 d	0.93 c	4.08 a	3.62 a	15.43 b	3.95 b	3.98 a
AC160	0.57 a	0.10 a	0.47 a	1.01 a	9.84 c	8.82 a	1.12 d	5.67 a	3.35 a	13.25 a	2.80 a	4.75 a
Mean	0.76	0.22	0.54	1.45	10.06	9.90	1.02	2.96	3.74	17.58	4.31	4.31
CV (%)	20.07	20.83	22.43	12.11	5.09	5.95	1.51	56.17	9.83	4.02	14.66	16.28
Standard Error	0.08	0.02	0.06	0.09	1.26	0.29	0.01	0.83	0.18	0.35	0.32	0.35

Mean values of four replicates of 32 fruits per access, expressed on a wet basis. Averages followed by the same letter in the column do not differ statistically by the Scott–Knott Test, at 5% probability. FW: fruit weight; SW: seed weight; PW: pulp weight; NS: number of seeds; LD: longitudinal diameter; TD: transverse diameter; Format: the relationship between LD/TD variables; pH: hydrogen potential; SS: soluble solids (Brix); TA: titratable acidity (% citric acid); SS/TA: the ratio between the two variables. Means on the same column followed by the same letter do not differ from each other by the Scott–Knott test at the 5% probability level.

The fruits of accesses AC67, AC92, and AC160 showed the lowest values in terms of fruit weight, pulp weight, and a number of seeds. The study by Rodrigues et al. [14] obtained an average for the weight of the fruit (2.87 g), pulp weight (1.29 g), and seed weight (0.49 g) higher than those reported in the present study, while the number of seeds remained equal, containing one to two seeds per fruit [1].

The accesses AC132 (11.64 mm of DL and 10.43 mm of DT) and AC156 (10.66 mm of DL and 11.50 mm of DT) presented the largest dimensions and the highest pulp yields (0.64 and 0.69 g, respectively). The fruits had a medium diameter (longitudinal and transversal) greater than those reported in red jambo (*Syzygium malaccensis*) [16] fruits and inferior to those reported for rumberry fruits (*Myrciaria floribunda*), in their orange and purple color—4.09 to 4.47 mm and 4.47 to 4.87, respectively [1].

The shape of the fruits showed an overall average of 1.02, characteristic of fruits with a more rounded shape (LD/TD = 1). The shape of the fruits is an index of ripeness since they are usually measured by the ratio of their diameter. This characteristic is also a factor of industrial quality since the industries prefer round-shaped fruits for easy cleaning and processing operations.

Statistically, there was not a significant difference regarding the firmness of the fruits. However, the fruits of the AC160 access showed greater firmness, with an average of 5.67 N, while the AC67 access showed the lowest average of 1.46 N. It is noteworthy that, to date, there are no reports in the literature regarding the firmness of rumberry fruits.

According to Becker et al. [17], firmness is a very important quality attribute in fruits because the greater the firmness, the greater the resistance to mechanical injuries during transport and commercialization of fruits responsible for flavor and aroma.

Rumberry fruits had a pH from 3.35 to 3.99, with an average value of 3.74, with no significant difference between the accesses studied (Table 1). These values were similar to those reported in fruits of the same species [1], whereas in fruits of the same genus, a pH of 2.41 was reported [18].

However, Almeida et al. [19] found pH values of 3.10 and 3.39 for the varieties of jabuticaba *Myrciaria jabuticaba* and *Myrciaria grandifolia*, respectively. According to the authors, the most acidic products are naturally more stable in terms of deterioration, and the relative proportion of organic acids present in fruits and vegetables might vary according to the degree of ripeness and growth conditions. This information is relevant for selecting accesses since more acidic fruits can be better used in the food industry.

The SS content varied significantly between accesses. AC67, AC92, AC132, and AC137 registered the highest values from 17.65 to 22.78 Brix, while accesses AC112, AC136, AC153, AC156, and AC160 registered values in a range of 13.25–16.88 Brix. These values were higher when compared to those reported by Vieira et al. [20] at different stages of fruit ripening of ubaia-azeda *(Eugenia azeda)*, whose values ranged from 3.13 to 4.35 °Brix, as well as those reported by Souza, Silva and Aguiar [18] when obtaining values of 9.02 °Brix in ripe fruits of camu-camu *(Myrciaria dubia)*.

As for AT, the lowest values found were in the AC137 (2.70), and AC160 (2.80) accesses. As for the accesses AC67, AC92, AC112, AC132, AC136, AC153, and AC156, there was no significant difference among them. Souza, Silva, Aguiar [18] reported values lower than those found when determining the chemical composition phases: unripe, semi-ripe, and ripe, finding averages of 2.75, 2.77, and 2.4 g of citric acid 100 g^{-1}, respectively.

Seraglio et al. [21], when analyzing the physicochemical parameters of mature fruits of three species of Myrtaceae, reported values of 0.19 g of citric acid 100 g^{-1} for jabuticaba fruits *(Myrciaria cauliflora)*, 0.02 g for guabiju fruits *(Myrcianthes pungens)*, and 0.04 g for jambolan fruits *(Syzygium cumini)*, respectively.

TA and SS are a crucial reference point for the ideal level of fruit ripeness, in addition to being an important parameter for assessing the conservation status of all foods, once through the decomposition process (hydrolysis, oxidation or fermentation), which changes the concentration of organic acids, thus altering the acidity of the food [21].

The SS/TA ratio of rumberry fruits was higher than that reported by Vieira et al. [20] and Neto, Silva and Dantas [22], who found mean values of 2.45 and 2.98, respectively. Lattuada et al. [23] observed higher values, ranging from 13.12 to 17.10, in whole and ripe fruits of five samples of jabuticaba trees *(Plinia peruviana* and *Plinia cauliflora)*.

For the fresh and/or processed fruit market, high SS values and SS/TA ratios are the most desirable and valued characteristics, both in natura and industrial use. In the case of consumers, they give preference to larger fresh fruits, with an appealing appearance, sweeter, and less acidic. Therefore, the best way to assess the fruit is through the SS/TA ratio, as it is more representative than the isolated analysis of SS or TA [6]. Thus, the fruits of the AC137 access would be the most suitable for presenting a higher SS/TA ratio (6.87).

2.2. Profile of Volatile Compounds

A total of 36 volatile organic compounds (VOCs) were identified by the polyacrylate fiber (PA) through solid-phase microextraction in the headspace mode (SPME-HS). The fruits of the different accesses of *Myrciaria floribunda* contained sesquiterpenes: 30.56% monoterpenes and 69.44% sesquiterpenes (Table 2).

Table 2. Volatile profile of fruits of *Myrciaria floribunda*, isolated by the polyacrylate fiber and SPME-HS/GC–MS.

No.	VOCs	CAS	% Area								
			AC67	AC92	AC112	AC132	AC136	AC137	AC153	AC156	AC160
1	α-pinene	80-56-8	0.59	0.24	0.19	1.09	-	0.54	-	0.08	-
2	Eucalyptus	470-82-6	10.6	0.29	1.26	4.36	4.04	2.45	8.74	6.56	0.7
3	3-carene	13466-78-9	5.0	1.06	0.09	1.09	2.48	1.66	1.69	-	-
4	Ocimene	502-99-8	1.17	1.58	0.2	4.76	0.21	0.63	0.74	0.03	0.18
5	α-terpineol	98-55-5	1	0.61	2.69	3.93	4.44	5.66	-	-	0.78
6	α-canfolenal	4501-58-0	0.2	0.71	0.11	0.21	0.5	-	0.48	0.37	-
7	Isopulegol acetate	57576-09-7	-	1.44	0.21	1.25	0.18	0.4	2.98	1.25	-
8	γ-terpineol	586-81-2	-	0.29	0.15	0.47	-	0.32	-	0.19	-
9	Acetate fenquila	13851-11-1	-	0.25	0.09	-	-	0.56	-	-	-
10	Borneol	507-70-0	-	-	-	0.4	-	1.1	-	-	-
11	Isobornyl format	1200-67-5	5.83	-	-	-	-	-	-	0.1	-

Table 2. Cont.

No.	VOCs	CAS	% Area								
			AC67	AC92	AC112	AC132	AC136	AC137	AC153	AC156	AC160
Monoterpenes			24.26	6.47	4.99	17.56	11.85	13.32	14.63	8.58	1.66
12	α-muurolene	10208-80-7	21.04	1.19	9.27	1.13	9.72	1.22	1.17	1.46	1.57
13	Cyclosativene	22469-52-9	3.95	9.29	6.78	4.37	0.64	0.96	0.99	0.48	2.51
14	β-guaiene		0.48	0.38	1	2.45	5.15	5.58	2.85	3.1	3.21
15	Caryophyllene		21.59	8.20	25.8	1.77	35.73	0.05	32.88	0.45	48.51
16	α-longipinene	5989-08-2	3.99	4.66	3.78	24.21	3.63	6.2	1.04	1.28	2.42
17	Longifolene	61262-67-7	6.35	3.50	4.58	4.24	0.36	-	0.4	5.18	0.48
18	α-selinene	473-13-2	0.86	11.62	2.39	2.45	1.23	1.4	-	-	0.3
19	Zonarene	41929-05-9	2.11	3.38	3.43	8.92	0.37	0.47	4.5	0.51	4.81
20	γ-selinene	515-17-3	-	4.64	11.2	0	16.05	58.18	0.4	63.1	0.63
21	Ledene	21747-46-6	0.4	0.03	4.38	13.21	0.17	7.7	0.52	-	2.44
22	Eudesma-3,7 (11) -diene	6813-21-4	0.15	0.16	0.07	0.3	0.16	0.62	2.27	8.14	0.14
23	α-gurjunene	489-40-7	6.6	4.05	4.45	0.3	3.61	0.34	0.16	0.32	0.23
24	Patchoulene	1405-16-9	1.62	32.56	1.87	1.74	3.07	0.48	0.52	0.42	0.92
25	Eremophila-1 (10), 11-diene	10219-75-7	2.6	2.67	4.88	5.85	2.64	-	1.69	0.05	3.94
26	γ-himachalene	53111-25-4	-	0.05	0.34	0.77	0.25	1	4.07	0.32	1.05
27	10s, 11s-himachala-3 (12), 4-diene	60909-28-6	0.2	0.68	0.78	7.6	1.18	0.49	8.89	0.65	6.94
28	Aristolen	88-84-6	0.2	0.23	1.37	-	1.05	0.71	18.22	0.52	16.7
29	γ-cadinene	39029-41-9	0.45	2.27	3.53	-	0.08	0.03	-	-	-
30	Cadina-3,9-diene	523-47-7	1	1.75	2.88	-	0.83	-	0.18	0.76	-
31	α-cubebene	17699-14-8	0.3	0.03	0.29	-	0.25	-	-	-	-
32	α-ylangene	14912-44-8	-	0.19	0.38	-	0.44	-	-	-	-
33	Guayana-1 (5), 11-diene	3691-12-1	0.11	0.52	0.29	-	-	0.05	-	0.03	-
34	δ-elemene	20307-84-0	-	0.04	0.13	-	-	-	-	-	-
35	Copena	3856-25-5	1.4	0.45	0.48	0.33	-	-	-	-	-
36	γ-muurolene	30021-74-0	-	-	-	-	-	0.04	-	-	-
Sesquiterpenes			75.42	92.54	94.35	79.64	86.61	85.52	80.75	86.77	96.8
Total identified			99.68	99.01	99.34	97.2	98.46	98.84	95.38	95.35	98.46

Accesses differentiated by the color characteristic of fruit: orange fruits (AC67, AC92, AC112, AC136, AC137, AC156, and AC157), red fruits (AC132 and AC153) and purple fruits (AC160). -: undetected compounds.

In oils derived from leaves and flowers of *Myrciaria floribunda*, monoterpenes have been reported as the main components, with sesquiterpenes predominating in stem oil [11]. Regarding the lyophilized fruits of the same species, the characterization was different, as they presented a higher composition of monoterpenes (59.4%) than sesquiterpenes (40.6%) [24]. By its turn, Kauffmann et al. [25] found oxygenated sesquiterpenes (82.66%) and sesquiterpene hydrocarbons (11.05%) in the composition of the *Myrciaria plinioides* essential oil.

For accesses AC67, AC132, AC92, AC160 and AC137, the main compounds were α-muurolene (21.04%), α-longipinene (24.21%), patchoulene (32.56%), caryophyllene (48.51%) and γ-selinene (58.18%), respectively. Similar results were reported by Silva Barbosa et al. [26] by identifying the chemical composition of the essential oil of the peel of fruits of *Myrciaria floribunda*. A total of 26 compounds were identified, most of them belonging to the sesquiterpenes class—notably γ-cadinene (15.69%), γ- muurolene (6.21%), α-selinene (6.11%), α-muurolene (6.11%), caryophyllene (5.54%) and α-copaene (5.02%), as they were the major compounds.

Comparing the results of the present study with those reported in fruits of the same genus (Myrciaria), there was similarity in the presence of several compounds, such as γ-muurolene, δ-elemene, α-cubebene, γ-cadinene, α-selinene, α -muurolene, aristolene, α-terpineol, eucalyptol, ocimene, and α-pinene, reported by Freitas et al. [27] and Rondán Sanabria et al. [28] in fruits of jabuticaba (*Myrciaria jabuticaba*).

In camu-camu fruits (*Myrciaria dubia*), the compound caryophyllene was identified, and presented greater abundance in the present study [18]. In the essential oil of *Eugenia involucrata* (Myrtaceae) leaves, the compounds δ-elemene, α-cubebene, γ-cadinene, α-gurjunene, and caryophyllene were reported [29].

The caryophyllene is a natural sesquiterpene found in the essential oils of various spices, fruits, and medicinal and ornamental plants [30], such as cloves (*Syzygium aromaticum*), belonging to the Myrtaceae family [31]. The caryophyllene and its derivatives have numerous properties—natural insecticide, acaricide, repellent, attractive and antifungal properties [32], anti-inflammatory, antitumor, antibacterial, antioxidant and spasmolytic [33].

In addition, the Food and Drug Administration (FDA) and the European Food Safety Authority (EFSA) have approved the use of caryophyllene in food products as a food additive, flavor enhancer, and a flavoring agent; and in the perfumery sector, as fragrance or fixative plants [30,31].

Figure 1 shows chromatograms with the largest number of compounds, differentiated by their color of orange (AC112), red (AC132) and purple (AC160), with ten of the compounds similar among them (eucalyptol, ocimene, α-muurolene, cyclosativene, α-longipinene, zonarene, eudesma-3,7(11)-diene, α-gurjunene, patchoulene and 10s, 11s-himachala-3 (12), 4 -diene).

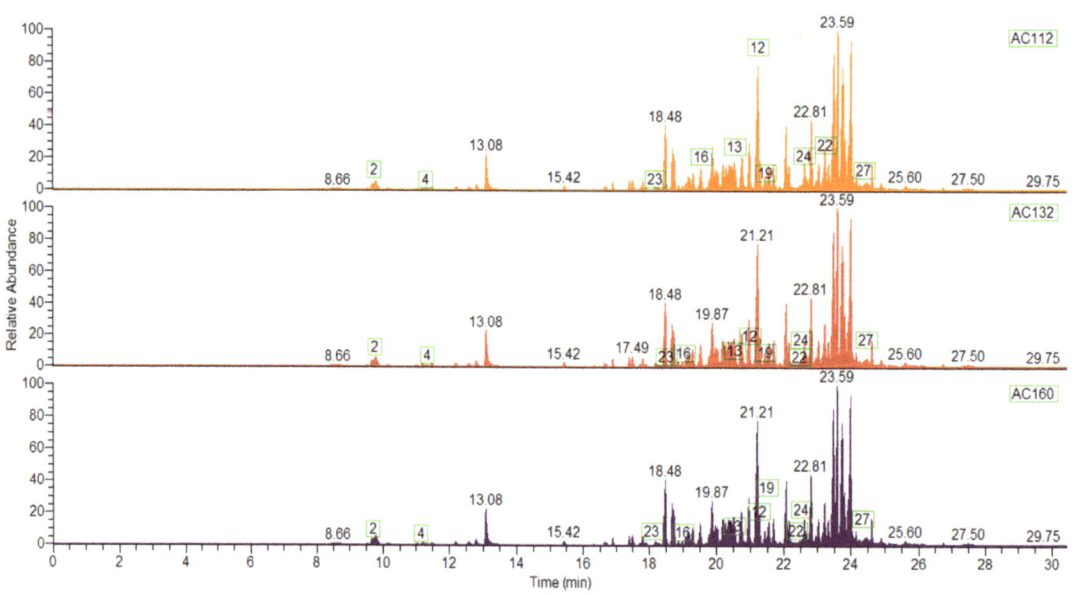

Figure 1. Chromatogram generated for rumberry fruits with the number of similar volatile compounds. The numbers presented in the graph corresponding to the volatile compounds arranged in Table 2 according to their numbering.

Overall, the orange and red fruits did not have substantial differences when it comes to the number of compounds detected, as both had monoterpenes and sesquiterpenes in their composition. However, the purple-colored fruits showed a higher composition of sesquiterpenes than monoterpenes, notably eucalyptol.

The 1,8-cineol or eucalyptol is a monoterpene with a high therapeutic index with antihypertensive, antiasthmatic, and analgesic effects, which makes it a potent drug candidate [34]. It has biological activities such as antibacterial, antifungal, anti-inflammatory, and anti-tumor [35]. In addition, it is used in the treatment of cough, rheumatism, neurosis, muscle pain, asthma, kidney stones, and in cosmetic products.

Currently, there are no previous reports on the volatile composition of rumberry fruits. However, it is known that the Myrtaceae family commonly presents the group of sesquiterpenes as the most prevalent chemical class, and to a lesser extent, the monoterpenes [29,36]—the latter with the fungicidal and attractive function of pollinators in plants [6]. The sesquiterpenes, in turn, have shown therapeutic potential with anti-inflammatory activity in the essential oils of several plant species. Furthermore, high levels of sesquiterpenes have antioxidant properties and a wide variety of biological and pharmaceutical activities [37].

The accesses of rumberry were differentiated into three groups—AC67, AC112, AC136, AC153, and AC160 (Group 1); AC92 and AC132 (Group 2); and AC137 and AC156 (Group 3) (Figure 2).

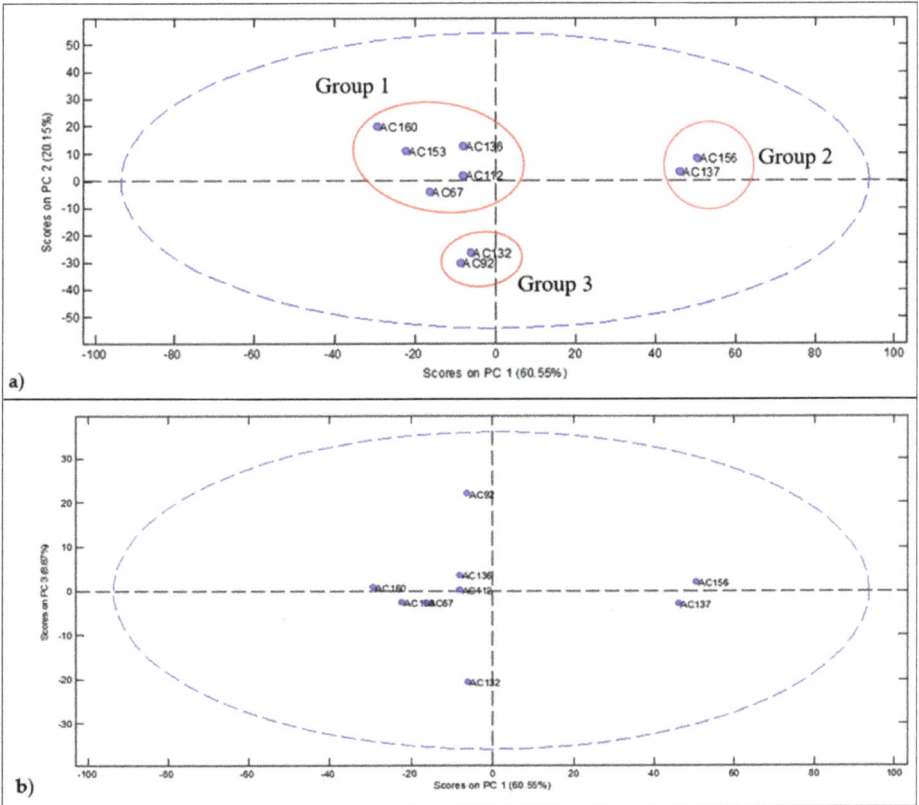

Figure 2. PCA scores plot obtained from the percentage of peak areas of the monoterpenes and sesquiterpenes of the volatile profile of fruits of *Myrciaria floribunda*: (**a**) groups formed with variability of 80.70% of areas; (**b**) groups formed with variability of 69.42% of the areas.

Group 1 (Figure 2a,b) was mainly characterized by the presence of caryophyllene (21.6% to 49.3%). However, it can be subdivided by the accesses AC112 and AC136, due to the similarity between the compounds α-muurolene (9.3% to 21.1) and γ-selinene (11.3% to 16.3), as seen in accesses AC153 and AC160, grouped by the compounds aristolene (17.0% to 19.1%) and 10s, 11s-himachala-3 (12), 4-diene (7.0% to 9.3%).

Group 2 behaved differently. In Figure 2b, the interaction between main component 1 (PC1) and main component 3 (PC3) is shown. This interaction was responsible for the separation of the group by the compounds patchoulene (32.9%) and α-longipinene (24.9%) (Figure 3), as these compounds had the highest percentage of VOCs area identified for such accesses (Table 2). This distinction between components may occur because they are fruits of a different color.

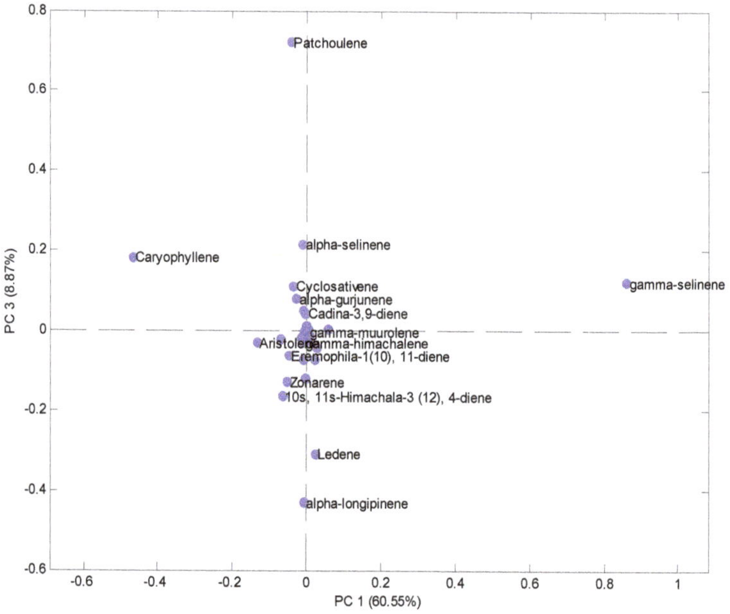

Figure 3. PCA loadings plot obtained from the monoterpenes and sesquiterpenes results of the volatile profile of fruits of *Myrciaria floribunda*.

Chalannavar et al. [38], when investigating the composition of the essential oil of *Psidium cattleianum* var. lucidum (Myrtaceae), reported patchoulene among the most predominant constituents. While Noudogbessi et al. [39] mentioned the presence of α-longipinene in the composition of the essential oil of dry leaves of *Syzygium guineense*. Li, Xuan and Shou [40] highlighted the therapeutic potential of α-longipinene against anxiety.

As for group 3, this includes the presence of the sesquiterpene γ-selinene, a compound with the highest percentage of area among the accesses (Table 2), which was similar to the subgroup composed of the accesses AC112 and AC136, as they are fruits of the same color.

Figure 3 shows the PCA loadings graph for the first three principal components. It is observed that the detailed analysis of the loadings showed in its greatest composition the class of sesquiterpenes. On the positive side of the axis of main component 1, the compounds ledene, γ-selinene, γ-muurolene, and γ-hymachalene stood out. In contrast, on the negative side of the axis were the following compounds: α-longipinene, 10s, 11s-himachala 3 (12), 4-diene, zonarene, eremophila-1 (10), 11-diene, aristolene, α-gurjunene, cyclosativene, α-selinene, cadina-3,9-diene, caryophyllene, and patchoulene.

It should be noted that four of the compounds present in the loadings were located at the ends of quadrant I (γ-selinene), quadrant III (α-longipinene), and quadrant VI (caryophyllene and patchoulene), due to the higher values percentage of total areas, considered as the most important compounds of the volatile profile of rumberry fruits.

The groups formed in the PCA (Figure 2) were confirmed with the application of the Hierarchical Cluster Analysis (HCA), considering Euclidean distances, which generated the dendrogram shown in Figure 4. The dendrogram of the Hierarchical Cluster Analysis (HCA) is presented, which grouped the rumberry accesses into three main groups with a distance = 50.

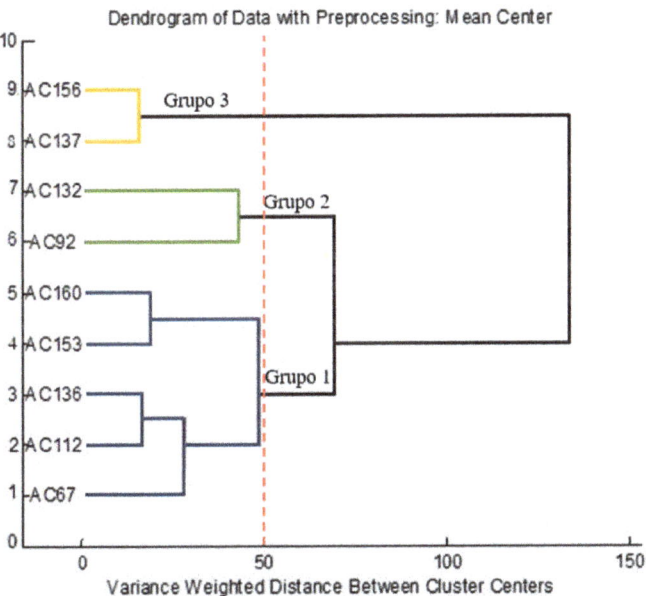

Figure 4. Dendrogram of the Hierarchical Cluster Analysis (HCA) of the different rumberry accesses, regarding the profile of volatile compounds.

The first group was formed by the purple (AC160), the red (AC153), and the three orange accesses (AC67, AC112, and AC136) due to the high percentages of the area obtained for the caryophyllene. The second group was formed by the accesses AC92 (orange fruits) and AC132 (red fruits). However, they were isolated in the graph of scores due to their different profiles. The third group was formed by the accesses AC137 and AC156 (orange fruits) grouped mainly by the compound γ-selinene, which had percentage total areas of 58.9% and 66.2%, respectively.

Assessing these groups, only two (caryophyllene and γ-selinene) of the identified VOCs stood out as the main compounds responsible for the volatile profile of rumberry fruits regardless of their color, as both compounds predominated in all accesses except AC67, which did not show the presence of γ-selinene. Despite this, no reports were found on the volatile profile of *Myrciaria floribunda* fruits, so this is believed to be the first description of their chemical composition, so the compounds mentioned above are the first record of the species.

3. Materials and Methods

3.1. Material

Fruits of nine accesses of rumberry orchards identified by the codes AC67, AC92, AC112, AC132, AC136, AC137, AC153, AC156, and AC160 were acquired from the Cambuí Active Germplasm Bank (BAG–Cambuí), located in the municipality of Rio Largo (latitude 09°28′42″ S, longitude 35°51′12″ W, altitude of 127 m), belonging to the Center of Agricultural Sciences, of the Federal University of Alagoas (CECA-UFAL). The accesses were differentiated by the color characteristics of the fruit: orange fruits (AC67, AC92, AC112, AC136, AC137, AC156, and AC157), red fruits (AC132 and AC153), and purple fruits (AC160).

Approximately 350 g of fruits were collected per access, bagged in polyethylene bags, labeled, and packed in thermally insulated boxes with ice while being transported to the Plant Biotechnology Laboratory (BIOVEG)/CECA-UFAL. The fruits were washed in running water and sanitized with 20 mL of 1% sodium hypochlorite for 5 min. Then, a second rinse was carried out under running water for 2 min to remove chlorine residues.

Subsequently, the fruits were manually pulped, discarding only the seed and, with the help of a mixer, the pulp and the peel were homogenized. The homogenate was placed in 50 mL Falcon flasks, kept in the freezer at −60 °C until they were transported to the Mass Spectrometry Laboratory of the Department of Chemistry at the Federal University of Minas Gerais (UFMG), Belo Horizonte-MG.

3.2. Methods

3.2.1. Physical and Physicochemical Characterization

In order to determine the physical characteristics, 128 fruits were separated by access. These fruits were individually evaluated using the weight of the fruit (FW) and the weight of the seed (SW), utilizing an analytical balance. Longitudinal diameter (LD) and transverse diameter (TD) were measured using a digital caliper. The pulp weight (PW) was obtained by the difference between the weight of the fruit and the seed. The number of seeds per fruit (NS) was determined by manual counting and the firmness obtained through a digital penetrometer with a 3.0 mm diameter tip, expressed in Newton (N).

For the physicochemical analysis, the samples (32 fruits) were crushed and homogenized with the aid of a mortar. The analyses were carried out in four replicates according to the "Analytical Norms of the Aldo Lutz Institute" [41]. The pH was obtained directly from a digital pH meter, duly calibrated with buffer solutions 4.0 and 7.0. Soluble solids (SS) were quantified in a digital refractometer, expressed in Brix, according to method 932.12 of the "Association of Official Analytical Chemists-AOAC" [42]. The titratable acidity (AT) was quantified by titrimetry; approximately 2.0 g of the pulp, diluted in 50 mL of distilled water with three drops of phenolphthalein, was titrated with a 0.1 N NaOH solution. The results were expressed in citric acid g/100 mL (%) [41]. The ratio (SS/AT) was calculated by the quotient between the values of soluble solids and titratable acidity.

3.2.2. Isolation of Volatile Organic Compounds (VOCs)

The solid-phase microextraction (SPME) technique was used in the headspace (HS) mode with a polyacrylate polar fiber (PA 85 µm, supelco). According to the manufacturer's instructions, the fiber was first conditioned in a gas chromatograph (Trace GC Ultra) with a temperature range of 220–300 °C, for 60 min.

Headspace bottles (20 mL), sealed with aluminum seal and rubber septum containing 0.5 g of sample were used. Subsequently, the vials were placed onto an aluminum block and heated to 85 °C over 5 min. After that, the fiber was exposed to the sample for 26 min in the headspace. Then, the fiber (PA) was removed from the sample, and manually inserted into the GC–MS injector.

3.2.3. Separation and Identification of Volatile Compounds (VOCs)

For the separation of VOCs, the gas chromatograph (Trace GC Ultra) was coupled to a mass spectrometer (Polaris Q) with an "ion-trap" analyzer (Thermo Scientific, San Jose, CA, USA.). A capillary column HP-5 MS (5% phenyl and 95% methylpolysiloxane), (30 m × 0.25 mm × 0.25 µm) was used for the separation. Helium was used as carrier gas with a constant flow of 1 mL/minutes (Agilent Techonolgies Inc, Waldbronn, Germany). The injector temperature was 250 °C in the splitless mode, time 5 min; the temperature of the ion source was 200 °C and the temperature of the interface temperature was 270 °C. For this, the following oven programming was used, starting at 40 °C for 5 min, with an increase of 5 °C/min until reaching 125 °C. Then, rising to 10 °C/min until reaching 245 °C, maintaining the isotherm for 3 min [43–45].

The chromatograms were extracted from the Xcalibur 1.4 software from Thermo Electron Corporation (San Jose, CA, USA.) [46]. The detection of the compounds was performed by electron impact ionization at 70 eV, in the full scan mode, identified according to their m/z ratio of 50 to 300, at a similarity level (RSI) greater than 500 [45,46]. The mass spectra were compared with the reference compounds from the NIST library (National Institute of Standards and Technology) and the data reported in the literature. For this, the percentages of the areas the VOCs obtained were analyzed by Excel version 2013 (Microsoft, Redmond, WA, USA) [47].

3.2.4. Statistics

For the physical and physical-chemical characterization of the samples, an experimental design was used entirely randomized (DIC), with four replicates of thirty-two fruits for each access, totaling 128 fruits per access. The data obtained were submitted to variance analysis and the means were compared by the Scott–Knott test, at 5% probability, using the Sisvar software, version 5.7.

The Principal Component Multivariate Analysis (PCA) and Hierarchical Cluster Analysis (HCA) were performed to verify the possible correlation between rumberry accesses and the identified VOCs. In this regard, the percentages of the total areas of the chromatographic peaks of the isolated VOCs were considered as variables, which were analyzed by the MatLab program version 7.9.0.529 (Mathworks, Natick, MA, USA) [48] with the aid of the PLS Toolbox version 5.2.2 (Eigenvectors Research, Manson, WA, USA) [49]. The data obtained were organized in a matrix composed of 9 columns (accesses) and 36 rows (VOCs), which were pre-processed by self-scaling so that each variable contributed with the same weight in the analysis. In the case of HCA, Euclidean distance was used as a dissimilarity coefficient.

4. Conclusions

In view of the values obtained in this work, it is concluded that the fruits of rumberry (*Myrciaria floribunda*) were characterized by being acidic, explained by the low pH levels. Among the evaluated accesses, the red fruits (AC132) stood out for presenting the highest weights and dimensions. However, the orange-colored fruits had the highest pulp yields (AC156), soluble solids (AC92), the ratio of soluble solids/titratable acidity (AC137). The purple fruits (AC160) showed greater firmness and low levels of titratable acidity.

As for the volatile profile, 36 compounds were identified, standing out from the sesquiterpenes class. The accesses AC67, AC132, AC92, AC160, and AC137 showed the highest percentages of areas corresponding to the following compounds: α-muurolene, α-longipinene, patchoulene, caryophyllene, and γ-selinene.

According to the PCA and HCA, there was chemical variability between the different accesses of *Myrciaria floribunda*, which was verified by the three groups formed. The compounds caryophyllene and γ-selinene stood out among the identified VOCs. Said compounds mainly contribute to the volatile profile of the fruits. The identification of new VOCs enables the choice and elaboration of quality products for both the industry and the consumer, thus reducing costs and bringing greater benefits, due to the demand for finding differentiated products in the market. Therefore, knowledge of the chemical composition of rumberry fruits can contribute to the expansion of chemotaxonomic studies of this genus species, as well as to the selection of new varieties.

Author Contributions: Y.M.G., E.E.P.d.L. and J.O.F.M. designed the experiments and analyzed the data. Y.M.G. and A.L.C.C.R. carried out the experiments, and M.H.d.N. contributed to the execution of the analysis. Y.M.G. wrote the manuscript and A.L.C.C.R. contributed to review and editing. R.A., R.L.B.d.A. and A.C.C.F.F.d.P. were responsible for the supervision and execution of the analysis. E.E.P.d.L. and J.O.F.M. reviewed and supervised this study. All authors have read and agreed to the published version of the manuscript.

Funding: This research was funded by Coordenação de Aperfeiçoamento de Pessoal de Nível Superior–CAPES (88887.229193/2018-00).

Institutional Review Board Statement: Not applicable.

Informed Consent Statement: Not applicable.

Data Availability Statement: All data are contained within the article.

Acknowledgments: The authors would like to thank Universidade Federal de Minas Gerais—UFMG, Universidade Federal de Alagoas—UFAL and Universidade Federal de São João Del Rei—UFSJ for the equipment loaned to carry out the analyzes and financial support. CAPES, CNPq, Pró-Reitoria de Pesquisa—PRPq-UFMG, Instituto Federal de Educação, Ciência e Tecnologia de Minas Gerais—Campus Bambuí, and FAPEAL for the financial support.

Conflicts of Interest: The authors declare that they have no conflict of interest.

Sample Availability: Samples of the *Myrciaria floribunda* are available from the authors.

References

1. Cruz Silva, A.V.; Sirqueira Nascimento, A.L.; Muniz, E.N. Fruiting and Quality Attributes of Cambui (*Myrciaria Floribunda* (West Ex Willd.) O. Berg in the Atlantic Forest of Northeast Brazil. *Revista Agro@ Mbiente Online* **2020**, *14*. [CrossRef]
2. Souza, M.C.; Morim, M.P. Subtribos Eugeniinae O. Berg e Myrtinae O. Berg (Myrtaceae) na Restinga da Marambaia, RJ, Brasil. *Acta Bot. Bras.* **2008**, *22*, 652–683. [CrossRef]
3. Santos, E.F.; Lemos, E.E.P.; Lima, S.T.; Araújo, R.R. Caracterização físico-química, compostos bioativos e atividade antioxidante total de frutos de cambuizeiro (*Myrciaria floribunda* O. Berg). *Rev. Ouricuri* **2017**, *7*, 64–79.
4. Araújo, R.R.; Santos, E.F.; Santos, E.D.; Lemos, E.E.P.; Endres, L. Quantificação de Compostos Fenólicos Em Diferentes Genótipos de Fruto de Cambuí (*Myrciaria Floribunda* O. Berg) Nativos Da Vegetação Litorânea de Alagoas. In Proceedings of the Congresso Brasileiro de Processamento Mínimo e Pós-Colheita de Frutas, Flores e Hortaliças, Anais 1, Aracaju, Brazil, 24–28 May 2015.
5. Narain, N.; Almeida, J.d.N.; Galvão, M.d.S.; Madruga, M.S.; de Brito, E.S. Compostos Voláteis Dos Frutos de Maracujá (*Passiflora Edulis* Forma Flavicarpa) e de Cajá (*Spondias Mombin* L.) Obtidos Pela Técnica de Headspace Dinâmico. *Ciênc. Tecnol. Aliment* **2004**, *24*, 212–216. [CrossRef]
6. Chitarra, M.I.F.; Chitarra, A. *Pós-Colheita de Frutos e Hortaliças; Fisiologia e Manuseio*, 2nd ed.; UFLA: Lavras, Brazil, 2005.
7. Kataoka, H.; Lord, H.L.; Pawliszyn, J. Applications of Solid-Phase Microextraction in Food Analysis. *J. Chromatogr. A* **2000**, *880*, 35–62. [CrossRef]
8. Franco, M.R.B.; Janzantti, N.S. Avanços Na Metodologia Instrumental Da Pesquisa Do Sabor. In *Aroma e Sabor de Alimentos: Temas Atuais*; Franco, M.R.B., Ed.; Editora Varela: São Paulo, Brazil, 2004; pp. 17–27.
9. Garruti, D.d.S. Composição de Volateis e Qualidade de Aroma Do Vinho de Caju. Ph.D. Thesis, Universidade Federal de Campinas, São Paulo, Brazil, 2001.
10. Tietbohl, L.A.C.; Barbosa, T.; Fernandes, C.P.; Santos, M.G.; Machado, F.P.; Santos, K.T.; Mello, C.B.; Araújo, H.P.; Gonzalez, M.S.; Feder, D.; et al. Laboratory Evaluation of the Effects of Essential Oil of *Myrciaria Floribunda* Leaves on the Development of *Dysdercus Peruvianus* and *Oncopeltus Fasciatus*. *Rev. Bras. Farmacogn.* **2014**, *24*, 316–321. [CrossRef]
11. Tietbohl, L.A.C.; Lima, B.G.; Fernandes, C.P.; Santos, M.G.; Silva, F.E.B.; Denardin, E.L.G.; Bachinski, R.; Alves, G.G.; Silva-Filho, M.V.; Rocha, L. Comparative Study and Anticholinesterasic Evaluation of Essential Oils from Leaves, Stems and Flowers of Myrciaria Floribunda (H. West Ex Willd.) O. Berg. *Lat. Am. J. Pharm.* **2012**, *31*, 637–641.
12. Apel, M.A.; Lima, M.E.L.; Souza, A.; Cordeiro, I.; Young, M.C.M.; Sobral, M.E.G.; Suffredini, I.B.; Moreno, P.R.H. Screening of the Biological Activity from Essential Oils of Native Species from the Atlantic Rain Forest (São Paulo–Brazil). *Pharmacologyonline* **2006**, *3*, 376–383.
13. Da Silva, A.V.C.; Rabbani, A.R.C.; Costa, T.S.; Clivati, D. Fruit and Seed Biometry of Cambuí (*Myciaria Tenella* O. Berg). *Rev. Agroambiente On-line* **2012**, *6*, 258. [CrossRef]
14. Rodrigues, M.A.; Guerra, D.; SEhn, T.T.S.; BOhrer, R.E.G.B.; da Silva, D.M. Caracterização Biométrica de Guabijuzeiros (*Myrcianthes Pungens* (O. Berg) d. Legrand). *Rev. Eletrôn. Científ. Uergs* **2020**, *6*, 83–91. [CrossRef]
15. de Abreu, L.A.F.; Paiva, R.; Mosqueira, J.G.A.; dos Reis, M.V.; Araújo, A.B.S.; Boas, E.V.d.B.V. Antioxidant Activity and Physico-Chemical Analysis of *Campomanesia Rufa* (O.Berg) Nied. Fruits. *Ciênc. Agrotec.* **2020**, *44*, e016720. [CrossRef]
16. Munhoz, C.L.; Ferreira, T.H.B.; Gomes, M.C.d.S. Caracterização Física de Frutos de Jambo Vermelho. *Cad. Agroecol.* **2018**, *13*, 7.
17. Becker, F.S.; Vilas Boas, A.C.; Sales, A.; Tavares, L.S.; de Siqueira, H.H.; Vilas Boas, E.V.D.B. Characterization of 'Sabará' Jabuticabas at Different Maturation Stages. *Acta Sci. Agron.* **2015**, *37*, 457. [CrossRef]
18. Souza, F.d.C.d.A.; Silva, E.P.; Aguiar, J.P.L. Vitamin Characterization and Volatile Composition of Camu-Camu (*Myrciaria Dubia* (HBK) McVaugh, Myrtaceae) at Different Maturation Stages. *Food Sci. Technol.* **2020**. [CrossRef]
19. De Almeida, E.S.; da Silva, R.N.; Gonçalves, E.M. Compostos fenólicos totais e características físico-químicas de frutos de jabuticaba. *Gaia Sci.* **2018**, *12*. [CrossRef]
20. Vieira, M.I.C.; Tavares, F.J.C.; Pinheiro, L.F.; Sampaio, V.d.S.; de Lucena, E.M.P. Alterações físico-químicas durante o crescimento dos frutos da ubaia-azeda. *Braz. J. Dev.* **2020**, *6*, 58707–58718. [CrossRef]
21. Seraglio, S.K.T.; Schulz, M.; Nehring, P.; Della Betta, F.; Valese, A.C.; Daguer, H.; Gonzaga, L.V.; Fett, R.; Costa, A.C.O. Nutritional and Bioactive Potential of Myrtaceae Fruits during Ripening. *Food Chem.* **2018**, *239*, 649–656. [CrossRef]
22. Chaves Neto, J.R.; Silva, S.D.M.; Dantas, R.L. Atributos de Qualidade, Compostos Bioativos e Atividade Antioxidante de Frutos de Uvaieira Durante a Maturação. *Agrarian* **2020**, *13*, 269–308. [CrossRef]

23. Lattuada, D.S.; Barros, N.; Hagemann, A.; de Souza, P.V.D. Caracterização Físico-Química e Desenvolvimento Pós-Colheita de Jabuticabas (*Plinia Peruviana* e *P. Cauliflora*). *Iheringia Série Bot.* **2020**, *75*, e2020015. [CrossRef]
24. de Oliveira, L.M.; Porte, A.; de Oliveira Godoy, R.L.; da Costa Souza, M.; Pacheco, S.; de Araujo Santiago, M.C.P.; Gouvêa, A.C.M.S.; da Silva de Mattos do Nascimento, L.; Borguini, R.G. Chemical Characterization of *Myrciaria Floribunda* (H. West Ex Willd) Fruit. *Food Chem.* **2018**, *248*, 247–252. [CrossRef]
25. Kauffmann, C.; Giacomin, A.C.; Arossi, K.; Pacheco, L.A.; Hoehne, L.; de Freitas, E.M.; de Carvalho Machado, G.M.; do Canto Cavalheiro, M.M.; Gnoatto, S.C.B.; Ethur, E.M. Antileishmanial in Vitro Activity of Essential Oil from *Myrciaria Plinioides*, a Native Species from Southern Brazil. *Braz. J. Pharm. Sci.* **2019**, *55*, e17584. [CrossRef]
26. da Silva Barbosa, D.C.; Holanda, V.N.; de Assis, C.R.D.; de Oliveira Farias de Aguiar, J.C.R.; DoNascimento, P.H.; da Silva, W.V.; do Amaral Ferraz Navarro, D.M.; da Silva, M.V.; de Menezes Lima, V.L.; dos Santos Correia, M.T. Chemical Composition and Acetylcholinesterase Inhibitory Potential, in Silico, of *Myrciaria Floribunda* (H. West Ex Willd.) O. Berg Fruit Peel Essential Oil. *Ind. Crops Prod.* **2020**, *151*, 112372. [CrossRef]
27. Freitas, T.P.; Taver, I.B.; Spricigo, P.C.; do Amaral, L.B.; Purgatto, E.; Jacomino, A.P. Volatile Compounds and Physicochemical Quality of Four Jabuticabas (*Plinia* Sp.). *Molecules* **2020**, *25*, 4543. [CrossRef] [PubMed]
28. Rondán, G.; Cabezas, A.; Oliveira, A.; Brousett-Minaya, M.; Narain, N. HS-SPME-GC-MS Detection of Volatile Compounds in *Myrciaria Jabuticaba* Fruit. *Sci. Agropecu.* **2018**, *9*, 319–327. [CrossRef]
29. Toledo, A.G.; de Souza, J.G.d.L.; da Silva, J.P.B.; Favreto, W.A.J.; da Costa, W.F.; Pinto, F.G.d.S. Chemical Composition, Antimicrobial and Antioxidant Activity of the Essential Oil of Leaves of *Eugenia Involucrata* DC. *Biosci. J.* **2020**, *36*. [CrossRef]
30. Sharma, C.; Al Kaabi, J.M.; Nurulain, S.M.; Goyal, S.N.; Kamal, M.A.; Ojha, S. Polypharmacological Properties and Therapeutic Potential of β-Caryophyllene: A Dietary Phytocannabinoid of Pharmaceutical Promise. *Curr. Pharm. Des.* **2016**, *22*, 3237–3264. [CrossRef]
31. Francomano, F.; Caruso, A.; Barbarossa, A.; Fazio, A.; La Torre, C.; Ceramella, J.; Mallamaci, R.; Saturnino, C.; Iacopetta, D.; Sinicropi, M.S. β-Caryophyllene: A Sesquiterpene with Countless Biological Properties. *Appl. Sci.* **2019**, *9*, 5420. [CrossRef]
32. da Silva, R.C.S.; Milet-Pinheiro, P.; Bezerra da Silva, P.C.; da Silva, A.G.; da Silva, M.V.; Navarro, D.M.d.A.F.; da Silva, N.H. (E)-Caryophyllene and α-Humulene: Aedes Aegypti Oviposition Deterrents Elucidated by Gas Chromatography-Electrophysiological Assay of *Commiphora Leptophloeos* Leaf Oil. *PLoS ONE* **2015**, *10*, e0144586. [CrossRef]
33. Carneiro, F.B.; Júnior, I.D.; Lopes, P.Q.; Macêdo, R.O. Variação Da Quantidade de β-Cariofileno Em Óleo Essencial de *Plectranthus Amboinicus* (Lour.) Spreng., Lamiaceae, Sob Diferentes Condições de Cultivo. *Rev. Bras. Farmacogn.* **2010**, *20*, 600–606. [CrossRef]
34. Souto, I.C.C.; Ferreira, J.L.S.; de Oliveira, H.M.B.F.; Alves, M.A.S.G.; de Oliveira Filho, A.A. Atividades Farmacológicas Do Monoterpeno 1,8-Cineol: Um Estudo in Silico. *Rev. Bras. Educ. Saúde* **2016**, *6*, 26–28. [CrossRef]
35. Rodenak Kladniew, B.E.; Castro, M.A.; Crespo, R.; García de Bravo, M.M. Eucaliptol (1, 8-Cineole) Inhibe La Proliferación de Celulas Tumorales Mediante Arresto Del Ciclo Celular, Estrés Oxidativo, Activación de MAPKs e Inhibición de AKT. *Terc. Época* **2017**, *7*. Available online: http://sedici.unlp.edu.ar/handle/10915/63248 (accessed on 1 November 2017).
36. Carneiro, N.S.; Alves, C.C.F.; Alves, J.M.; Egea, M.B.; Martins, C.H.G.; Silva, T.S.; Bretanha, L.C.; Balleste, M.P.; Micke, G.A.; Silveira, E.V.; et al. Chemical Composition, Antioxidant and Antibacterial Activities of Essential Oils from Leaves and Flowers of *Eugenia Klotzschiana* Berg (Myrtaceae). *An. Acad. Bras. Cienc.* **2017**, *89*, 1907–1915. [CrossRef]
37. de Cássia Da Silveira e Sá, R.; Andrade, L.N.; de Sousa, D.P. Sesquiterpenes from Essential Oils and Anti-Inflammatory Activity. *Nat. Prod. Commun.* **2015**, *10*, 1767–1774. [CrossRef]
38. Chalannavar, R.K.; Narayanaswamy, V.K.; Baijnath, H.; Odhav, B. Chemical Composition of Essential Oil of *Psidium Cattleianum* Var. Lucidum (Myrtaceae). *Afr. J. Biotechnol.* **2012**, *11*, 8341–8347. [CrossRef]
39. Noudogbessi, J.P.; Yédomonhan, P.; Sohounhloué, D.C.; Chalchat, J.C.; Figuérédo, G. Chemical Composition of Essential Oil of *Syzygium Guineense* (Willd.) DC. Var. Guineense (Myrtaceae) from Benin. *Rec. Nat. Prod.* **2008**, *2*, 33–38.
40. Li, Y.-J.; Xuan, H.-Z.; Shou, Q.-Y.; Zhan, Z.-G.; Lu, X.; Hu, F.-L. Therapeutic Effects of Propolis Essential Oil on Anxiety of Restraint-Stressed Mice. *Hum. Exp. Toxicol.* **2012**, *31*, 157–165. [CrossRef] [PubMed]
41. IAL, Instituto Adolf Lutz. Métodos Físico-Químicos Para Análises de Alimentos. In *Normas Analíticas do Instituto Adolfo Lutz*; Instituto Adolf Lutz: São Paulo, Brazil, 2008; p. 1020.
42. AOAC, Association of Official Analytical Chemists. *Official Methods of Analysis of AOAC International*, 18th ed.; Association of Official Analytical Chemists: Washington, DC, USA, 2005.
43. García, Y.; Rufini, J.; Campos, M.; Guedes, M.; Augusti, R.; Melo, J. SPME Fiber Evaluation for Volatile Organic Compounds Extraction from Acerola. *J. Braz. Chem. Soc.* **2019**, *30*, 247–255. [CrossRef]
44. García, Y.M.; de Lemos, E.E.P.; Augusti, R.; Melo, J.O.F. Optimization of Extraction and Identification of Volatile Compounds from *Myrciaria floribunda*. *Rev. Ciênc. Agron.* **2021**, *52*, e20207199. [CrossRef]
45. Garcia, Y.M.; Guedes, M.N.S.; Rufini, J.C.M.; Souza, A.G.; Augusti, R.; Melo, J.O.F. Volatile Compounds Identified in Barbados Cherry 'BRS-366 Jaburú'. *Sci. Electron. Arch.* **2016**, *9*, 67. [CrossRef]
46. *Xcalibur*; Versión 1.4; Thermo Scientific: San Jose, CA, USA, 2011.
47. *Excel 2013*; Microsoft: Redmond, WA, USA, 2013.
48. *MatLab*; Version 7.10.0.499; MathWorks: Natick, MA, USA, 2009.
49. *PLS Toolbox*; Version 5.2.2; Eigenvectors Research: Manson, WA, USA, 2009.

Article

Optimization of Extraction Conditions and Characterization of Volatile Organic Compounds of *Eugenia klotzschiana* O. Berg Fruit Pulp

Ana P. X. Mariano [1], Ana L. C. C. Ramos [1], Afonso H. de Oliveira Júnior [2], Yesenia M. García [2], Ana C. F. F. de Paula [3], Mauro R. Silva [4], Rodinei Augusti [5], Raquel L. B. de Araújo [1] and Júlio O. F. Melo [2,*]

1. Departamento de Alimentos, Faculdade de Farmácia, Campus Belo Horizonte, Universidade Federal de Minas Gerais, Belo Horizonte 31270-901, Brazil; anapaula.xavier.mariano@hotmail.com (A.P.X.M.); analuizacoeli@gmail.com (A.L.C.C.R.); raquel@bromatologiaufmg.com.br (R.L.B.d.A.)
2. Departamento de Ciências Exatas e Biológicas, Campus Sete Lagoas, Universidade Federal de São João Del-Rei, Sete Lagoas 36307-352, Brazil; afonsohoj@gmail.com (A.H.d.O.J.); jenny_thesiba@hotmail.com (Y.M.G.)
3. Departamento de Ciências Agrárias, Instituto Federal de Educação, Ciência e Tecnologia de Minas Gerais, Campus Bambuí, Bambui 38900-000, Brazil; ana.paula@ifmg.edu.br
4. Departamento de Nutrição, Pontifícia Universidade Católica de Minas Gerais, Belo Horizonte 30640-070, Brazil; mauroramalhosilva@yahoo.com.br
5. Departamento de Química, Campus Belo Horizonte, Universidade Federal de Minas Gerais, Belo Horizonte 35702-031, Brazil; augusti.rodinei@gmail.com
* Correspondence: onesiomelo@gmail.com

Abstract: *Eugenia klotzschiana* O. Berg is a native species to the Cerrado biome with significant nutritional value. However, its volatile organic compounds (VOCs) chemical profile is not reported in the scientific literature. VOCs are low molecular weight chemical compounds capable of conferring aroma to fruit, constituting quality markers, and participating in the maintenance and preservation of fruit species. This work studied and determined the best conditions for extraction and analysis of VOCs from the pulp of *Eugenia klotzschiana* O. Berg fruit and identified and characterized its aroma. Headspace solid-phase microextraction (HS-SPME) was employed using different fiber sorbents: DVB/CAR/PDMS, PDMS/DVB, and PA. Gas chromatography and mass spectrometry (GC-MS) were employed to separate, detect, and identify VOCs. Variables of time and temperature of extraction and sample weight distinctly influenced the extraction of volatiles for each fiber. PDMS/DVB was the most efficient, followed by PA and CAR/PDMS/DVB. Thirty-eight compounds that comprise the aroma were identified among sesquiterpenes (56.4%) and monoterpenes (30.8%), such as α-fenchene, guaiol, globulol, α-muurolene, γ-himachalene, α-pinene, γ-elemene, and patchoulene.

Keywords: Myrtaceae; headspace solid-phase microextraction; Cerrado; aroma; volatile organic compounds

1. Introduction

Cerrado is an ecosystem that concentrates one of the greatest biodiversities on the planet, occupying about 22% of the Brazilian territory and typified by a variety of plant species such as herbs, sub-shrubs, shrubs, trees, and vines, which together add up to more than 6000 identified species [1]. It is formed by different ecoregions due to its great latitudinal and longitudinal variation, comprising open pastures, savanna woods, forests, and perennial riparian forests [2]. Fruit trees stand out mainly due to their economic potential and the high nutrient content that they present in their compositions, in addition to the distinct flavor and aroma characteristics of their fruits [3].

Cerrado fruits such as araticum [4], grumixama [5,6], acerola [7], cagaita [8], murici [9,10], and pequi [11,12] have stood out commercially due to their unique nutritional properties, flavors, and aromas. The expansion of cultivation and advances in research on these fruits have fostered the development of new products for the food industry, providing income generation to communities of small producers and fruit collectors in the Cerrado [13–15].

Eugenia klotzschiana O. Berg belongs to the Myrtaceae family and has important nutritional value, including the content of polyphenolic compounds (566.3 mg Gallic Acid Equivalent 100 g^{-1}) and flavonoids (550.0 mg Quercetin Equivalent/100 g), in addition to a considerable content of ascorbic acid (8.66 mg 100 g^{-1} fresh weight) and dietary fiber (6.45 g 100 g^{-1}) [16,17]. The present study represents an advance in optimizing extraction conditions of volatile compounds from the fruits of *Eugenia klotzschiana* by the solid-phase microextraction (SPME) method. It is the first time that the influence of the fibers PA, DVB/CAR/PDMS, and PDMS/DVB, as well as temperature, time, and agitation employed in the extraction of VOCs present in these fruits were evaluated. In a previous study, one experimental condition used PDMS/DVB SPME fiber, allowing isolation of only 11 compounds [18].

Volatile compounds constitute the aroma of fruits, which is understood as the process of releasing low molecular weight compounds that impart odors and are captured by the human sensory system through olfaction [19]. It is noteworthy that the fruit aroma associated with the characteristics of color, texture, and size is fundamental in the perception and subsequent predilection of the consumer to a particular food. Furthermore, volatile compounds can be used to differentiate varieties of the same plant species [9], as important chemical signals in the cultivation and maintenance of the species in nature [20], and also allow the characterization of the fruit maturation stages and its quality as a whole [21].

Analysis of volatile compounds covers the process of extraction or recovery, desorption, separation, detection, and identification. Several extraction methods have been developed. Currently, headspace (HS) solid-phase microextraction (SPME) (HS-SPME) is used for the analysis of samples with complex matrices, such as food, due to the excellent results obtained and no employment of organic solvents or destruction of the analyzed sample. In addition to its repeatability and possibility to adjust the conditions and types of fibers used. Fibers available on the market vary in the type of coating and the thickness. Coatings represent the phase that establishes affinity for the analytes in the HS-SPME system, allowing their extraction, such as divinylbenzene/carboxen/polydimethylsiloxane (DVB/CAR/PDMS), polydimethylsiloxane/divinylbenzene (PDMS/DVB), and polyacrylate (PA). Desorption, separation, and detection steps can be performed using gas chromatography coupled with mass spectrometry (GC-MS) or high-performance liquid chromatography (HPLC) [9,22–24].

Therefore, the objective of this work was to study the optimal conditions for extraction, detection, and determination of volatiles from the pulp of *Eugenia klotzschiana* O. Berg fruits, as well as to define the best SPME fiber and discriminate the volatile chemical constituents employing HS-SPME and GC-MS.

2. Results and Discussion

Major components of the aroma found in the pulp of *Eugenia klotzschiana* are sesquiterpenes (55.3%) and monoterpenes (31.6%). Esters, amines, and other compounds are present in smaller amounts (13.2%) (Figure 1). It is noteworthy that the volatile chemical composition is characteristic of each fruit due to the characteristics of the food matrix and the influence of external factors, such as environmental and cultivation conditions, to which the plant species was exposed [25].

In acerola, for example, volatiles are mainly made up of esters and alcohols [7], whereas for cambuí, sesquiterpenes correspond to 71% of the identified volatiles [9]. A study with cagaita (*Eugenia dysenterica*) showed monoterpenes representing 34.6% of the volatile profile and esters, 36.3% [26]. In contrast, grumixama (*Eugenia brasiliensis*) comprised 94.7% sesquiterpenes and 5.3% monoterpenes [5].

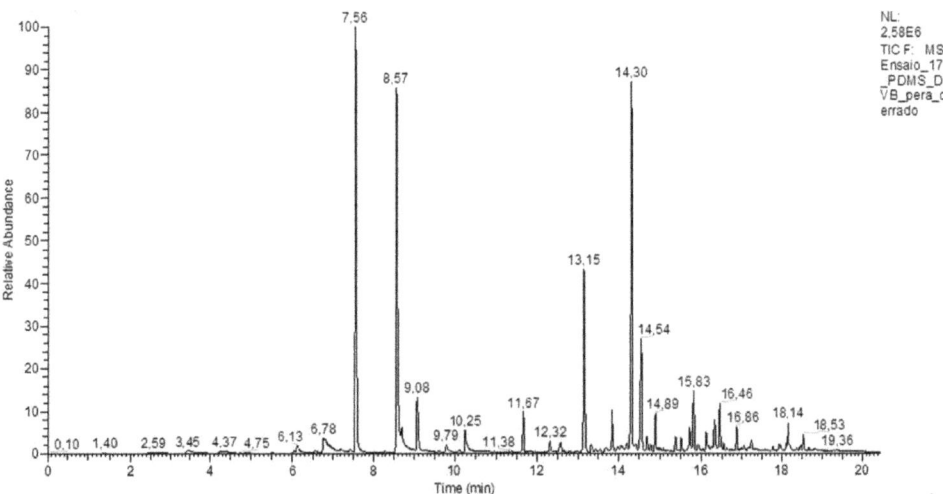

Figure 1. Peaks detected by GC-MS represent volatile compounds extracted by the PDMS/DVB fiber.

Thirty-eight volatile compounds were identified by employing three extraction fibers. Table A1 (Appendix A) shows the breakdown of volatile chemical constituents detected in the fruit pulp. The arranged columns represent identified compounds and the respective type(s) of fiber(s) that extracted them. The letter "X" marks the fiber that extracted the compound for identification purposes.

What is sensorially perceived as the aroma is the set of synergistic effects of volatile compounds present in fruits. The composition of the compounds becomes specific to each species [10,26]. However, it is possible to identify similar compounds in *E. klotzschiana* and cambuí, which have eight common volatiles in the aromatic profile: α-fenchene, guaiol, globulol, α-muurolene, γ-himachalene, α-pinene, γ-elemene, and patchoulene [27].

α-Pinene confers a characteristic odor of pine [28], also characteristic of the presence of camphene, which emits a woody and citrusy odor [29]. Myrtenol contributes to the odor of flowers and mint, while isogeraniol contributes to flowers and jasmine [30]. Linalyl acetate is responsible for the fruit and lavender notes, softening the fennel and lavanduol, the green lavender aroma [31]. α-Humulene is present in high amounts in hops (*Humulus lupulus*), which is used in beer production due to its flavoring potential. It provides a characteristic aroma to the beverage and other compounds and woody notes [29]. It is noteworthy that *E. klotzschiana* is consumed by the regional population in juices and is known as cervejinha do campo, "field beer", due to the characteristic aroma of the fruit. Finally, guaiene and benzyl acetate contribute to the perception of wood and balsamic, apple, floral, and fruity notes, respectively [32].

Figure 2 shows the percentage of the area of volatiles extracted from the chromatograms using each SPME fiber. The largest volatile areas were detected in the chromatograms of the tests performed with PDMS/DVB fiber, followed by PA fiber and with the smallest area, DVB/CAR/PDMS fiber, corresponding to 73.4%, 15.9% and 10.7% of the entire identified area, respectively.

A possible interpretation of the fibers' behavior due to the volatiles present in the pulp of *E. klotzschiana* is the affinity of the type of coating of each fiber for specific analytes since the aromatic profile of *E. klotzschiana* is mainly composed of sesquiterpenes (56.41%), which are compounds characterized by medium polarity. The PDMS/DVB coating is semipolar, making it chemically more susceptible to adsorb substances with similar polarity [10]. While esters, which have higher polarity, represent less than 13.60% of the volatiles, corroborating the values of the area extracted by the fiber with the polyacrylate coating.

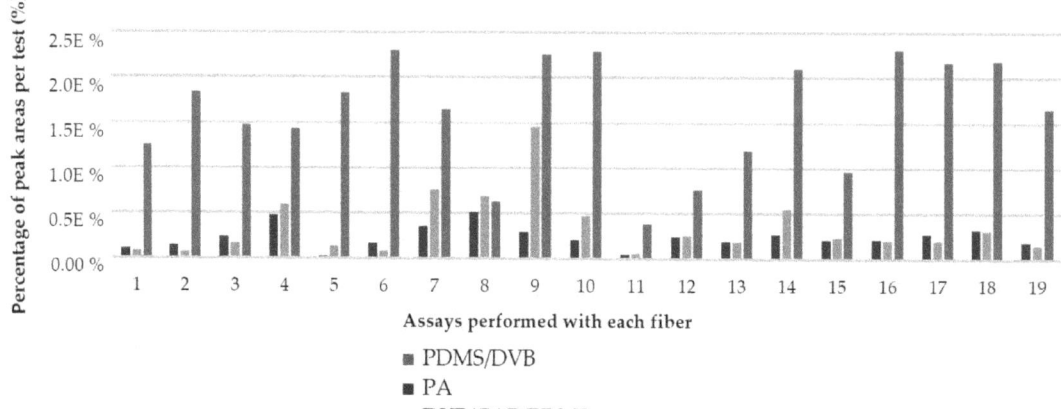

Figure 2. Percentage of peak areas of volatile chemical compounds extracted by SPME-HS using different types of coating.

The DVB/CAR/PDMS fiber had the smallest area extracted from volatiles. However, five compounds were only isolated in assays carried out with this coating: cadina-3,9-diene, (±)-α-pinene, γ-himachalene, (±)-camphene, and sabinene hydrate trans acetate.

2.1. Evaluation of the Effect of Independent Variables on the Extraction of VOCs

2.1.1. Extraction Temperature

Temperature is one of the main factors influencing the yield of volatile compounds' extraction, especially in plant matrices [7,9,26]. This condition is due to its ability to maximize or reduce the extracted volatiles due to the potential degradation of analytes that exacerbated temperature values can trigger [33]. For the levels used in this study, the temperature significantly affected the experiment performed with DVB/CAR/PDMS and PDMS/DVB fibers.

As for DVB/CAR/PDMS fibers, the temperature was the only variable that exerted a significant effect, maximizing the area of extracted volatiles (Figure 3). It was observed that the largest extraction areas were obtained at temperatures from 80 °C. Related to the region with the most significant area, the use of larger amounts of pulp is noted, indicating that, in order to obtain significant values for the volatile area, it is necessary to use high temperatures and a larger quantity of pulp. A possible explanation is the lower affinity of the compounds with the fiber coating, making it demand high values of the system variables to obtain a better response [22].

Observing the graph of the response surface of the PDMS/DVB fiber (Figure 3), which depicts the behavior of temperature in the extraction of volatiles, it appears that temperature had a positive effect on the area of extracted volatiles. Thus, the response was maximized in the presence of heat, especially from 50 to 107 °C, where the region of the response surface was identified with a larger area. However, temperatures higher than 107 °C reduced the volatile area, indicating that although it is effective and important for the HS-SPME system in the analyzed sample, exacerbating values can negatively affect volatile extraction. A possible explanation for this is the occurrence of degradation of volatiles due to the use of high temperatures [33], which in the studied system was temperatures higher than 107 °C. From 50 °C, a positive effect on the extraction yield is already observed, indicating that applying high temperatures is unnecessary.

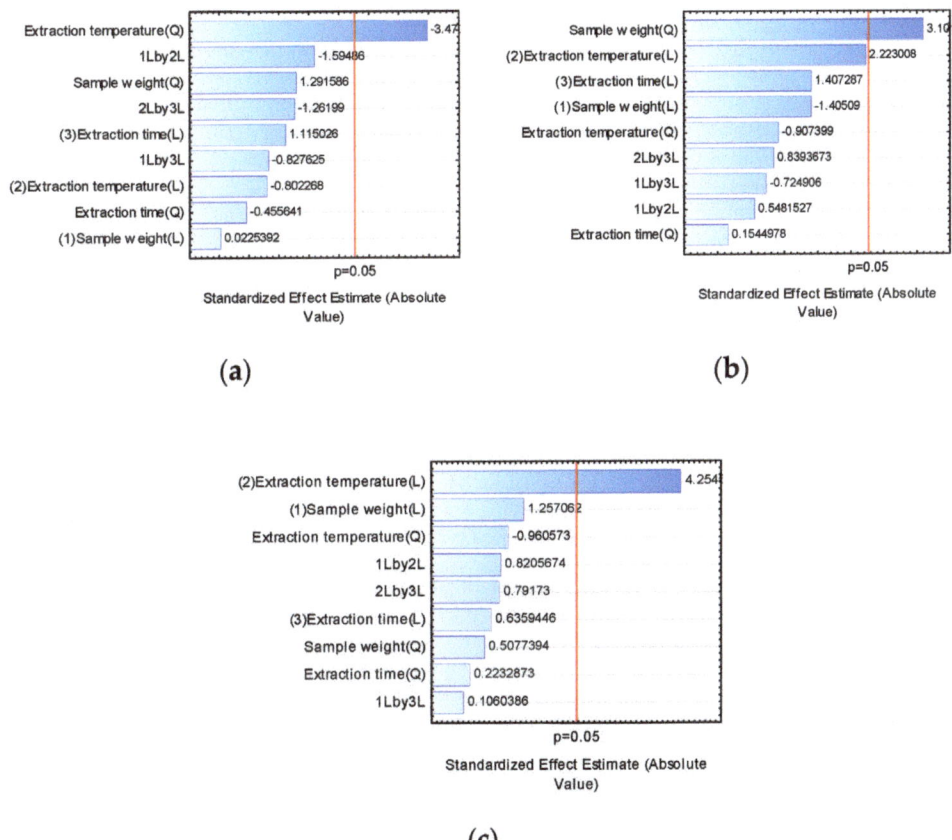

Figure 3. Effects of time and temperature of extraction and quantity of pulp variables on the extraction of volatiles using different fiber coatings for HS-SPME: (**a**) PDMS/DVB, (**b**) PA, and (**c**) CAR/PDMS/DVB.

In assays carried out with PA fibers, the temperature conditions had no significant effect on the extraction of volatiles. The lowest temperature values were enough to release analytes from the plant matrix and their adsorption by compatibility with the polyacrylate coating from the headspace. This reinforces that it is not necessary to use high-temperature values to maximize the yield of volatile extraction from *E. klotzschiana* in the studied systems.

2.1.2. Extraction Time

Extraction time is an essential parameter for the effective use of the HS-SPME technique. It must be sufficient to release the analytes from the analyzed matrix until they settle in equilibrium in the headspace vial and are adsorbed or absorbed by the fiber coating [23,33,34]. In the experiments carried out, ranging from 10 to 30 min, the time did not significantly affect the volatile extraction yield for the studied fibers.

Observing the DVB/CAR/PDMS fiber response surface graph (Figure 4), it is possible to confirm no influence of weather conditions on the extracted volatile area. The most significant areas were obtained from 24 min of fiber exposure. It was observed that there was no increase in the efficiency of extracting volatile compounds from the sample, even in tests with more extended time values at a constant sample quantity.

On the other hand, the volatile areas extracted by PDMS/DVB were greater when using time values greater than 20 min, even when associated with a smaller amount of sample, and intrinsically, a lower concentration of VOCs. In regions with sample amounts greater than 2.0 g, it is also possible to obtain higher extraction yields. However, they were not superior to those obtained with 0.5 g of *E. klotzschiana* pulp at a time longer than 20 min. The optimal extraction time calculated by the model was 19.71 min, confirming what was demonstrated by the response surface.

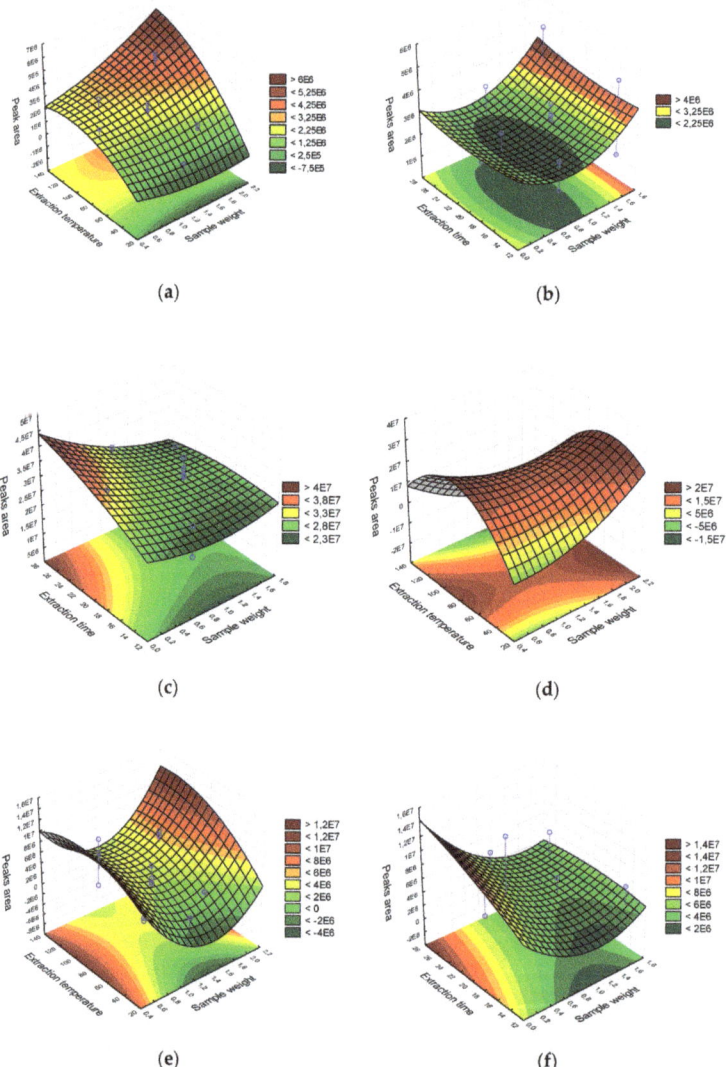

Figure 4. Three-dimensional response surface (RSM) graphs of the variables time and temperature of extraction and amount of pulp in the extraction of volatiles using different fiber coatings for HS-SPME: (**a**) DVB/CAR/PDMS extraction temperature vs. sample weight, (**b**) DVB/CAR/PDMS extraction time vs. sample weight, (**c**) PDMS/DVB extraction time vs. sample weight, (**d**) PDMS/DVB extraction temperature vs. sample weight, (**e**) PA extraction temperature vs. sample weight, and (**f**) PA extraction time vs. sample weight.

As for the PA fiber, it was observed that the highest extraction yield values were detected in a time greater than 20 min. However, similar values were not observed when larger amounts of sample were used, as occurred in the PDMS/DVB fiber. A possible explanation is that the fiber sorbent reached its maximum adsorption capacity of the analytes to which it was exposed [35].

2.1.3. Sample Weight

Sample weight in the experiments ranged from 0.5 to 2.0 g of *E. klotzschiana* pulp. The volume contained in the headspace flask must be sufficient for the isolation of volatiles and their adsorption and/or absorption by the sorbent to occur [9].

It was found that sample weight did not influence volatile extraction yield when using DVB/CAR/PDMS and PDMS/DVB fibers. However, it is noteworthy that, due to the larger sample volume, it was possible to detect an increase in yield in PDMS/DVB assays that used more than 2.0 g of pulp with extraction times shorter than 16 min. This result portrays that a greater quantity of analytes were eluted into the vapor phase (headspace) due to the more significant amount of pulp, and the fiber could adsorb this greater fraction, maximizing the extracted area.

When analyzing results with DVB/CAR/PDMS fibers, the most considerable areas were obtained only from 1.4 g of pulp, regardless of the temperature or time employed.

As for PA fibers, the variation in pulp amount positively affected the yield of extraction of volatiles, increasing the area extracted when the amount of pulp was more significant than 2.0 g. However, it is essential to emphasize that 0.5 g was enough to obtain the same volatile areas. What happens when using 2.0 g of pulp or more is the greater accumulation of analytes in the headspace due to the more significant amount of substrate, so that the content was able to be absorbed by the polyacrylate sorbent, for the volatiles present in the pulp of *E. klotzschiana* at the levels and parameters studied [34].

3. Materials and Methods

3.1. Sample Acquisition

Fruit samples of *Eugenia klotzschiana* O. Berg were obtained through a donation from trees located in the region of Turmalina, State of Minas Gerais, Brazil (Lat-17.287410, Long-42.718210). Fully ripened fruits (greenish-yellow skin color) were collected directly from trees from January to February 2019, amounting to about 3 kg.

After collection, the fruits were washed under running water to remove impurities and sanitized with a chlorinated solution (200 ppm) for 15 min, followed by rinsing in running water. The fruits were then fully packed in polyethylene bags and stored at a temperature of $-18\ °C$ until analysis. Samples were transported in a secondary package containing ice to the Mass Spectrometry Laboratory of the Department of Chemistry at the Federal University of Minas Gerais (UFMG), Belo Horizonte, Minas Gerais, Brazil.

Analyses were performed using the fruit pulp, obtained by manual pulp removal, followed by homogenization with a mixer (Mixer Mondial Versatile Black M-08).

3.2. Experimental Design

The central composite design was used with five repetitions at the central point and six axial points, with three independent variables: extraction time (minutes), extraction temperature (Celsius), and sample weight (grams), with two levels each, namely: for extraction time a minimum level of 10 min, a maximum of 30 min, and a center point of 20 min, for extraction temperature a minimum of $30\ °C$, a maximum of $120\ °C$, and a center point of $75\ °C$, and for sample weight a minimum of 0.5 g, a maximum of 2.0 g, and a center point of 1.25 g.

Three types of fibers were used, DVB/CAR/PDMS (50/30 μm), PDMS/DVB (65 μm), and PA (85 μm) (Sigma-Aldrich, St. Louis, MO, USA) [24]; thus, three experiments were performed with nineteen repetitions each.

Peaks included were those with a relative abundance greater than 2% of the total area of the chromatogram. The total relative area (%) of the peaks: the sum of the percentage abundance of the area of valid peaks obtained in the chromatograms, was determined as the dependent variable using Microsoft Office Excel® 2010. Results were analyzed according to the area extracted by each fiber and the behavior of the independent variables through the response surface methodology using Statisticv.10 (Stat-Soft Inc., Tulsa, OK, USA).

3.3. Headspace Solid-Phase Microextraction (HS-SPME)

Eugenia klotzschiana O. Berg pulp was transferred to identified glass vials. Pulp amount was weighed according to the experimental planning, and the flasks were sealed with an aluminum seal and septum rubber [7,10].

Subsequently, the vials containing the pulp were subjected to time, temperature, fiber type, and extraction conditions. For this purpose, a system was structured containing a heating plate, an aluminum block to contain the vials with the sample, and an iron support with a clamp to ensure the proper positioning of the SPME device containing the fiber when exposed. After extraction, the fiber was retracted and taken to the chromatograph (GC-MS) for desorption of the compound [9,23,34,36].

3.4. Gas Chromatography Coupled with Mass Spectrometry Analysis and Identification of Volatile Compounds

Volatile compounds were analyzed using a gas chromatograph (Trace CG Ultra) coupled to the mass spectrometer (Polaris Q) with an ion-trap analyzer (Thermo Scientific, San Jose, CA, USA). Compounds were separated using a capillary column HP−5 MS (5% phenyl and 95% methylpolysiloxane) of 30 m long, 0.25 mm internal diameter, 0.25 μm film thickness, and helium gas with a constant flow rate of 1 mL min^{-1} (Agilent Techonolgies Inc., Waldbronn, Germany). The injector (splitless mode) was maintained for 5 min at a temperature of 250 °C, the ion source at 200 °C, and the interface at 270 °C. The oven was programmed at 40 °C for 1 min, followed by an increase in temperature at a rate of 12 °C min^{-1} until it reached 120 °C, maintaining it for 2 min. Then, at 15 °C min^{-1} to 150 °C and finally, at 20 °C min^{-1} to 245 °C, maintaining for 2 min.

The identification of volatile compounds was carried out using the National Institute of Standards and Technology Research Library (NIST). This identification was also based on articles that determined volatile compounds in *Eugenia klotzschiana* O. Berg. The total peak area was obtained in Xcalibur1.4 from Thermo Electron Corporation (Thermo Electron, San Jose, CA, USA) and analyzed in Microsoft Office Excel 2010®.

4. Conclusions

The profile of volatile compounds from the pulp of *Eugenia klotzschiana* O. Berg is varied and complex, containing 38 volatiles constituted by 55.3% of sesquiterpenes and 31.6% of monoterpenes. The PDMS/DVB fiber allowed the identification of 23 volatiles, the PA fiber 17, and the DVB/CAR/PDMS fiber only 8. Variables of time, temperature, and weight of the sample behaved differently for the three fibers studied. However, the HS-SPME method proved to be effective in extracting volatile compounds present in the pulp of *Eugenia klotzschiana* O. Berg, with the best performance occurring when using the PDMS/DVB fiber, under conditions of 56 °C, 2.6 g of pulp, and 20 min of extraction. Consequently, this study presented new information about a fruit species of great importance in the Cerrado.

Author Contributions: Conceptualization, J.O.F.M., A.P.X.M., R.L.B.d.A., Y.M.G., and R.A.; methodology, R.L.B.d.A., J.O.F.M., A.P.X.M., R.A., and M.R.S.; software, Y.M.G., A.H.d.O.J., and A.L.C.C.R.; validation, A.L.C.C.R., M.R.S., and A.H.d.O.J.; formal analysis, A.P.X.M., M.R.S., and A.L.C.C.R.; investigation, A.P.X.M.; resources, R.L.B.d.A., R.A., A.C.C.F.F.d.P., and J.O.F.M.; data curation, A.P.X.M. and R.L.B.d.A.; writing—original draft preparation, A.P.X.M.; writing—review and editing, A.P.X.M., A.H.d.O.J., and A.L.C.C.R.; visualization, R.L.B.d.A., and A.C.C.F.F.d.P.; supervision, R.L.B.d.A. and J.O.F.M.; project administration, J.O.F.M. and A.P.X.M.; funding acquisition, R.A., J.O.F.M., A.P.X.M.,

R.L.B.d.A., and A.C.C.F.F.d.P. All authors have read and agreed to the published version of the manuscript.

Funding: This research was funded by CAPES, CNPq (304051/2018), FAPEMIG, IFMG, UFMG, and UFSJ, to whom the authors would like to extend their gratitude. This research is developed within the framework of the Sustainable Rural Project - Cerrado, funded by Technical Cooperation approved by the Inter-American Development Bank (IDB), with resources from the UK Government's International Climate Finance, with the Ministry of Agriculture, Livestock, and Supply (MAPA) as the institutional beneficiary. The Brazilian Institute for Development and Sustainability (IABS—08026000510/2003-51) is responsible for the project's execution and administration, and the ILPF Network Association, through Embrapa, is responsible for the scientific coordination and technical support.

Institutional Review Board Statement: Not applicable.

Informed Consent Statement: Not applicable.

Data Availability Statement: All data are contained within the article.

Acknowledgments: The authors would like to thank the Universidade Federal de Minas Gerais (UFMG), Instituto Federal de Educação Ciência e Tecnologia de Minas Gerais (IFMG), and Universidade Federal de São João del-Rei (UFSJ) for the infrastructure to carry out the analyses and for the financial support. The authors also thank CAPES, CNPq, Pró-Reitoria de Pesquisa - PRPq - FMG, and FAPEMIG for the financial support.

Conflicts of Interest: The authors declare no conflict of interest.

Sample Availability: Samples of *Eugenia klotzschiana* O. Berg are available from the authors.

Appendix A

Table A1. Volatile organic compounds identified by three different types of SPME fibers and GC-MS in the pulp of *Eugenia klotzschiana* O. Berg.

N.	Name	Formula	CAS	Reference	PA	PDMS/DVB	CAR/PDMS/DVB
Monoterpenes							
01	(±)-α-pinene	$C_{10}H_{16}$	80-56-8	[10]			X
02	p-mentha-1,8-dien-7-ol	$C_{10}H_{14}O$	536-59-4	[37]		X	
03	(2S, 5S)-2-methyl-5-propan-2-ylbicyclo [3.1.0] hexan-2-ol	$C_{10}H_{18}O$	-			X	
04	myrtenol	$C_{10}H_{16}O$	515-00-4	[30]		X	
05	p-mentha-3,8-diene	$C_{10}H_{16}$	586-67-4	[38]		X	
06	(±)-camphene	$C_{10}H_{16}$	79-92-5	[29]			X
07	α-fenchene	$C_{10}H_{16}$	471-84-1	[10]	X	X	
08	linalyl acetate	$C_{12}H_{20}O_2$	115-95-7	[31]			X
09	ocimene	$C_{10}H_{16}$	29714-87-2	[31]	X	X	
10	(−)-lavandulol	$C_{10}H_{18}O$	498-16-8	[31]	X		
11	sabinene hydrate trans acetate	$C_{12}H_{20}O_2$	-	-			X
12	isogeraniol	$C_{10}H_{18}O$	5944-20-7	[30]	X		
Sesquiterpenes							
13	δ-elemene	$C_{15}H_{24}$	20307-84-0	[10]	X	X	
14	isoledene	$C_{15}H_{24}$	95910-36-4	[21]		X	
15	β-guaiene	$C_{15}H_{24}$	88-84-6	[39]		X	
16	α-humulene	$C_{15}H_{24}$	-	[29]	X	X	
17	(+)-calarene	$C_{15}H_{24}$	17334-55-3	[40]		X	X
18	cadinene	$C_{15}H_{24}$	29350-73-0	[21]		X	
19	viridiflorene	$C_{15}H_{24}$	21747-46-6	[10]		X	

Table A1. Cont.

N.	Name	Formula	CAS	Reference	PA	PDMS/DVB	CAR/PDMS/DVB
20	α-guaiene	$C_{15}H_{24}$	3691-12-1	[10]	X	X	
21	γ-himachalene	$C_{15}H_{24}$	53111-25-4	[10]			X
22	(−)-γ-muurolene	$C_{15}H_{24}$	24268-39-1	[10]	X	X	
23	acoradiene	$C_{15}H_{24}$	24048-44-0	[41]	X	X	
24	(+)-cyclosativene	$C_{15}H_{24}$	22469-52-9	[21]	X	X	
25	(−)-caryophyllene	$C_{15}H_{24}$	87-44-5	[21]			
26	cadina-3,9-diene	$C_{15}H_{24}$	523-47-7	[42]			X
27	globulol	$C_{15}H_{26}O$	489-41-8	[10]	X	X	
28	patchoulene	$C_{15}H_{24}$	-	[10]	X		
29	α-caryophyllene alcohol	$C_{15}H_{26}O$	4586-22-5	[43]	X		
30	(−)-guaiol	$C_{15}H_{26}O$	489-86-1	[10]	X		
31	cedryl acetate	$C_{17}H_{28}O_2$	77-54-3	[44]	X		
32	2-naphthalenol,2,3,4,4alpha,5,6,7-octahydro-1,4alpha-dimethyl-7-(2-hydroxi-1-methylethyl)	$C_{15}H_{26}O_2$	-	-	X		
33	spiro [4.5]decan-7-one, 1,8-dimethyl-8,9-epoxi-4-isopropyl	$C_{15}H_{24}O_2$	61050-91-7	[20]		X	
Other							
34	N-(2-hydroxipropyl) ethylenediamine	$C_{15}H_{14}N_2O$	123-84-2	-		X	
35	ethyl hexanoate	$C_8H_{16}O_2$	123-66-0	[45]		X	
36	benzyl alcohol	C_7H_8O	100-51-6	[45]	X		
37	benzyl acetate	$C_9H_{10}O_2$	140-11-4	[46]	X	X	X
38	(E)-3,7-dimetlnon-6-enal	$C_{11}H_{20}O$	-	-		X	
	Total compounds identified by each fiber				18	23	08

CAS (American Chemical Society), divinylbenzene/carboxen/polydimethylsiloxane (DVB/CAR/PDMS), divinylbenzene/polydimethylsiloxane (PDMS/DVB), polyacrylate (PA).

References

1. Junqueira, N.T.V.; Junqueira, K.P.; Pereira, A.; Pereira, E.; Braga, M.; Conceição, L.D.; Faleiro, F. Frutíferas nativas do cerrado: O extrativismo e a busca da domesticação. In Proceedings of the Embrapa Cerrados-Artigo em Anais de Congresso (ALICE), Congresso Brasileiro de Fruticultura, Bento Gonçalves, Brazil, 22–26 October 2012.
2. Sano, E.E.; Bettiol, G.M.; Martins, E.D.S.; Couto Júnior, A.F.; Vasconcelos, V.; Bolfe, E.L.; Victoria, D.D.C. Características gerais da paisagem do Cerrado. In *Embrapa Informática Agropecuária-Capítulo em livro científico (ALICE)*; Embrapa: Brasília, Brazil, 2020.
3. Vieira, R.F.; Agostini-Costa, T.; Silva, D.B.; Ferreira, F.R.; Sano, S.M. *Frutas Nativas da Região Centro-Oeste do Brasil*; Embrapa Recursos Genéticos e Biotecnologia: Brasília, Brasil, 2006.
4. Arruda, H.S.; Pereira, G.A.; de Morais, D.R.; Eberlin, M.N.; Pastore, G.M. Determination of free, esterified, glycosylated and insoluble-bound phenolics composition in the edible part of araticum fruit (*Annona crassiflora* Mart.) and its by-products by HPLC-ESI-MS/MS. *Food Chem.* 2018, 245, 738–749. [CrossRef] [PubMed]
5. Ramos, A.L.C.C.; Mendes, D.D.; Silva, M.R.; Augusti, R.; Melo, J.O.F.; de Araújo, R.L.B.; Lacerda, I.C.A. Chemical profile of *Eugenia brasiliensis* (Grumixama) pulp by PS/MS paper spray and SPME-GC/MS solid-phase microextraction. *Res. Soc. Dev.* 2020, 9, e318974008. [CrossRef]
6. Teixeira, L.D.L.; Bertoldi, F.C.; Lajolo, F.M.; Hassimotto, N.M.A. Identification of ellagitannins and flavonoids from *Eugenia brasilienses* Lam. (Grumixama) by HPLC-ESI-MS/MS. *J. Agric. Food Chem.* 2015, 63, 5417–5427. [CrossRef]
7. García, Y.M.; Rufini, J.; Campos, M.P.; Guedes, M.N.; Augusti, R.; Melo, J.O. SPME fiber evaluation for volatile organic compounds extraction from acerola. *J. Braz. Chem. Soc.* 2019, 30, 247–255. [CrossRef]
8. Silva, M.R.; De Souza, A.G.; De Araújo, R.L.B.; Lacerda, I.C.A.; Augusti, R.; Melo, J.O.F.; Mendonça, H.D.O.P. Análise metabolômica de cagaitas utilizando a espectrometria de massas com ionização por paper spray. *Avanços Ciência E Tecnol. Alimentos. Científica* 2020, 1, 25–41. [CrossRef]
9. García, Y.M.; Ramos, A.L.C.C.; de Paula, A.C.C.F.F.; do Nascimento, M.H.; Augusti, R.; de Araújo, R.L.B.; Melo, J.O.F. Chemical Physical Characterization and Profile of Fruit Volatile Compounds from Different Accesses of Myrciaria floribunda (H. West Ex Wild.) O. Berg through Polyacrylate Fiber. *Molecules* 2021, 26, 5281. [CrossRef]
10. García, Y.M.; Ramos, A.L.C.C.; de Oliveira Júnior, A.H.; de Paula, A.C.C.F.F.; de Melo, A.C.; Andrino, M.A.; Melo, J.O.F. Physicochemical Characterization and Paper Spray Mass Spectrometry Analysis of *Myrciaria Floribunda* (H. West ex Willd.) O. Berg Accessions. *Molecules* 2021, 26, 7206. [CrossRef]
11. Bess, J.; Araújo, S.S.; Grellmann, B.P.; Cruz, D.C.; Alamino, M.; Melo, J.O.F.; Reina, L.D.C.B. Biometric and physical-chemical analysis of pequis collected in the state of Mato Grosso. *Sci. Electron. Arch.* 2020, 13, 24–30. [CrossRef]
12. Santos, B.O.; Augusti, R.; Melo, J.O.F.; Takahashi, J.A.; de Araújo, R.L.B. Optimization of extraction conditions of volatile compounds from pequi peel (*Caryocar brasiliense* Camb.) using HS-SPME. *Res. Soc. Dev.* 2020, 9, e919974893. [CrossRef]
13. Ramos, S.A.; Silva, M.R.; Jacobino, A.R.; Damasceno, I.A.N.; Rodrigues, S.M.; Carlos, G.A.; Capobiango, M. Caracterização físico-química, microbiológica e da atividade antioxidante de farinhas de casca e amêndoa de manga (*Mangifera indica*) e sua aplicação em brownie. *Res. Soc. Dev.* 2021, 10, e22210212436. [CrossRef]
14. Figueiredo, Y.G.; Bueno, F.C.; Júnior, A.H.D.O.; Mazzinghy, A.C.D.C.; Mendonça, H.D.O.P.; de Oliveira, A.F.; De Melo, A.C.; Reina, L.D.C.B.; Augusti, R.; Melo, J.O.F. Profile of the volatile organic compounds of pink pepper and black pepper. *Sci. Electron. Arch.* 2021, 14. [CrossRef]
15. Silva, M.R.; Freitas, L.G.; Mendonça, H.D.O.; Souza, A.G.; Pereira, H.V.; Augusti, R.; Melo, J.O.F.; Araújo, R.L. determination of chemical profile of *Eugenia dysenterica* ice cream using PS-MS and HS-SPME/GC-MS. *Química Nova* 2021, 44, 129–136. [CrossRef]
16. Carneiro, N.S. Caracterização Química e Zvidade Antioxidante da Polpa e Óleo Essencial da Pera do Cerrado (*Eugenia klotzschiana* Berg.). 2016. Available online: https://repositorio.ifgoiano.edu.br/handle/prefix/42 (accessed on 20 October 2020).
17. Carneiro, N.S.; Alves, J.M.; Alves, C.C.F.; Esperandim, V.R.; Miranda, M.L.D. Essential oil of flowers from *Eugenia klotzschiana* (myrtaceae): Chemical composition and in vitro trypanocidal and cytotoxic activities. *Rev. Virtual Quim.* 2017, 9, 1381–1392. [CrossRef]
18. Mariano, A.P.X.; Ramos, A.L.C.C.; Augusti, R.; Araújo, R.L.B.; Melo, J.O.F. Analysis of the chemical profile of cerrado pear fixed compounds by mass spectrometry with paper spray and volatile ionization by SPME-HS CG-MS. *Res. Soc. Dev.* 2020, 9, e949998219. [CrossRef]
19. da Rocha, R.F.J.; da Silva Araújo, Í.M.; de Freitas, S.M.; dos Santos Garruti, D. Optimization of headspace solid phase microextraction of volatile compounds from papaya fruit assisted by GC–olfactometry. *J. Food Sci. Technol.* 2017, 54, 4042–4050. [CrossRef] [PubMed]
20. Sánchez-Hernández, E.; Buzón-Durán, L.; Andrés-Juan, C.; Lorenzo-Vidal, B.; Martín-Gil, J.; Martín-Ramos, P. Physicochemical characterization of *Crithmum maritimum* L. and *Daucus carota* subsp. gummifer (Syme) Hook. fil. and their antimicrobial activity against apple tree and grapevine phytopathogens. *Agronomy* 2021, 11, 886. [CrossRef]
21. Wei, C.; Ma, Z.; Liu, Y.; Qiao, J.; Sun, G. Effect of boron on fruit quality in pineapple. In *AIP Conference Proceedings*; AIP Publishing LLC: Melville, NY, USA, 2018.
22. Biajoli, A.F.P. Fibras para SPME (microextração em fase solida) Recobertas com Novos Ormosils Sol-gel. (Doctoral Dissertation, Universidade Estadual de Campinas (UNICAMP). Instituto de Química). 2008. Available online: http://repositorio.unicamp.br/jspui/handle/REPOSIP/250195 (accessed on 20 July 2021).

23. Pawliszyn, J. Theory of solid-phase microextraction. In *Handbook of Solid Phase Microextraction*; Elsevier: Amsterdam, The Netherlands, 2012; pp. 13–59. [CrossRef]
24. Sigma, S. *SPME for GC Analysis: Getting Started with Solid Phase Microextraction*; Merck: Darmstadt, Germany, 2018.
25. Canuto, K.; Garruti DD, S.; Magalhaes, H. Microextração em Fase Sólida: Métodos Analíticos Práticos Para Extração de Compostos Voláteis de Frutas. Embrapa Agroindústria Tropical-Comunicado Técnico (INFOTECA-E). 2011. Available online: https://ainfo.cnptia.embrapa.br/digital/bitstream/item/42119/1/COT11003.pdf (accessed on 20 October 2020).
26. Silva, M.R.; Bueno, G.H.; Araújo, R.L.; Lacerda, I.C.; Freitas, L.G.; Morais, H.A.; Augusti, R.; Melo, J.O. Evaluation of the influence of extraction conditions on the isolation and identification of volatile compounds from cagaita (*Eugenia dysenterica*) using HS-SPME/GC-MS. *J. Braz. Chem. Soc.* **2019**, *30*, 379–387. [CrossRef]
27. García, Y.M.; Lemos, E.E.P.D.; Augusti, R.; Melo, J.O.F. Optimization of extraction and identification of volatile compounds from *Myrciaria Floribunda*. *Rev. Ciência Agronômica* **2021**, *52*, 1–8. [CrossRef]
28. Hui, Y.H.; Chen, F.; Nollet, L.M.; Guiné, R.P.; Martín-Belloso, O.; Mínguez-Mosquera, M.I.; Stanfield, P. (Eds.) *Handbook of Fruit and Vegetable Flavors*; John Wiley and Sons: Hoboken, NJ, USA, 2010.
29. Gasiński, A.; Kawa-Rygielska, J.; Szumny, A.; Czubaszek, A.; Gąsior, J.; Pietrzak, W. Volatile compounds content, physicochemical parameters, and antioxidant activity of beers with addition of mango fruit (*Mangifera Indica*). *Molecules* **2020**, *25*, 3033. [CrossRef]
30. Wu, Y.; Zhang, W.; Yu, W.; Zhao, L.; Song, S.; Xu, W.; Zhang, C.; Ma, C.; Wang, L.; Wang, S. Study on the volatile composition of table grapes of three aroma types. *LWT* **2019**, *115*, 108450. [CrossRef]
31. Xiao, Z.; Li, Q.; Niu, Y.; Zhou, X.; Liu, J.; Xu, Y.; Xu, Z. Odor-active compounds of different lavender essential oils and their correlation with sensory attributes. *Ind. Crops Prod.* **2017**, *108*, 748–755. [CrossRef]
32. Freitas, T.P.; Taver, I.B.; Spricigo, P.C.; do Amaral, L.B.; Purgatto, E.; Jacomino, A.P. Volatile Compounds and Physicochemical Quality of Four Jabuticabas (*Plinia* sp.). *Molecules* **2020**, *25*, 4543. [CrossRef] [PubMed]
33. Yang, L.; Zhu, Y.; He, Z.; Zhang, T.; Xiao, Z.; Xu, R.; He, J. Plantanone D, a new rare methyl-flavonoid from the flowers of *Hosta plantaginea* with anti-inflammatory and antioxidant activities. *Nat. Prod. Res.* **2021**, *35*, 4331–4337. [CrossRef] [PubMed]
34. Pawliszyn, J. *Solid Phase Microextraction: Theory and Practice*; John Wiley & Sons: Hoboken, NJ, USA, 1997.
35. Krokou, A.; Kokkinofta, R.; Stylianou, M.; Agapiou, A. Decoding carob flavor aroma using HS–SPME–GC–MS and chemometrics. *Eur. Food Res. Technol.* **2020**, *246*, 1419–1428. [CrossRef]
36. Kataoka, H.; Lord, H.L.; Pawliszyn, J. Applications of solid-phase microextraction in food analysis. *J. Chromatogr. A* **2000**, *880*, 35–62. [CrossRef]
37. Garcia, C.V.; Stevenson, R.J.; Atkinson, R.G.; Winz, R.A.; Quek, S.Y. Changes in the bound aroma profiles of 'Hayward' and 'Hort16A' kiwifruit (*Actinidia* spp.) during ripening and GC-olfactometry analysis. *Food Chem.* **2013**, *137*, 45–54. [CrossRef]
38. Do Nascimento, A.F. Atividade do Óleo Essencial de Frutos de *Schinus Terebinthifolius* Raddi (Anacardiaceae) em *Tetranychus urticae* Koch (Acari: Tetranychidae) e *Rhyzopherta dominica* Fabricius (Coleoptera: Bostrichidae). 2012. Available online: http://www.tede2.ufrpe.br:8080/tede2/handle/tede2/5951 (accessed on 20 January 2021).
39. Rondán Sanabria, G.G.; Cabezas Garcia, A.J.; Oliveira Lima, A.W.; Brousett-Minaya, M.A.; Narain, N. HS-SPME-GC-MS detection of volatile compounds in *Myrciaria jabuticaba* Fruit. *Sci. Agropecu.* **2018**, *9*, 319–327. [CrossRef]
40. Núñez-Carmona, E.; Abbatangelo, M.; Zottele, I.; Piccoli, P.; Tamanini, A.; Comini, E.; Sberveglieri, G.; Sberveglieri, V. Nanomaterial gas sensors for online monitoring system of fruit jams. *Foods* **2019**, *8*, 632. [CrossRef]
41. Asadi, K.; Abbasi-Maleki, S.; Hashjin, G.S. Antidepressant-like effect of *Cuminum cyminum* essential oil on the forced swim and tail suspension tests in male mice. *J. Shahrekord Univ. Med. Sci.* **2020**, *22*, 167–172. [CrossRef]
42. Boachon, B.; Burdloff, Y.; Ruan, J.-X.; Rojo, R.; Junker, R.R.; Vincent, B.; Nicolè, F.; Bringel, F.; Lesot, A.; Henry, L.; et al. A promiscuous CYP706A3 reduces terpene volatile emission from *Arabidopsis* flowers, affecting florivores and the floral microbiome. *Plant Cell* **2019**, *31*, 2947–2972. [CrossRef]
43. Thongkham, E.; Aiemsaard, J.; Kaenjampa, P. Antioxidant and Antimicrobial Properties of Ethanolic Extract of Asam Gelugor Fruit (*Garcinia atroviridis*). *Burapha Sci. J.* **2021**, *26*, 1293–1307.
44. Mei, L.; Shi, L.; Song, X.; Liu, S.; Cheng, Q.; Zhu, K.; Zhuge, R. Characterization of Carboxymethyl Cellulose Films Incorporated with Chinese Fir Essential Oil and Their Application to Quality Improvement of Shine Muscat Grape. *Coatings* **2021**, *11*, 97. [CrossRef]
45. Fang, H.; Chen, J.; Tian, Y.; Liu, Y.; Li, H.; Cheng, H. Chemometrics characterization of volatile changes in processed bayberry juice versus intact fruit during storage by headspace solid-phase micro-extraction combined with GC–MS. *J. Food Processing Preserv.* **2020**, *44*, e14444. [CrossRef]
46. Cano-Lamadrid, M.; Galindo, A.; Collado-González, J.; Rodríguez, P.; Cruz, Z.N.; Legua, P.; Burló, F.; Morales, D.; Carbonell-Barrachina, Á.A.; Hernández, F. Influence of deficit irrigation and crop load on the yield and fruit quality in Wonderful and Mollar de Elche pomegranates. *J. Sci. Food Agric.* **2018**, *98*, 3098–3108. [CrossRef]

Article

Profile of *Myracrodruon urundeuva* Volatile Compounds Ease of Extraction and Biodegradability and In Silico Evaluation of Their Interactions with COX-1 and iNOS

Yuri G. Figueiredo [1], Eduardo A. Corrêa [2,3], Afonso H. de Oliveira Junior [1], Ana C. d. C. Mazzinghy [1], Henrique d. O. P. Mendonça [1], Yan J. G. Lobo [2], Yesenia M. García [1], Marcelo A. d. S. Gouvêia [4], Ana C. C. F. F. de Paula [4], Rodinei Augusti [5], Luisa D. C. B. Reina [6], Carlos H. da Silveira [7], Leonardo H. F. de Lima [1] and Júlio O. F. Melo [1,*]

[1] Departamento de Ciências Exatas e Biológicas, Campus Sete Lagoas, Universidade Federal de São João Del-Rei, Sete Lagoas 35700-000, MG, Brazil; yuri.gfigueiredo@hotmail.com (Y.G.F.); afonsohoj@gmail.com (A.H.d.O.J.); anamazzinghy@yahoo.com.br (A.C.d.C.M.); hp.quimico@hotmail.com (H.d.O.P.M.); jenny_thesiba@hotmail.com (Y.M.G.); leofrancalima@ufsj.edu.br (L.H.F.d.L.)

[2] Campus Dona Lindu, Universidade Federal de São João Del-Rei, Divinópolis 35501-296, MG, Brazil; eduardo@epamig.br (E.A.C.); yanjeronimo88@gmail.com (Y.J.G.L.)

[3] Empresa de Pesquisa Agropecuária de Minas Gerais, Unidade EPAMIG ITAC, Pitangui 35650-000, MG, Brazil

[4] Departamento de Ciências Agrárias, Instituto Federal de Educação, Ciência e Tecnologia de Minas Gerais, Campus Bambuí, Bambuí 38900-000, MG, Brazil; marcelogouveiatc@gmail.com (M.A.d.S.G.); ana.paula@ifmg.edu.br (A.C.C.F.F.d.P.)

[5] Departamento de Química, Campus Pampulha, Universidade Federal de Minas Gerais, Belo Horizonte 35702-031, MG, Brazil; augusti.rodinei@gmail.com

[6] Instituto de Ciências Naturais, Humanas e Sociais, Universidade Federal de Minas Gerais, Belo Horizonte 35702-031, MG, Brazil; luisabarrett@gmail.com

[7] Instituto de Ciências Tecnológicas, Campus Itabira, Universidade Federal de Itajubá, Itabira 35903-087, MG, Brazil; carlos.silveira@gmail.com

* Correspondence: onesiomelo@gmail.com

Abstract: *Myracrodruon urundeuva* Fr. Allem. (Anacardiaceae) is a tree popularly known as the "aroeira-do-sertão", native to the caatinga and cerrado biomes, with a natural dispersion ranging from the Northeast, Midwest, to Southeast Brazil. Its wood is highly valued and overexploited, due to its characteristics such as durability and resistance to decaying. The diversity of chemical constituents in aroeira seed has shown biological properties against microorganisms and helminths. As such, this work aimed to identify the profile of volatile compounds present in aroeira seeds. Headspace solid phase microextraction was employed (HS-SPME) using semi-polar polydimethylsiloxane-divinylbenzene fiber (PDMS/DVB) for the extraction of VOCs. 22 volatile organic compounds were identified: nine monoterpenes and eight sesquiterpenes, in addition to six compounds belonging to different chemical classes such as fatty acids, terpenoids, salicylates and others. Those that stood out were p-mentha-1,4, 4(8)-diene, 3-carene (found in all samples), caryophyllene and cis-geranylacetone. A virtual docking analysis suggested that around 65% of the VOCs molar content from the aroeiras seeds present moderate a strong ability to bind to cyclooxygenase I (COX-I) active site, oxide nitric synthase (iNOS) active site (iNOSas) or to iNOS cofactor site (iNOScs), corroborating an anti-inflamatory potential. A pharmacophoric descriptor analysis allowed to infer the more determinant characteristics of these compounds' conferring affinity to each site. Taken together, our results illustrate the high applicability for the integrated use of SPME, in silico virtual screening and chemoinformatics tools at the profiling of the biotechnological and pharmaceutical potential of natural sources.

Keywords: cerrado; Anacardiaceae; aroeira; biological properties; virtual screening; chemoinformatics

1. Introduction

Aroeira (*Myracrodruon urundeuva*), commonly known in Brazil as "aroeira do sertão" is a dioecious species. Its fruits have an oval shape, and a firm calyx, considered fruit-seed [1,2]. This species comes from cerrado and semiarid regions [3] and can be found in Brazil from the Northeast to the South of the country, presenting significant economic value [4–6].

Aroeira has been widely studied due to its therapeutic properties, with emphasis on anti-inflammatory, antioxidant, antimicrobial and healing activities. Its peel, leaves and fruits are used [7]. A set of recent studies have also pointed to the medicinal properties of the whole plant, particularly the extracts and essential oils from the seeds [8–10]. The complex mixture of aroeira seed organic compounds has also shown anthelmintic and antimicrobial activity [11].

Certain bioactive compounds present in the aroeira species, such as flavonoids, terpenes and tannins, have already been probed to exert anti-inflammatory effects with great proficiency, for instance in selectively inhibiting the phospholipase A2, an enzyme that has pro-inflammatory activity [12–14]. In the literature, there are several reports that the use of aroeira's hydroalcoholic extract improves healing of several variations of inflammatory conditions and therefore, provides better healing aspects [15].

In particular, essential oils (EOs) are mainly made up of terpenes, terpenoids and their derivatives; for which evolution has selected a myriad of chemical groups and cross interactions with an equal number of biological targets in plants and animals; presenting these targets' varied functions [16]. Allied to this, the EOs molecular constituents are usually small and lipophilic plant metabolites, with a natural ability to overcome biological barriers, in order to presenting ease of extraction and biodegradability [17,18]. This set of attributes means these plant extracts have been used in natural medicine and in various technological and biotechnological applications since ancient times [19–24]. Of the different biological activities reported for EOs in humans, anti-inflammatory properties appear with a prominent role, presenting potential for a less invasive aid to more orthodox drugs against classical and current illnesses (such as COVID-19) [25–29]. For the above reasons, the development of an inexpensive, environmentally safe and sufficiently accurate method to profile the molecular content and its relative variances of the EOs from *M. urundeuva* seeds and a first approach to obtain insights about such activity are both valuable goals.

Solid-phase microextraction (SPME) is a technique that has been substantially successful in profiling volatile compounds with a low economic and environmental burden [30,31]. SPME consists of the analyte partitioning between the sample and an extracting microcomponent, constituting a polymeric phase that can be solid or liquid that involves a fused silica fiber [32]. Relatively to the advantages of this technique, it is noteworthy that it does not demand sophisticated analytical tools, in addition to organic solvents dispensing and enabling the reuse of the extraction fibers [33–36]. This technique also presents itself widely dependent on the type of fiber. It was shown in the work of [36] that the PDMS/DVB fiber allowed the identification of more than twice the volatile compounds than the DVB/CAR/PDMS fiber for the pulp of *Eugenia klotzchiana* O.

Regarding the research of anti-inflammatory targets for natural compounds, cyclooxygenases (COX) and inducible nitric oxide synthase (iNOS) are two canonical targets [37,38]. The cyclooxygenase enzyme has two isoforms called COX-1 and COX-2, reported in the literature [39–41]. Both cyclooxygenases have been already proved as targets of Eos and other natural compounds with anti-inflammatory properties [42–45]. In fact, the COX natural substrate, arachidonic acid, is a 20-carbon polyunsaturated fatty acid (Figure 1A,B). It is hence expected that the natural adaptation of the enzyme active site to this huge and hydrophobic substrate can also promote a substantial fit to a set of equally bulk and hydrophobic terpene derivatives (overall sesquiterpenes and sesquiterpenoids) present in EOs. Constantly expressed, Cyclooxygenase 1 (COX-1) is constitutive in most tissues and is extremely important to maintain their normal physiological state [46].

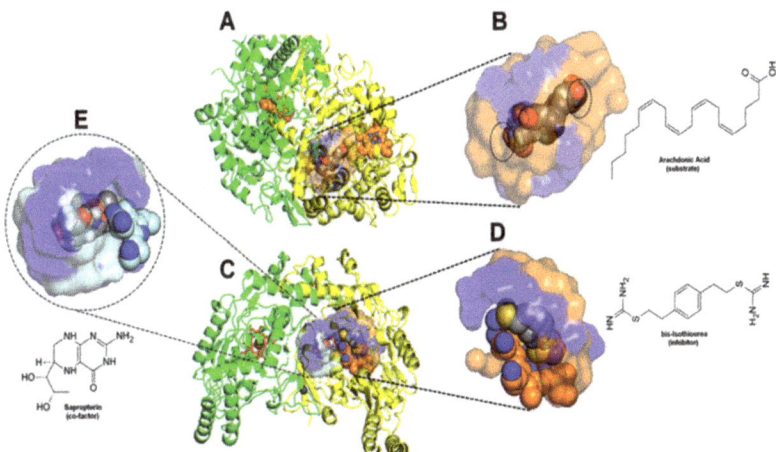

Figure 1. Physical-chemical characteristics for each docking site and their complementarity to the respective higher affinity controls. (**A**)—Full vision of the COX-1 with their two monomers (cartoons in *green* and *yellow*), cofactors (spheres and sticks) and the docked active site in surface and spheres. (**B**)—inlet vision of the same active site containing the crystallographic and the three first docking poses of the more positive control (substrate arachidonic acid represented in spheres, colored in *grey* and *red* for carbons and oxygens, respectively), as well its skeletal formula. The polar residues at the active site (a.s.) surface are shown in *blue*, while the nonpolar in *orange*. Dashed circles show the two a.s. entrances. (**C**)—Similar full vision of the two monomers from iNOS. (**D**)—Inlet of the iNOS active site (iNOSas) containing the more positive control bis-isothiourea (competitive inhibitor). Colors and schemes similar to B, with the heme group from the a.s. also shown in spheres, as well the ligand's sulfurs and nitrogens, respectively in *yellow* and *blue*. (**E**)—inlet of the iNOS cofactor site (iNOScs) containing the more positive control sapropterin (cofactor). Colors and schemes similar to D, but with the nonpolar regions of the cofactor site (c.s.) in *silver*.

The iNOS enzyme produces nitric oxide (NO) in a sustained manner and is expressed in many inflammatory conditions [47,48]. In particular, the participation of this enzyme on the inflammatory and angiogenic mechanisms implicated on tumor growth and carcinogenesis (with both enhancing as inhibiting paradoxical behaviors) has made it a promising and intriguing therapeutic target [49–52]. To find efficient and safe iNOS inhibitors persists as a difficult task, with just the considerably cytotoxic bis-isothiourea showing significant inhibitory potency currently (Figure 1C,D) [53]. However, a set of natural substances, between them phytochemicals and EO compounds, have shown direct and/or indirect inhibitory effects on this enzyme [54,55].

In addition, iNOS function is possible to modulate both by direct binding at its active site (iNOSa.s.) and by allosteric binding to its coactivator site (iNOSc.s.), i.e., by some substance competing to the binding to this site with the enzyme cofactor tetrahydroneopterin, as shown in Figure 1C,E [56]. This makes this enzyme an interesting and promising double target for EOs compounds.

Computational and chemoinformatics screening approaches have proved to be useful in natural bioactive compounds selection, as well as in the elucidation of which physicochemical characteristics are most related to their structure–activity relationship [57,58]. In particular, such applications have been used successfully for COX-1 and iNOS [58–60].

In the present work, we used the SPME technique to identify the volatile compounds (VOCs) present in EOs from aroeira seed samples from different regions of the state of Minas Gerais (Brazil). The bibliographical research for the present work did not find works that had the aroeira seeds as a target for the extraction and evaluation of volatile compounds, whereas works that prioritized the leaves, stems, and bark for this purpose were found. Therefore, we strategically chose the seeds as the object of study due to the lack of information based on the seeds chemical profile. Moreover, we complement them with virtual screening techniques to verify promising compounds to bind to human COX-1 and iNOS, followed by chemical descriptors and a chemoinformatics analysis to understand which characteristics could make them more likely to interact with these canonical anti-inflammatory targets. For iNOS, both affinity with the active site (iNOSa.s.) and with the cofactor site (iNOSc.s.) were verified. The results point to a high incidence of promising compounds for both enzymes. For COX-1 a higher direct correlation between the in silico recovered affinity and the molecular volume and hydrophobicity was found, making the sesquiterpene fraction stand out on the binding. For the iNOS binding at the iNOSa.s., some correlation is still observed between the affinity and molar volume/hydrophobicity, although with considerably less significance compared to COX-1, both sesquiterpenes and monoterpenes, as well a minor fraction of oxygenated compounds, figurating between the higher affinity hits. Finally, for the iNOSc.s. binding, no significant correlation is observed with any simple chemical descriptor, indicating a more complex structure–activity relationship. A relatively weak correlation was found, however, between the affinity and the molecular length and extensibility. Beside this, no sesquiterpenes were found between the higher affinity compounds, the top hits being equally distributed between monoterpenes and oxygenated molecules. This indicates a less permissive site to substantially voluminous and highly hydrophobic molecules. The computational analysis of the more relevant contacts with each respective active site shed light on the above mentioned chemical descriptor features, in addition to the support found on the comparison with experimental structures of the targets bound to canonical ligands.

Taken together, the results of our study illustrate the applicability of SPME profiling added to chemoinformatics analysis to obtain fast, environmentally friendly and non-expansive information about the medicinal and biotechnological potential of natural sources. It is hoped that the results presented here will be useful in directing future prospecting approaches for natural bioactive compounds, as well as rational planning for new ligands.

2. Results

2.1. Solid Phase Microextraction Allows Fine Profiling of the Aroeira Seeds Volatiles

Figure 2 shows the chromatograms of the seed samples of all aroeira trees analyzed.

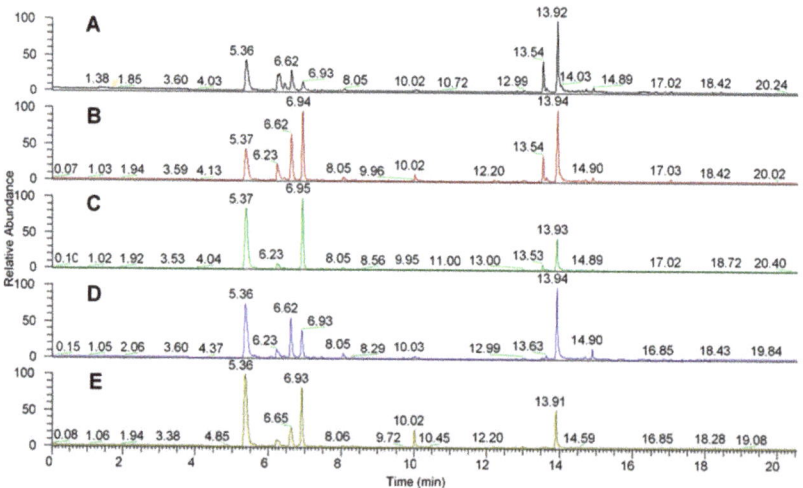

Figure 2. Chromatograms showing DVB/PDMS fiber VOC extraction results. Each chromatogram listed from (**A–E**) represents, respectively, matrices 1 to 5.

The peaks show the volatile organic compounds identified in a m/z ratio.

As it can be seen, all chromatograms show similarities between the peaks in relation to retention time, because even though they are chromatograms of different trees, they are still the same species which do not differentiate significantly in the synthesis of secondary metabolites, since these are determined primarily by the genotype.

Table 1 shows the HS-SPME/GC-MS identified compounds through the chromatograms of the samples.

Table 1. Volatile compounds extracted from Aroeira (*Myracrodruon urundeuva*) seeds by HS-SPME/GC-MS.

N°	Volatile Organic Compound	CAS	Formula	Sampled Trees				
				A.1	A.2	A.3	A.4	A.5
			MONOTERPENES					
1	α-Pinene [a,b,c,d,e,f,g,i,j,l,m,n]	7785-70-8	$C_{10}H_{16}$	22.6%	ND	ND	ND	ND
2	3-Carene [a,b,c,d,f,g,i,k,l]	13466-78-9	$C_{10}H_{16}$	11.0%	34.8%	46.5%	49.9%	55.2%
3	Camphene [f,i,l,m,n]	79-92-5	$C_{10}H_{16}$	ND	ND	0.5%	0.6%	0.5%
4	α-Campholenal [f]	4501-58-0	$C_{10}H_{16}O$	ND	7.7%	ND	4.6%	ND
5	Camphenol,6-	3570-04-5	$C_{10}H_{16}O$	14.9%	ND	4.2%	ND	5.7%
6	D-Limonene [a,b,c,d,e,f,g,h,l,n]	5989-27-5	$C_{10}H_{16}$	ND	23.0%	ND	ND	ND
7	m-Mentha-6,8-diene®	1461-27-4	$C_{10}H_{16}$	ND	ND	30.7%	ND	ND
8	p-Mentha-1,4(8)—diene [d,f,k]	586-62-9	$C_{10}H_{16}$	1.0%	1.3%	0.9%	2.9%	ND
9	p-1,8-diene,(S)	5989-54-8	$C_{10}H_{16}$	5%	ND	ND	10.4%	21.1%
			SESQUITERPENES					
10	β-Guaiene [k]	88-84-6	$C_{15}H_{24}$		0.8%			0.2%
11	Caryophyllene [b,c,d,e,f,g,i,j,m,n]	87-44-5	$C_{15}H_{24}$	11.4%	6.0%	1.9%	0.6%	0.3%
12	Cedr-8(15) ene	11028-42-5	$C_{15}H_{24}$	1.7%	1.2%	0.6%	2.6%	ND
13	Elemene [d,f,m,n]	339154-9	$C_{15}H_{24}$	0.2%	ND	ND	ND	ND
14	β-Chamigrene	18431-82-8	$C_{15}H_{24}$	ND	ND	ND	ND	0.4%
15	Guaia-1(5),11-diene	3691-12-1	$C_{15}H_{24}$	ND	ND	0.4%	ND	ND
16	Patchoulene	1405-16-9	$C_{15}H_{24}$	ND	0.6%	ND	ND	ND
17	1H-Benzocycloheptene,2,4aα,5,6,7,8,9aα-octahydro-3,5,5-trimethyl-9-methylene-	3853-83-6	$C_{15}H_{24}$	1%	0.8%	ND	0.8%	ND
			OTHER CLASSES					
18	Ethylcaproate	123-66-0	$C_8H_{16}O_2$	4.0%	ND	0.8%	ND	ND
19	*Cis-* Geranylacetone	3879-26-3	$C_{13}H_{22}O$	27.2%	ND	25.4%	24.3%	10.3%
20	o-Anisic acid, methylester	606-45-1	$C_9H_{10}O_3$	ND	0.5%	ND	ND	0.7%
21	Salicylic acid, methyl, methylester [e]	119-36-8	$C_8H_8O_3$	ND	2.3%	ND	2.0%	5.6%
22	*Trans*-Geranylacetone	3796-70-1	$C_{13}H_{22}O$	ND	21.6%	ND	ND	ND
23	Hexanoic acid 1-cyclopentylethylester	NA	$C_{13}H_{24}O_2$	ND	ND	ND	0.6%	ND

Identified Volatile compounds: letters indicate compounds found by other authors in different species of the Anarcadiceae family.

Among the compounds found in the aroeira samples using DVB-PDMS fibers, 22 volatile organic compounds of different chemical classes were identified, including terpenoids, carboxylic acids, and ketones. With 17 compounds identified, terpenoides were the most found in all samples, especially caryophyllene and 3-carene. When analyzing the literature of similar works in the characterization of aroeira's chemical profile, chemical compounds such as 1 R-α-pinene, 3-carene, caryophyllene, camphene, and limonene were also frequently found, such as in the work of [12]. In addition to being identified in all five samples, 3-carene was the major constituent in samples 3, 4 and 5, with an abundance of 46.5%, 49.9%, and 55.2%, respectively, and the second most abundant in sample 2 (34.8%). [38] also identified 3-carene as the most common monoterpene (30.4%) when the activity of the essential oil of the "red aroeira" (*Schinus terebinthifolius* Radd) was evaluated as an antibacterial agent.

In the work of [61], two samples of aroeira regarding the profile of their essential oils were analyzed. One sample was from Mato Grosso, and the other from Tocantins, and both presented 3-carene as a major constituent (78.1%; 56.3%), which is in line with the present work, where all the samples comprised this compound, which was also a major contributor in the composition of most samples (samples 3, 4, and 5). However, in the work by [30] where the chemical profile of aroeira and "red aroeira" (*Schinus terebinthifolius* Radd) was evaluated, the constituent 3-Carene was not found, and trans-geranylacetone was the major compound found. Therefore, it is evident that although samples from different regions show differences in the composition of volatile compounds, there is still a predominance of certain terpenes that characterize the species. Regarding monoterpenes found in this work, it is worth mentioning that the essential oils of the aroeira tree are pointed out in other papers as antimicrobial agents due to the action of some compounds, such as 3-carene, a monoterpene involved in the bactericide action on wild-type hospital strains [62]. Ref [63] found the monoterpenes α-pinene and D-limonene as compounds with a higher antibacterial activity against gram-negative and gram-positive bacteria in the constitution of the essential oil of two species of the genus Schinus (Anacardiaceae).

2.2. Comparative Virtual Docking Points for Significant Distinctions between Hits for COX-1 and iNOScs, with iNOSas Presenting an "Intermediary" Behavior between Both Sites

These compounds were compared against the compounds recovered by the experimental assays (Figure 3—blue bars). As expected, the negative controls presented an intermediary or low binding affinity in all target sites. The positive and negative controls are clearly distinguishable for COX-1and iNOScs, but iNOSas presented a mixing behavior, except bis-isothiourea (a known competitive inhibitor) and hexane which presented respectively a high and low affinity to the active site compared to the other controls. Thus, we sorted the tested compounds into three groups based on how close their docking scores were from the best positive control score (top compounds group), from the worst negative control score (worst compounds group) or compounds with intermediary behavior (middle compounds group). Once the groups were defined independently for each target site, we looked closer to the molecular structure of these compounds searching for possible structural patterns (Figure 4). Perhaps predictably, no superposition of compound structures was found between COX-1 and iNOScs once they were at quite distinct sites. Otherwise, iNOSas selected multiple top compounds able to bind favorably both in COX-1 and iNOScs, suggesting possible multi-front anti-inflammatory compounds. Between these, stand out compounds as caryophyllene and patchoulene (respective compounds **4** and **6** in Figure 4), already with substantially reported anti-inflammatory activities [25–29]. Similarly, a set of monoterpenes and oxygenated compounds were recovered as top hits for iNOS by binding to both sites, the active and the cofactor one. In this way, the iNOSas pocket has shown an intermediary behavior concerning compound selectivity compared to the other two sites here studied. The same was observed for the worst groups, highlighting the compound **23** which was the only one to present a low affinity to COX-1 and iNOSas and a high affinity to iNOScs, suggesting a potential specific inhibitor.

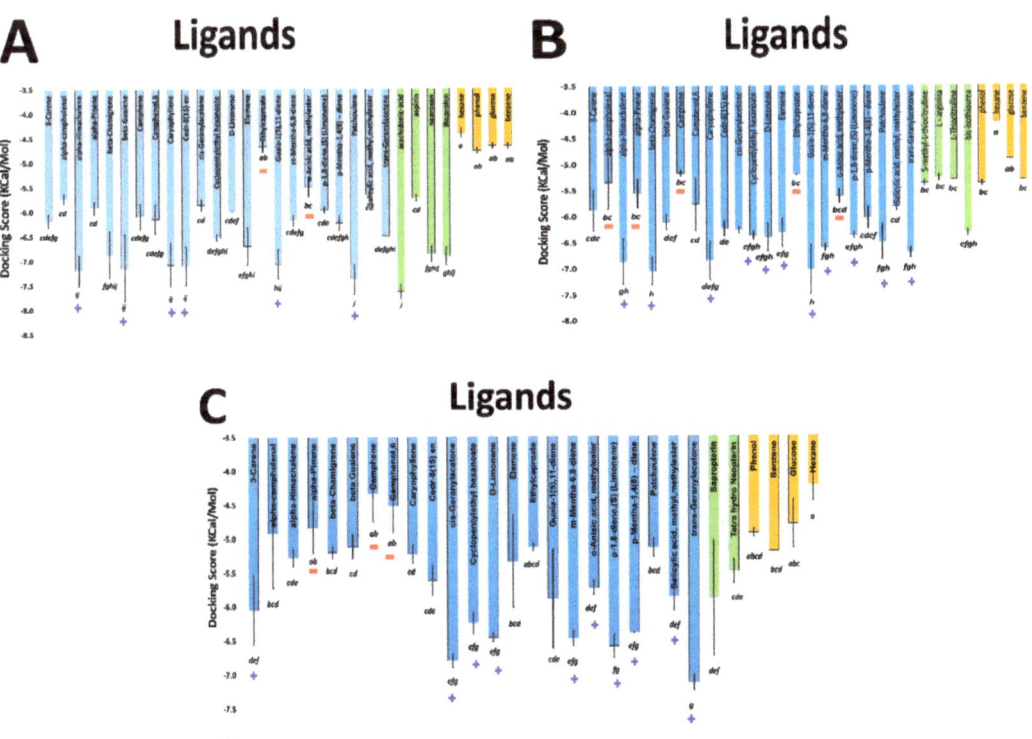

Figure 3. Average docking scores for the three first poses of the respective aroeira compounds and controls at the three targeted sites. (**A**)—COX-1; (**B**)—iNOS active site (iNOSas); (**C**)—iNOS cofactor site (iNOScs). Bars are colored in blue for the aroeira compounds, green for positive controls and yellow for negative controls. Letters below the bars indicate the grouping of the scores according to ANOVA. A blue "plus" (+) and a red "minus" (−) symbols indicate, respectively, top and worst hits between the aroeira volatiles for each site at each enzyme.

2.3. Volume and Hydrophobicity as Major Components for the COX-1 Affinity, Followed by Inos and with iNOScs Presenting a More Complex and Less Predictable Behavior

The in silico recovered affinity for aroeira compounds to the COX-1 active site suggests to be directly dependent on volume and hydrophobicity. This can be inferred both by the fact that its top hits are totally composed of voluminous sesquiterpenes, as by the substantial anti-correlation between a set of chemical descriptors related to volume and hydrophobicity (as molar volume, polarizability, molar refractivity, LogP, parachor) and their docking free energy (Figures 4–6). Also, the considerable positive correlation between the docking free energy and descriptors related to polarity and hydrogen bonds, as total polar surface area (TPSA) and the number of hydrogen bond acceptors, corroborates this trend (Figure 5).

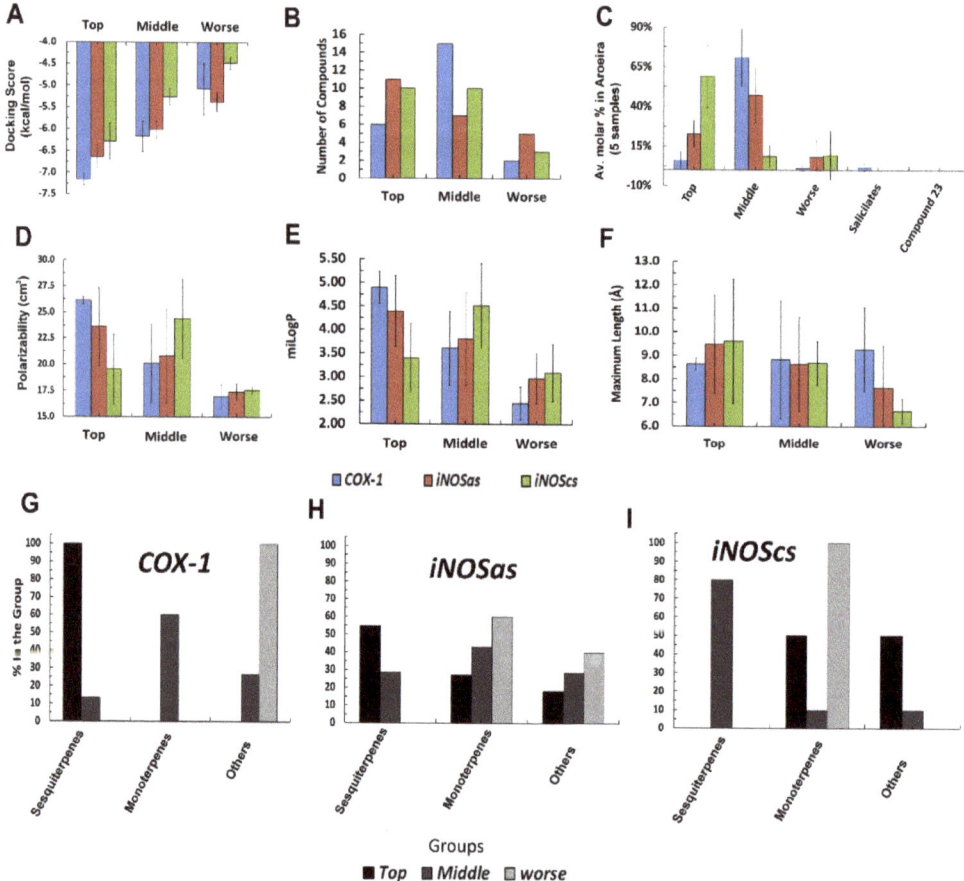

Figure 6. Statistics and numbers concerning the aroeira compounds classified according to their respective virtual affinities for the three targeted sites. (**A–F**)-respective plots of the docking scores, number of compounds (between the 23 profiled), average percent of the compounds at the five samples, average polarizability, miLogP and maximal length, according to the compounds belong to the *top*, *middle*, or *worst* groups for COX-1, iNOSas and iNOScs. (**G–I**)-Respective plots for COX-1, iNOSas and iNOScs of the percent of sesquiterpenes, monoterpenes and other classes at each one of the *top*, *middle*, and *worst* docking score classifications.

2.4. Structural Interpretation of the Differential Selectivity for COX-1 and iNOS Active/Cofactor Sites for the Different Ligand Classes on Aroeira Seeds

In order to better understand the correlations between the target selectivity and the aroeira VOCs physical-chemical attributes, we carried structural analysis on the contact patterns between the top hits at each active site. We also compared the contact patterns from the docked VOCs and from the respective first ranked positive controls (from experimental and docked structures) in order to best validate our results.

A primary point to be discussed here is that the first ranked docking positive control for COX-1 (the substrate arachidonic acid) does not reproduce the poses of the crystallographic models at the two available PDB structures, the PDB-ID:1DIY (with 3.00 Å resolution) and the PDB-ID:1U67 (with 3.10 Å resolution) (Figure 7). This is not surprising, however, when both the resolution and the specific electronic density map inside the enzyme active site of each crystal structure are taken into accounts.

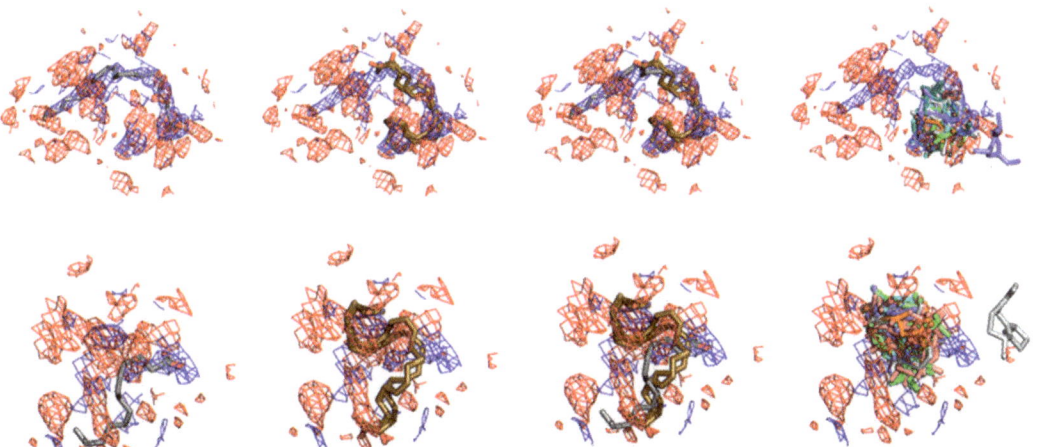

Figure 7. Different binding modes for the ligands at the COX-1 crystallographic and docked structures and their fits to crystal electronic density. At the *Top* the electronic densities and different molecular model superpositions for the human COX-1 complexed with arachidonic acid structure at the PDB-ID:1DIY is shown. At the *bottom*, the superpositions with the electronic density of the same complex as solved at the PDB-ID:1U67 are depicted. In both cases, the *red* and *blue meshes* depict the respective F_oF_c electronic density (i.e., electronic densities for which the authors model have found atomic fitting) and $2F_oF_c$ maps (i.e., electronic densities for which the author's molecular model has not found atomic superposition). In both cases also the respective electronic density maps are considered with a σ factor of 1.0. From *left* to *right* it can be seen the respective superposition at both densities for the models deposited at the *protein data bank*; for our Arachidonic acid docked structures (positive control); for both (crystallographic and docked); for the set of the three first docked poses of the top hits for this enzyme between the VOCs profiled from the *M. urundeuva* seeds.

In Figure 8, three-dimensional maps considering the intensity (the sphere size) and frequency between different ligands and poses (the font size) for the major contacts at each target are comparatively depicted. The contacts are compared (from left to right) between crystallographic poses (arachidonic acid at the PDB-ID:1DIY for COX-1; L-arginine at PDB-ID:3NOS and isothiourea at PDB-ID:4NOS for iNOSas; sapropterin at PDB-ID:4NOS for iNOScs), the docked three first poses of the respective most positive controls (arachidonic acid for COX-1; bis-isothiourea for iNOSas; sapropterin for iNOScs) and the set of three first poses for all the top compounds for each target.

2.5. Two Salicylate Derivatives on Aroeira Seeds Seems to Be Promissor for Direct, or after Modification, Suicide Inhibition of the COX-1 Enzyme

Two salicylate derivatives (compounds **16** and **23** in Figure 4) are present in minor concentrations at the VOCs from M. urundeuva seeds (Figure 6). Although they are present in substantially low concentrations and even absent in some samples, the docking poses recovered by these two compounds trend to approximate the respective methoxy and hydroxyl groups from the COX-1 catalytic serine on similar way that the suicide acetyl in aspirin^{TM} (Figure 9).

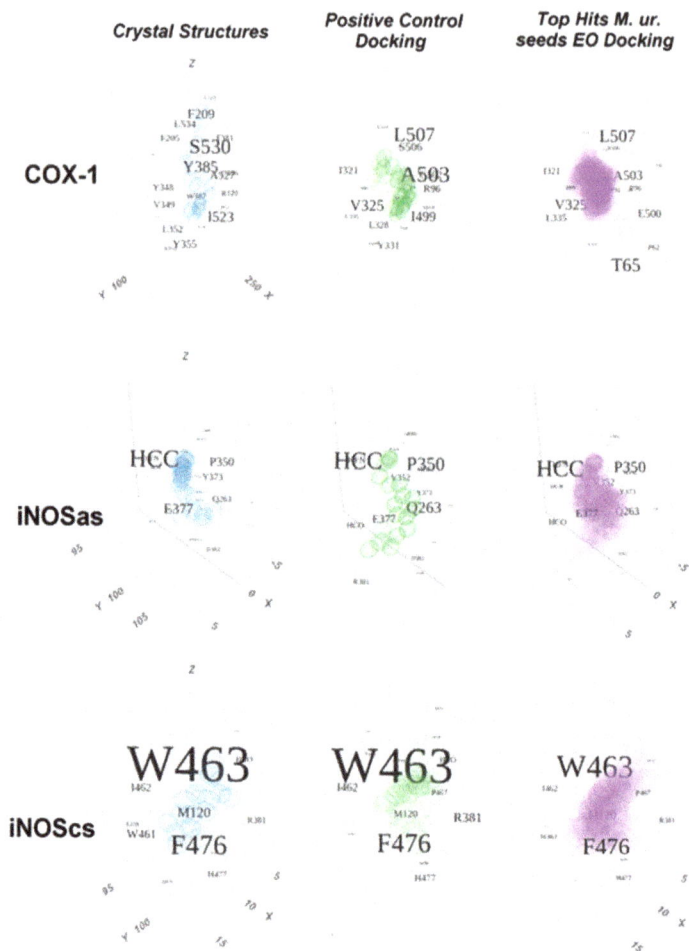

Figure 8. Tridimensional graphic representation of the major contacts at the active site for positive controls and top compounds from essential oils from M. urundeuva seeds. The tridimensional representation (at the x, y, z axes) of each respective active site is arbitrarily centralized at the geometrical center of the set of atoms involved in contacts around all the analysis. The grid spacement at the three-dimensional space in each graphic is always of 5 Å at each dimnension. The density of the spheres on the image depicts how many superposed atoms there are at that region between different docks with the same ligand (the three first poses from the docking procedure were considered) and between different compounds. The font size depicting the residue identity and number increases according to the average intensity of the contacts involving it (i.e., the average contact area superposition between the ligands and the residue). The subdivision of the large Heme group was carried out as mentioned in the Materials and Methods section. From top to bottom it can be noticed the contacts for the COX-1, iNOSa.s. and iNOSc.s. pockets. From the left to the right the respective contact intensities for the crystallographic controls (the arachidonic acid contacts at the PDB-ID:1DIY for COX-1, the L-Arginine-iNOS contacts at the PDB-ID:3NOS plus the isothiourea-iNOS contacts at the PDB-ID:4NOS for iNOSa.s., the tetrahydroneopterin-iNOS contacts at the PDB-ID:4NOS for iNOSc.s.); for the docked three first poses for the most positive controls (arachidonic acid for COX-1, bis-isothiourea for iNOSa.s. and tetrahydropterin for iNOSc.s.); as well the three first poses of the top hits for each site are depicted. M. ur. = Myracrondruon urundeuva. EO = Essential oil.

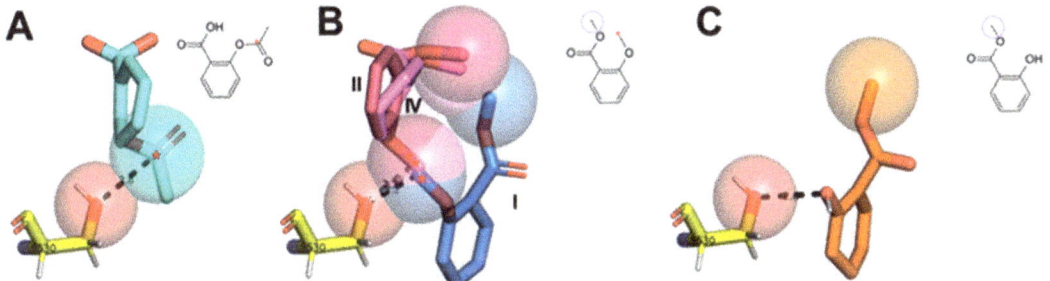

Figure 9. Docking poses for acetylsalicylic acid and salicylate related compounds from aroeira seeds at COX-1 active site. (**A**)-Acetylsalycilic acid (aspirin^{TM}, positive control); (**B**)-Aroeira compound **23** (the respective docking poses I, II and IV are depicted); (**C**)-Aroeira compound **16**. The contacts between the S530 residue (subjected to suicide inhibition by aspirin^{TM}) and the compound's nucleophilic attack sensitive center (when present) or hydroxyl radical are both shown as dashed lines. Both at the skeletal formulas as at the docked structures, the respective groups susceptible to S530 nucleophilic attack (*red* asterisk/sphere) and esterified to the original acid (dashed circle/sphere) are, all of them, highlighted when present.

3. Material and Methods

Mature seeds of *Myracrodruon urundeuva* from 5 adult matrices were collected in October and November 2019 in different areas of Sete Lagoas-MG, Brazil, located at coordinates 19°28′29.0″ S 44°11′39.9″ W, at an altitude of 751 m. According to Köeppen, the regional climate is Cwa, i.e., typical savanna climate, with dry winters and wet and rainy summers [61]. The seeds of each matrix were transferred separately to the chemistry laboratory, where the manual process of removing dirt particles and undesirable parts of the plant was carried out. The next step consisted of grinding the seeds of each matrix separately in an A 11 IKA analytical mill, followed by weighing each sample on a Marbeg balance and storing it in headspace flasks.

Headspace solid phase microextraction (HS-SPME) was employed for the extraction of volatile compounds, using a semi-polar polymeric film, polydimethylsiloxane-divinylbenzene (PDMS/DVB). In the extraction of the VOCs, 2 g of the previously ground seeds were used, placed in a 20 mL headspace vial, the containers were closed with an aluminum seal and a rubber septum. The 20-mL headspace vial was then placed on an aluminum block and heated to 60 °C. After 5 min of heating, the SPME polymeric film (PDMS/DVB) was exposed to the sample for 20 min, and then the holder containing the polymeric film was retracted and manually inserted into the injector of the gas chromatograph coupled to the mass spectrometer, exposing the polymeric film during 5 min for the desorption of the extracted volatile organic compounds. The two figures below show aroeira seeds in early stages of development.

3.1. Gas Chromatography—Mass Spectrometry

Aroeira seed samples were analyzed by a gas chromatograph (Trace GC Ultra) coupled to a mass spectrometer (Polaris Q model, Thermo Scientific, San Jose, CA, USA), with an ion trap type analyzer, located in the Mass Spectrometry Laboratory of the UFMG's Chemistry Department. The samples were analyzed in the following settings: injector temperature of 250 °C; splitless mode injection, desorption time of 5 min; injector temperature of 200 °C; interface temperature of 275 °C. The column heating temperature was set up starting at 40 °C and remaining at the temperature for 1 min followed by gradual temperature increase of 10 °C/min up to 100 °C keeping the isotherm for 1 min, 12 °C/min up to 150 °C, keeping the isotherm for 1 min and then 15 °C/min up to 245 °C, temperature at which the isotherm was kept for 1 min. The detector was kept in scanning mode (fullscan, from 35 to 300 *m/z*), using the Electron Impact Ionization (EI) technique, at an energy of 70 electron-volt (eV). Throughout the process an HP-5 MS capillary chromatographic column (5% phenyl and 95% methylpolisiloxane) was used, of the following dimensions: 30 m in length, 0.25 mm (mm) internal diameter and 0.25 µm film thickness [33,35].

3.2. Volatile Compound Identification

For the identification of the volatile compounds, the mass-to-charge ratio (m/z) corresponding to each peak generated by the chromatogram was compared with the mass spectra obtained through ionization by EI, using energy of 70 eV, and the fullscan range from 35 to 300 m/z As such, the mass spectra of the analytes found were compared with the mass spectra data obtained from the NIST (National Institute of Standards and Technology) library, using as an auxiliary tool the data recorded in the literature for the confirmation of the volatile compounds found in the seed samples. The RSI index consists of a numerical comparison factor, where the higher the value, the closer the compound is to the one found in the NIST library. However, only peaks with values of relative standard intensity (RSI) higher than 600, and a signal-to-noise ratio (S/N) above 50 decibels were selected. Intensity values of the peaks obtained and the S/N ratio were obtained using Thermo Electron Corporation's XCalibur 1.4 program and the data were transferred to Microsoft Office Excel 2013, where the peak selections were made according to the S/N ratio in the UFSJ/CSL Chemistry Laboratory.

3.3. Virtual Docking Assays

Structure- based virtual screening applying docking simulations was performed using the AutoDock Vina tool [64]. The respective structures for each target were obtained from the protein data bank [65] or modeled by homology. The human COX-1 was modeled from the correspondent human sequence obtained from UniProt [66] using the tool Swiss-Model [67] and the ovine structure at the PDB-ID:1DIY (originally complexed to the arachidonic acid substrate). The human iNOS structure was obtained from the PDB-ID:4NOS (originally complexed the inhibitor isothiourea and containing the tetrahydroneopterin cofactor) using the Swiss-Model tool to fill small gaps at the protein crystal construction. For this enzyme, two docking procedures were carried out for each ligand: one at the enzyme active site (here called iNOSa.s.) and the other at the cofactor site (here called iNOSc.s.). In all the three cases, the enzyme dimmeric structure was used and the crystallographic ligand was removed before the docking procedures, conserving just the protein dimmer and the respective co-factors (heme groups for both enzymes, in addition to zinc ion and both tetrahydroneopterin for iNOS at the iNOSa.s. docking and just heme, zinc ion and the tetrahydroneopterin at the non-docked monomer for this enzyme at the iNOSc.s. docking procedure).

The docking boxes for both targets were centered at the position originally occupied by the respective crystallographic ligands (the arachidonic acid from the PDB-ID:1DIY for COX-1; the isothiourea ligand for the iNOSa.s. docking in iNOS and the tetrahydroneopterin cofactor for the iNOSc.s. docking at this enzyme). The box dimensions were planned in order to preserve enough of the crystallographic context about 8 Å around different crystall ligands previously inspected by the present authors for each enzyme, finishing on a common x, y, z set of dimensions of 17, 25, 17 Å, respectively.

A ligand virtual dataset was composed of volatile compounds of aroeira determined by the mass spectrometry method above. MOL2 files from the ligands were downloaded from PubChem [68]. As negative controls for the virtual docking procedures, we chose four non-druggable molecules concerning these enzymes (glucose, hexane, benzene, and phenol) for all the three pockets. Specific positive controls were chosen for each pocket: the substrate arachidonic acid, the suicide inhibitor acetyl-salicylic acid and the competitive inhibitors naproxen and ibuprofen for COX-1; the substrate L-Arginine and the competitive inhibitors S-methyl-L-thiocitruline, L-thiocitruline and bis-isothiourea for the iNOSa.s. pocket in iNOS; the cofactor tetrahydroneopterin and its derivative sapropterin for the iNOSc.s. pocket in this same enzyme. This virtual ligand dataset was prepared using MGLTools [69] and automatically docked through AutoDockVina tool usign Python housemade scripts. The docking procedures were carried with an exhaustiveness parameter of 128, in order maximize the sampling and accuracy. For each system, the first three docked poses were taken to score and structural analysis. The docking results were analyzed using housemade scripts and the tool PyMOL [70].

The three first poses considering the docking affinity scores were taken from each ligand and ANOVA was used for statistical analysis and mean comparison as determined by Tukey's HSD (honestly significant difference) test. The R language and computational environment was used for all the statistical analyses [71].

3.4. Molecular Descriptors and Statistical Analysis

Chemoinformatics molecular descriptors for the volatile compounds of aroeira were obtained from the free tools: ACD/ChemSketch [72], Molinspiration platform [73] and 3D-QSAR.com platform [74]. Correlogram among different molecular descriptors and docking score affinities were calculated using R. The Pearson correlation coefficient was determined.

3.5. Estimation of the Major Contacts Involved in Ligand-Target Interactions

The contact areas were computed according to the methodology described in [75]. Basically, this area is computed for each pair of atoms at the ligand-target complexes. It is equivalent to the excluded or untouched area of a rolling water molecule (probe). Thus, a null contact area indicates that one (or more) water molecule could be interposed between two atoms, defining a potential cavity in terms of this heuristics. It also indicates how tightly packed a group of atoms is in space, given that the larger the contact area, the better spatially clustered they will be. As a consequence, atoms with non-zero contact areas delimit an interface region between any sets of different biomolecules in close contact.

We represented three-dimensionally the interaction intensity at each residue/subsite at the pocket with dot spheres designating superimposed ligand atoms as a result of the successive docking into the respective targets. Letters and numbers represent target residues positioned according to the geometric center of its atoms in the ligand-target complex. The font size indicates the average contact area of the residues with such ligands.

As a larger prosthetic group, the HEM group at the iNOS a.s. and c.s. respective pockets (in which the ligand interacts or is able to interact with this group) was divided into parts, with the following names: HFN = the Fe^{2+} and the N from the porphyrins at the center; HCC = SP2 carbons of the porphyrin ring; HCR = SP3 carbons branched from the porphyrin ring that support the carboxylic groups; HCO = oxygens of carboxylic groups at the branch ends. The HEM groups were considered as part of the targets.

4. Conclusions

It was possible to observe that HS-SPME, a method considered green for not using organic solvents and requiring minimum sample preparation, was efficient in the extraction of volatile compounds from aroeira seed samples collected in five trees of different areas in Sete Lagoas- MG, where it was possible to find and identify 22 volatile organic compounds of different chemical classes, among terpenoids, carboxylic acids, and ketones. 3-carene was present in all samples as the main constituent in three of the five samples.

The chromatography technique used was efficient in the separation process of volatile organic compounds, coupled with the analysis in mass spectrometry which allowed the identification of these compounds.

Author Contributions: Conceptualization, J.O.F.M., Y.G.F., A.H.d.O.J., A.C.d.C.M., H.d.O.P.M., L.H.F.d.L. and R.A.; methodology, J.O.F.M., Y.G.F., A.H.d.O.J., A.C.d.C.M., H.d.O.P.M., L.H.F.d.L., E.A.C., Y.J.G.L., Y.M.G., L.D.C.B.R. and C.H.d.S.; software, L.H.F.d.L., E.A.C., Y.J.G.L. and C.H.d.S.; validation, J.O.F.M., Y.G.F., A.H.d.O.J., M.A.d.S.G., A.C.d.C.M., H.d.O.P.M., L.H.F.d.L., E.A.C.; Y.J.G.L., Y.M.G., L.D.C.B.R. and C.H.d.S.; formal analysis, J.O.F.M., Y.G.F., A.C.d.C.M., L.H.F.d.L. and E.A.C.; investigation, Y.G.F., A.C.d.C.M. and E.A.C.; resources, Y.G.F., A.C.d.C.M., R.A., E.A.C., L.H.F.d.L. and J.O.F.M.; data curation, Y.G.F., A.C.d.C.M., L.H.F.d.L. and J.O.F.M.; writing—original draft preparation, Y.G.F., A.C.d.C.M., E.A.C., L.H.F.d.L. and J.O.F.M.; writing—review and editing, Y.G.F., A.C.d.C.M., E.A.C., L.H.F.d.L., L.D.C.B.R., M.A.d.S.G., A.C.C.F.F.d.P., C.H.d.S. and J.O.F.M. visualization, Y.G.F., M.A.d.S.G., A.C.d.C.M., L.H.F.d.L. and J.O.F.M.; supervision, L.H.F.d.L. and J.O.F.M.; project administration, J.O.F.M. and L.H.F.d.L. funding acquisition, R.A., J.O.F.M., L.H.F.d.L., M.A.d.S.G. and A.C.C.F.F.d.P. All authors have read and agreed to the published version of the manuscript.

Funding: This research was funded by CAPES, CNPq, FAPEMIG, IFMG, UFMG, UFMT, UNIFEI and UFSJ to whom the authors would like to extend their gratitude. This research is developed within the framework of the Sustainable Rural Project—Cerrado, funded by Technical Cooperation approved by the Inter-American Development Bank (IDB), with resources from the UK Government's International Climate Finance, with the Ministry of Agriculture, Livestock and Supply (MAPA) as institutional beneficiary. The Brazilian Institute for Development and Sustainability (IABS) is responsible for the execution and administration of the project and the ILPF Network Association, through Embrapa, is responsible for the scientific coordination and technical support.

Institutional Review Board Statement: Not applicable.

Informed Consent Statement: Not applicable.

Data Availability Statement: All data is contained within the article.

Acknowledgments: The authors would like to thank the Universidade Federal de Minas Gerais (UFMG), Instituto Federal de Educação Ciência e Tecnologia de Minas Gerais (IFMG), Universidade Federal do Mato Grosso (UFMT), Universidade Federal de Itajubá (UNIFEI) and Universidade Federal de São João del-Rei (UFSJ) for the infrastructure to carry out the analyses and for the financial support and thank CAPES, CNPq, Pró-Reitoria de Pesquisa—PRPq—UFMG, The Brazilian Institute for Development and Sustainability (IADS), and FAPEMIG for the financial support.

Conflicts of Interest: The authors declare no conflict of interest.

References

1. Nunes, Y.R.F.; Fagundes, M.; Almeida, H.S.; Veloso, M.D.M. Aspectos ecológicos da aroeira (*Myracrodruon urundeuva* Allemão-Anacardiaceae): Fenologia e germinação de sementes. *Rev. Árvore Viçosa* **2008**, *32*, 233–243. [CrossRef]
2. Feliciano, A.L.P.; Maragon, L.C.; Holanda, A.C. Morfologia de sementes, de plântulas e de plantas jovens de aroeira (*Myracrodruon urundeuva* Allemão). *Rev. Biol. Ciências Terra Paraíba* **2008**, *8*, 110–118.
3. Virgens, I.O.; Castro, R.D.; Fernandez, L.G.; Pelacani, C.R. Comportamento fisiológico de sementes de *Myracrodruon urundeuva* fr. all. (Anacardiaceae) submetidas a fatores abióticos. *Ciência Florest. St. Maria RS* **2012**, *22*, 681–692. [CrossRef]
4. Araújo, R.D.I. Atividade Antimicrobiana e Citotóxica de Óleo Essencial e Extratos Orgânicos Provenientes da Myracrodruon urundeuva Fr. Allem. (Aroeira-do-Sertão). Dissertação de Mestrado, Universidade Federal do Rio Grande do Norte, Natal, RN, Brasil, 2017.
5. Aquino, N.C.; Araújo, M.R.; Silveira, R.E. Intraspecific Variation of the Volatile Chemical Composition of *Myracrodruon urundeuva* Fr. Allem. ("Aroeira-do-Sertão"): Characterization of Six Chemotypes. *J. Braz. Chem. Soc.* **2017**, *28*, 907–912. [CrossRef]
6. Medeiros, A.C.S.; Smith, R.; Probert, R.J.; Sader, R. Comportamento fisiológico de sementes de aroeira (*Myracrodruon urundeuva* Fr. All.), em condições de armazenamento. *Bol. Pesq. Fl. Colombo* **2000**, *40*, 77.
7. Silva, P.D.D. Caracterização de Compostos Fenólicos por Espectrometria de Massas e Potencial Antioxidante das Cascas de Myracrodruon urundeuva (Aroeira-do-Sertão) do Cariri Paraibano. Dissertação de Mestrado, Universidade Federal da Paraíba, Sumé, PB, Brasil, 2018.
8. Ferreira, P.M.P.; Farias, D.F.; Viana, M.P.; Souza, P.M.; Vasconcelos, I.M.; Soares, B.M.; Pessoa, C.; Costa-Lotufo, L.V.; Moraes, M.O.; Carvalho, A.F.U. Study of the antiproliferative potential of seed extracts from Northeastern Brazilian plants. *Biomed. Sci. An. Acad. Bras. Cienc.* **2011**, *83*, 1045–1058. [CrossRef] [PubMed]
9. Soares, A.M.S.; Oliveira, J.T.A.; Rocha, C.Q.; Ferreira, A.T.S.; Perales, J.; Zanatta, A.C.; Vilegas, W.; Silva, C.R.; Costa-Junior, L.M. *Myracrodruon urundeuva* seed exudates proteome and anthelmintic activity against Haemonchus contortus. *PLoS ONE* **2018**, *13*, e0200848. [CrossRef] [PubMed]
10. Machado, A.C.; Oliveira, R.C. Phytotherapy medicines in dentistry: Evidence and perspectives on the use of "Aroeira-do-sertão" (*Myracrodruon Urundeuva* Allemão). *Rev. Bras. Plantas Med.* **2014**, *16*, 283–289. [CrossRef]
11. Lica, I.C.L.; Soares, A.; de Mesquita, L.S.S.; Malik, S. Biological properties and pharmacological potential of plant exudates. *Food Res. Int.* **2018**, *105*, 1039–1053. [CrossRef]
12. Carvalho, M.G.; Melo, A.G.N.; Aragão, C.F.S.; Raffin, F.N.; Moura, T.F.A.L. *Schinus terebinthifolius* Raddi: Chemical composition, biological properties and toxicity. *Rev. Bras. Pl. Med. Botucatu* **2013**, *15*, 158–169. [CrossRef]
13. Ceruks, M.; Romoff, P.; Favero, A.O.; Lago, J.H.G. Constituintes fenólicos polares de *Schinus terebinthifolius* Raddi (Anacardiacea). *Rev. Química Nova São Paulo* **2007**, *30*, 597–599. [CrossRef]

14. Jain, M.K.; Yu, B.Z.; Rogers, J.M.; Smith, A.E.; Boger, E.T.; Ostrander, R.L.; Rheingold, A.L. Specifi c competiti ve inhibitor of secreted phospholipase A2 from berries of Schinus terebinthifolius. *Phytochemistry* **1995**, *39*, 537–547. [CrossRef]
15. Santos, B.O.; Augusti, R.; Melo, J.O.F.; Takahashi, J.A.; Araújo, R.L.B. Optimization of extraction conditions of volatile compounds from pequi peel (*Caryocar brasiliense* Camb.) using HS-SPME. *Res. Soc. Dev.* **2020**, *9*, e919974893. [CrossRef]
16. Bergman, M.E.; Davis, B.; Phillips, M.A. Medically Useful Plant Terpenoids: Biosynthesis, Occurrence, and Mechanism of Action. *Molecules* **2019**, *24*, 3961. [CrossRef]
17. Aponso, M.; Patti, A.; Bennett, L.E. Dose-related effects of inhaled essential oils on behavioural measures of anxiety and depression and biomarkers of oxidative stress. *J. Ethnopharmacol.* **2020**, *250*, 112469. [CrossRef] [PubMed]
18. Aziz, Z.A.A.; Ahmad, A.; Setapar, S.H.M.; Karakucuk, A.; Azim, M.M.; Lokhat, D.; Rafatullah, M.; Ganash, M.; Kamal, M.A.; Ashraf, G. Essential Oils: Extraction Techniques, Pharmaceutical and Therapeutic potential Review. *Curr. Drug Metab.* **2018**, *19*, 1100–1110. [CrossRef] [PubMed]
19. Da Silva, L.L.; De Almeida, R.; Verícimo, M.A.; De Macedo, H.W.; Castro, H.C. Atividades terapêuticas do óleo essencial de melaleuca (*Melaleuca alternifolia*) Uma revisão de literatura. *Braz. J. Health Rev.* **2019**, *2*, 6011–6021. [CrossRef]
20. Dhakad, A.K.; Pandey, V.V.; Beg, S.; Rawat, J.M.; Singh, A. Biological, medicinal and toxicological significance ofEucalyptusleaf essential oil: A review. *J. Sci. Food Agric.* **2017**, *98*, 833–848. [CrossRef]
21. Farrar, A.J.; Farrar, F.C. Clinical Aromatherapy. *Nurs. Clin. N. Am.* **2020**, *55*, 489–504. [CrossRef]
22. Hosseinzadeh, S.; Jafarikukhdan, A.; Hosseini, A.; Armand, R. The Application of Medicinal Plants in Traditional and Modern Medicine: A Review of Thymus vulgaris. *Int. J. Clin. Med.* **2015**, *6*, 635–642. [CrossRef]
23. Yuan, R.; Zhang, D.; Yang, J.; Wu, Z.; Luo, C.; Han, L.; Yang, F.; Lin, J.; Yang, M. Review of aromatherapy essential oils and their mechanism of action against migraines. *J. Ethnopharmacol.* **2020**, *265*, 113326. [CrossRef] [PubMed]
24. Ramsey, J.T.; Shropshire, B.C.; Nagy, T.R.; Chambers, K.D.; Li, Y.; Korach, K.S. Essential Oils and Health. *Yale J. Biol. Med.* **2020**, *93*, 291–305. [PubMed]
25. Asif, M.; Saleem, M.; Saadullah, M.; Yaseen, H.S.; Al Zarzour, R. COVID-19 and therapy with essential oils having antiviral, anti-inflammatory, and immunomodulatory properties. *Inflammopharmacology* **2020**, *28*, 1153–1161. [CrossRef] [PubMed]
26. Zuo, X.; Gu, Y.; Wang, C.; Zhang, J.; Zhang, J.; Wang, G.; Wang, F. A Systematic Review of the Anti-Inflammatory and Immunomodulatory Properties of 16 Essential Oils of Herbs. *Evid. Based Complement. Altern. Med.* **2020**, *2020*, 8878927. [CrossRef]
27. Jurenka, J.S. Anti-inflammatory properties of curcumin, a major constituent of Curcuma longa: A review of preclinical and clinical research. *Altern. Med. Rev. J. Clin. Ther.* **2009**, *14*, 141–143.
28. Borges, R.S.; Ortiz, B.L.S.; Pereira, A.C.M.; Keita, H.; Carvalho, J.C.T. Rosmarinus officinalis essential oil: A review of its phytochemistry, anti-inflammatory activity, and mechanisms of action involved. *J. Ethnopharmacol.* **2018**, *229*, 29–45. [CrossRef]
29. Peterfalvi, A.; Miko, E.; Nagy, T.; Reger, B.; Simon, D.; Miseta, A.; Czéh, B.; Szereday, L. Much More Than a Pleasant Scent: A Review on Essential Oils Supporting the Immune System. *Molecules* **2019**, *24*, 4530. [CrossRef]
30. Figueiredo, Y.G.; Bueno, F.C.; Júnior, A.H.D.O.; Mazzinghy, A.C.D.C.; Mendonça, H.D.O.P.; de Oliveira, A.F.; De Melo, A.C.; Reina, L.D.C.B.; Augusti, R.; Melo, J.O.F. Profile of the volatile organic compounds of pink pepper and black pepper. *Sci. Electron. Arch.* **2021**, *14*, 39–43. [CrossRef]
31. Silva, M.R.; Bueno, G.H.; Araujo, R.L.B.; Lacerda, I.C.A.; Freitas, L.G.; Morais, H.A.; Augustini, R.; Melo, J.O.F. Evaluation of the Influence of Extraction Conditions on the Isolation and Identification of Volatile Compounds from Cagaita (*Eugenia dysenterica*) Using HS SPME/GC-MS. *J. Braz. Chem. Soc.* **2019**, *30*, 379–387. [CrossRef]
32. Nascimento, P.T.; Fadini, M.A.M.; Rocha, M.S.; Souza, C.S.F.; Barros, B.A.; Melo, J.O.F.; Von Pinho, R.G.; Valicente, F.H. Olfactory response of Trichogramma pretiosum (Hymenoptera: Trichogrammatidae) to volatiles induced by transgenic maize. *Bull. Entomol. Res.* **2021**, *111*, 674–687. [CrossRef]
33. García, Y.M.; Ramos, A.L.C.C.; de Oliveira, A.H., Jr.; de Paula, A.C.C.F.F.; de Melo, A.C.; Andrino, M.A.; Silva, M.R.; Augusti, R.; de Araújo, R.L.B.; de Lemos, E.E.P.; et al. Physicochemical Characterization and Paper Spray Mass Spectrometry Analysis of *Myrciaria floribunda* (H. West ex Willd.) O. Berg Accessions. *Molecules* **2021**, *26*, 7206. [CrossRef] [PubMed]
34. García, Y.M.; de Lemos, E.E.P.; Augusti, R.; Melo, J.O.F. Otimização da extração e identificação dos compostos voláteis de Myrciaria floribunda. *Rev. Ciência Agronômica* **2021**, *52*, e20207199. [CrossRef]
35. García, Y.M.; Ramos, A.L.C.C.; de Paula, A.C.C.F.F.; do Nascimento, M.H.; Augusti, R.; de Araújo, R.L.B.; de Lemos, E.E.P.; Melo, J.O.F. Chemical Physical Characterization and Profile of Fruit Volatile Compounds from Different Accesses of Myriciaria floribunda (H. West Ex Wild.) O. Berg through Polyacrylate Fiber. *Molecules* **2021**, *26*, 5281. [CrossRef] [PubMed]
36. Mariano, A.P.X.; Ramos, A.L.C.C.; de Oliveira Júnior, A.H.; García, Y.M.; de Paula, A.C.C.F.F.; Silva, M.R.; Augusti, R.; de Araújo, R.L.B.; Melo, J.O.F. Optimization of Extraction Conditions and Characterization of Volatile Organic Compounds of Eugenia klotzschiana O. Berg Fruit Pulp. *Molecules* **2022**, *27*, 935. [CrossRef] [PubMed]
37. Kim, S.F.; Huri, D.A.; Snyder, S.H. Inducible Nitric Oxide Synthase Binds, S-Nitrosylates, and Activates Cyclooxygenase-2. *Science* **2005**, *310*, 1966–1970. [CrossRef]
38. Attiq, A.; Jalil, J.; Husain, K.; Ahmad, W. Raging the War Against Inflammation With Natural Products. *Front. Pharmacol.* **2018**, *9*, 1–27. [CrossRef]
39. Zidar, N.; Odar, K.; Glavac, D.; Jerse, M.; Zupanc, T.; Stajer, D. Cyclooxygenase in normal human tissues—Is COX-1 really a constitutive isoform, and COX-2 an inducible isoform? *J. Cell. Mol. Med.* **2008**, *13*, 3753–3763. [CrossRef]
40. Rouzer, C.A.; Marnett, L.J. Cyclooxygenases: Structural and functional insights. *J. Lipid Res.* **2009**, *50*, S29–S34. [CrossRef]

41. Moro, M.G.; Sanchez, P.K.V.; Lupepsa, A.C.; Baller, E.M.; Franco, G.C.N. Biología de la ciclooxigenasa en la función renal–Revisión de la literatura. *Rev. Colomb. Nefrol.* **2017**, *4*, 27. [CrossRef]
42. Jaradat, N.; Al-Lahham, S.; Abualhasan, M.N.; Bakri, A.; Zaide, H.; Hammad, J.; Hussein, F.; Issa, L.; Mousa, A.; Speih, R. Chemical Constituents, Antioxidant, Cyclooxygenase Inhibitor, and Cytotoxic Activities of Teucrium pruinosum Boiss. Essential Oil. *BioMed. Res. Int.* **2018**, *2018*, 1–9. [CrossRef]
43. Salaria, D.; Rolta, R.; Sharma, N.; Patel, C.N.; Ghosh, A.; Dev, K.; Sourirajan, A.; Kumar, V. In Vitro and in silico antioxidant and anti-inflammatory potential of essential oil of *Cymbopogon citratus* (DC.) Stapf. of North-Western Himalaya. *J. Biomol. Struct. Dyn* **2021**, 1–5. [CrossRef] [PubMed]
44. Al-Maharik, N.; Jaradat, N.; Qneibi, M.; Abualhasan, M.N.; Emwas, N. Glechoma curviflora Volatile Oil from Palestine: Chemical Composition and Neuroprotective, Antimicrobial, and Cyclooxygenase Inhibitory Activities. *Evid.-Based Complement. Altern. Med.* **2020**, *2020*, 4195272. [CrossRef] [PubMed]
45. Rollinger, J.M.; Haupt, S.; Stuppner, H.; Langer, T. Combining Ethnopharmacology and Virtual Screening for Lead Structure Discovery: COX-Inhibitors as Application Example. *J. Chem. Inf. Comput. Sci.* **2004**, *44*, 480–488. [CrossRef]
46. Batlouni, M. Anti-inflamatórios não esteroides: Efeitos cardiovasculares, cérebro-vasculares e renais. *Arq. Bras. Cardiol.* **2010**, *94*, 556–563. [CrossRef]
47. Cerqueira, N.F.; Yoshida, W.B. Óxido nÌtrico: Revisão. *Acta Cir. Bras.* **2002**, *17*, 417–423. Available online: http://www.scielo.br/acb (accessed on 28 December 2021). [CrossRef]
48. Da Silva, C.B.; Ceron, C.S.; Mendes, A.S.; de Martinis, B.S.; Castro, M.M.; Tirapelli, C.R. Inducible nitric oxide synthase (iNOS) mediates ethanol-induced redox imbalance and upregulation of inflammatory cytokines in the kidney. *Can. J. Physiol. Pharmacol.* **2021**, *99*, 1016–1025. [CrossRef]
49. De Oliveira, G.A.; Cheng, R.Y.; Ridnour, L.A.; Basudhar, D.; Somasundaram, V.; McVicar, D.W.; Monteiro, H.P.; Wink, D.A. Inducible Nitric Oxide Synthase in the Carcinogenesis of Gastrointestinal Cancers. *Antioxid. Redox Signal.* **2017**, *26*, 1059–1077. [CrossRef] [PubMed]
50. Bahadoran, Z.; Mirmiran, P.; Ghasemi, A.; Kashfi, K. Type 2 Diabetes and Cancer: The Nitric Oxide Connection. *Crit. Rev. Oncog.* **2019**, *24*, 235–242. [CrossRef] [PubMed]
51. Singh, S.; Gupta, A.K. Nitric oxide: Role in tumour biology and iNOS/NO-based anticancer therapics. *Cancer Chemother. Pharmacol.* **2011**, *67*, 1211–1224. [CrossRef] [PubMed]
52. Sahebnasagh, A.; Saghafi, F.; Negintaji, S.; Hu, T.; Shabani-Boroujeni, M.; Safdari, M.; Ghaleno, H.R.; Miao, L.; Qi, Y.; Wang, M.; et al. Nitric Oxide and Immune Responses in Cancer: Searching for New Therapeutic Strategies. *Curr. Med. Chem.* **2021**, *28*, 1. [CrossRef] [PubMed]
53. Cinelli, M.A.; Do, H.T.; Miley, G.P.; Silverman, R.B. Inducible nitric oxide synthase: Regulation, structure, and inhibition. *Med. Res. Rev.* **2019**, *40*, 158–189. [CrossRef] [PubMed]
54. Özenver, N.; Efferth, T. Small molecule inhibitors and stimulators of inducible nitric oxide synthase in cancer cells from natural origin (phytochemicals, marine compounds, antibiotics). *Biochem. Pharmacol.* **2020**, *176*, 113792. [CrossRef] [PubMed]
55. Tabanca, N.; Nalbantsoy, A.; Kendra, P.E.; Demirci, F.; Demirci, B. Chemical Characterization and Biological Activity of the Mastic Gum Essential Oils of Pistacia lentiscus var. chia from Turkey. *Molecules* **2020**, *25*, 2136. [CrossRef]
56. Prabhakar, S.S. Tetrahydrobiopterin reverses the inhibition of nitric oxide by high glucose in cultured murine mesangial cells. *Am. J. Physiol. Physiol.* **2001**, *281*, F179–F188. [CrossRef]
57. Maruca, A.; Moraca, F.; Rocca, R.; Molisani, F.; Alcaro, F.; Gidaro, M.C.; Alcaro, S.; Costa, G.; Ortuso, F. Chemoinformatic Database Building and in Silico Hit-Identification of Potential Multi-Targeting Bioactive Compounds Extracted from Mushroom Species. *Molecules* **2017**, *22*, 1571. [CrossRef]
58. Lagunin, A.A.; Goel, R.K.; Gawande, D.Y.; Pahwa, P.; Gloriozova, T.A.; Dmitriev, A.V.; Ivanov, S.M.; Rudik, A.V.; Konova, V.I.; Pogodin, P.V.; et al. Chemo- and bioinformatics resources for in silico drug discovery from medicinal plants beyond their traditional use: A critical review. *Nat. Prod. Rep.* **2014**, *31*, 1585–1611. [CrossRef] [PubMed]
59. Salvemini, D.; Kim, S.F.; Mollace, V. Reciprocal regulation of the nitric oxide and cyclooxygenase pathway in pathophysiology: Relevance and clinical implications. *Am. J. Physiol. Integr. Comp. Physiol.* **2013**, *304*, R473–R487. [CrossRef] [PubMed]
60. Maldonado-Rojas, W.; Olivero-Verbel, J. Food-Related Compounds That Modulate Expression of Inducible Nitric Oxide Synthase May Act as Its Inhibitors. *Molecules* **2012**, *17*, 8118–8135. [CrossRef]
61. Maia, J.G.S.; Silva, M.H.L.; Andrade, E.H.A.; Zoghbi, M.G.B.; Carreira, L.M.M. Essential oils from *Astronium urundeuva* (Allemao) Engl. And, A. fraxinifolium Schott ex Spreng. *Flavour Fragr. J.* **2002**, *17*, 72–74. [CrossRef]
62. Cole, E.; Dos Santos, R.; Lacerda, V., Jr.; Martins, J.; Greco, S.; Neto, A.C. Chemical composition of essential oil from ripe fruit of Schinus terebinthifolius Raddi and evaluation of its activity against wild strains of hospital origin. *Braz. J. Microbiol.* **2014**, *45*, 821–828. [CrossRef]
63. Elshafie, H.S.; Ghanney, N.; Mang, S.M.; Ferchichi, A.; Camele, I. An In Vitro attempt for controlling severe phytopathogens and human pathogens using essential oils from mediterranean plants of genus Schinus. *J. Med. Food* **2016**, *19*, 266–273. [CrossRef] [PubMed]
64. Trott, O.; Olson, A.J. AutoDock Vina: Improving the speed and accuracy of docking with a new scoring function, efficient optimization, and multithreading. *J. Comput. Chem.* **2010**, *31*, 455–461. [CrossRef] [PubMed]

65. Zardecki, C.; Dutta, S.; Goodsell, D.S.; Lowe, R.; Voigt, M.; Burley, S.K. PDB-101: Educational resources supporting molecular explorations through biology and medicine. *Protein Sci.* **2022**, *31*, 129–140. [CrossRef] [PubMed]
66. The UniProt Consortium UniProt: A hub for protein information. *Nucleic Acids Res.* **2014**, *43*, D204–D212. [CrossRef]
67. Guex, N.; Peitsch, M.C.; Schewde, P.T. Automated comparative protein structure modeling with SWISS-MODEL and Swiss-PdbViewer: A historical perspective. *Electrophoresis* **2009**, *30*, S162–S173. [CrossRef]
68. Kim, S.; Chen, J.; Cheng, T.; Gindulyte, A.; He, J.; He, S.; Li, Q.; Shoemaker, B.A.; Thiessen, P.A.; Yu, B.; et al. PubChem in 2021: New data content and improved web interfaces. *Nucleic Acids Res.* **2021**, *49*, D1388–D1395. [CrossRef]
69. Morris, G.M.; Huey, R.; Lindstrom, W.; Sanner, M.F.; Belew, R.K.; Goodsell, D.S.; Olson, A.J. Olson AutoDock4 and AutoDockTools4: Automated docking with selective receptor flexibility. *J. Comput. Chem.* **2009**, *30*, 2785–2791. [CrossRef]
70. *The PyMOL Molecular Graphics System*; Version 2.0; Schrödinger, LLC.: New York, NY, USA, 2010.
71. Morandat, F.; Hill, B.; Osvald, L.; Vitek, J. *Evaluating the Design of the R Language*; Springer: Berlin/Heidelberg, Germany, 2012; Volume 7313, pp. 104–131. [CrossRef]
72. Mweene, P.; Muzaza, G. Implementation of Interactive Learning Media on Chemical Materials. *J. Educ. Verkenn.* **2020**, *1*, 8–13. [CrossRef]
73. Marpaung, D.N.; Pongkendek, J.J.; Azzajjad, M.F.; Sukirno, S. Analysis of Student Motivation using Chemsketch on Hydrocarbon Topic in SMA Negeri 2 Merauke. *J. Appl. Sci. Eng. Technol. Educ.* **2021**, *3*, 69–73. [CrossRef]
74. Ragno, R. www.3d-qsar.com: A web portal that brings 3-D QSAR to all electronic devices—The Py-CoMFA web application as tool to build models from pre-aligned datasets. *J. Comput. Mol. Des.* **2019**, *33*, 855–864. [CrossRef]
75. Araújo, B.M.; Coelho, A.L.; Silveira, S.A.; Romanelli, J.P.; de Melo-Minardi, R.; Silveira, C.H. GAPIN: Grouped and Aligned Protein Interface Networks. *bioRxiv* **2019**, *3*, 520833. [CrossRef]

Article

The Difference of Volatile Compounds in Female and Male Buds of *Herpetospermum pedunculosum* Based on HS-SPME-GC-MS and Multivariate Statistical Analysis

Zhenying Liu [1], Ye Fang [2], Cui Wu [1], Xian Hai [2], Bo Xu [1], Zhuojun Li [1], Pingping Song [1], Huijun Wang [1] and Zhimao Chao [1,*]

Citation: Liu, Z.; Fang, Y.; Wu, C.; Hai, X.; Xu, B.; Li, Z.; Song, P.; Wang, H.; Chao, Z. The Difference of Volatile Compounds in Female and Male Buds of *Herpetospermum pedunculosum* Based on HS-SPME-GC-MS and Multivariate Statistical Analysis. *Molecules* **2022**, *27*, 1288. https://doi.org/10.3390/molecules27041288

Academic Editor: Hiroyuki Kataoka

Received: 23 January 2022
Accepted: 11 February 2022
Published: 14 February 2022

Publisher's Note: MDPI stays neutral with regard to jurisdictional claims in published maps and institutional affiliations.

Copyright: © 2022 by the authors. Licensee MDPI, Basel, Switzerland. This article is an open access article distributed under the terms and conditions of the Creative Commons Attribution (CC BY) license (https://creativecommons.org/licenses/by/4.0/).

[1] Institute of Chinese Materia Medica, China Academy of Chinese Medical Sciences, Beijing 100700, China; liuzy9607@163.com (Z.L.); wucuidalian@163.com (C.W.); xubo_345@163.com (B.X.); 18811385399@163.com (Z.L.); songpingping122@163.com (P.S.); huijunde163@163.com (H.W.)
[2] Shangri-La Alpine Botanical Garden, Diqing 674400, China; sabg001@163.com (Y.F.); haixian9310@126.com (X.H.)
* Correspondence: chaozhimao@163.com or zmchao@icmm.ac.cn; Tel.: +86-135-2270-5161

Abstract: *Herpetospermum pedunculosum* (Ser.) C. B. Clarke (Family Cucurbitaceae) is a dioecious plant and has been used as a traditional Tibetan medicine for the treatment of hepatobiliary diseases. The component, content, and difference in volatile compounds in the female and male buds of *H. pedunculosum* were explored by using headspace solid-phase microextraction-gas chromatography-mass spectrometry (HS-SPME-GC-MS) technology and multivariate statistical analysis. The results showed that isoamyl alcohol was the main compound in both female and male buds and its content in males was higher than that in females; 18 compounds were identified in female buds including 6 unique compounds such as (*E*)-4-hexenol and isoamyl acetate, and 32 compounds were identified in male buds, including 20 unique compounds such as (*Z*)-3-methylbutyraldehyde oxime and benzyl alcohol. (*Z*)-3-methylbutyraldehyde oxime and (*E*)-3-methylbutyraldehyde oxime were found in male buds, which only occurred in night-flowering plants. In total, 9 differential volatile compounds between female and male buds were screened out, including isoamyl alcohol, (*Z*)-3-methylbutanal oxime, and 1-nitropentane based on multivariate statistical analysis such as principal component analysis (PCA) and orthogonal partial least squares discrimination analysis (OPLS-DA). This is the first time to report the volatile components of *H. pedunculosum*, which not only find characteristic difference between female and male buds, but also point out the correlation between volatile compounds, floral odor, and plant physiology. This study enriches the basic theory of dioecious plants and has guiding significance for the production and development of *H. pedunculosum* germplasm resources.

Keywords: *Herpetospermum pedunculosum*; HS-SPME-GC-MS; dioecious plant; bud; volatile compound; multivariate statistical analysis

1. Introduction

Herpetospermum pedunculosum (Ser.) C. B. Clarke (Family Cucurbitaceae) is an annual climbing herb, distributed in Tibet, Yunnan, Sichuan, and other high-altitude areas in China, Nepal, and northeastern India [1–3]. As a traditional Tibetan medicine, it has the functions of clearing away heat and detoxification, and removing the gallbladder and internal heat for its bitter taste and cool potency [3,4], and is widely used for the treatment of hepatobiliary diseases [5].

H. pedunculosum is a dioecious plant, whose bisexual flower is shown in Figure 1. Its flowering period is June to September, and it blooms at night. Its flowers are golden-yellow with five petals, trumpet-shaped, eventually tubular, and covered with fine hairs. The male flower is usually solitary with sparsely pubescent hairs or concomitant with the same raceme; pedicels with 2.0–6.0 cm length and sparse villous hairs; calyx tube with 2.0–2.5 cm length, enlarged similarly to a funnel, tubular lower part, and lanceolate lobes; yellow corolla with 2.0–2.2 cm in length and 1.2–1.3 cm in width, and with elliptic lobes and sharp tips. The female flower is solitary with the same perianth as male flowers and with three staminodes or none, and oblong ovary with three rooms. The mature fruit is 7–8 cm in length and 3–4 cm in width with long pubescent hairs.

Figure 1. The flowers of *H. pedunculosum* (**Left**: female; **right**: male).

There were many studies of *H. pedunculosum*. Its seeds had some pharmacological activities of hepatoprotection, anti-hepatitis B virus, and anti-liver fibrosis [6,7], contained chemical constitutes of fatty acid, polysaccharides, coumarin, and spinasterol glycoside [6–9], and had 156,531 bp length of the whole genome [10]. Because the seeds have medicinal, economic, and sowing values, the female plant is more popular with farmers [11,12]. Fifty metabolites including carbohydrates, amino acids, organic acids, lipids, and polyamine were identified in its leaves [13]. Twelve compounds including *n*-benzyltyramine, 1H-indol-3-carboxylic acid, rhodiocyanoside B, and matteflavoside A were isolated from its stems [14]. The moisture, total ash, acid-insoluble ash, and alcohol-soluble extract were determined as quality indexes of its flower [15–17]. However, there were a few studies on the gender difference in the dioecious plant of *H. pedunculosum*.

In this study, the female and male buds of *H. pedunculosum* were collected. Headspace solid-phase microextraction (HS-SPME) was used to extract their volatile components. Gas chromatography-mass spectrometry (GC-MS) was used to separate and identify these volatile compounds. The multivariate statistical analysis was chosen to screen the differential compounds, aiming to explore the difference in volatile compounds between female and male buds of this dioecious plant.

2. Results

2.1. Volatile Compounds of GC-MS Analysis

The total ion chromatogram obtained by GC-MS analysis was shown in Figure 2. The results showed that there were no significant differences among the three female buds (f1, f2, and f3) nor among the three male buds (m1, m2, and m3), but there were significant differences between female and male buds. It was indicated that there were no significant differences of buds in the same sex, but there were essential differences between different sexes regarding their volatile constitutes.

The results of volatile compounds by GC-MS analysis were shown in Table 1. The volatile compounds were identified by comparing retention time and mass spectra. For example, compound 1 (4.044 min) yielded a parent ion at m/z 88 and fragment ions at m/z 77, 55, and 42, and was identified as isoamyl alcohol according to NIST database. Compound 3 (8.675 min) displayed a parent ion at m/z 101 and fragment ions at m/z 86, 59, and 41, and was identified as (Z)-3-methylbutanal oxime. Compound 9 (10.578 min) produced a parent ion at m/z 117 and fragment ions at m/z 71, 43, and 29, and was identified as 1-nitropentane.

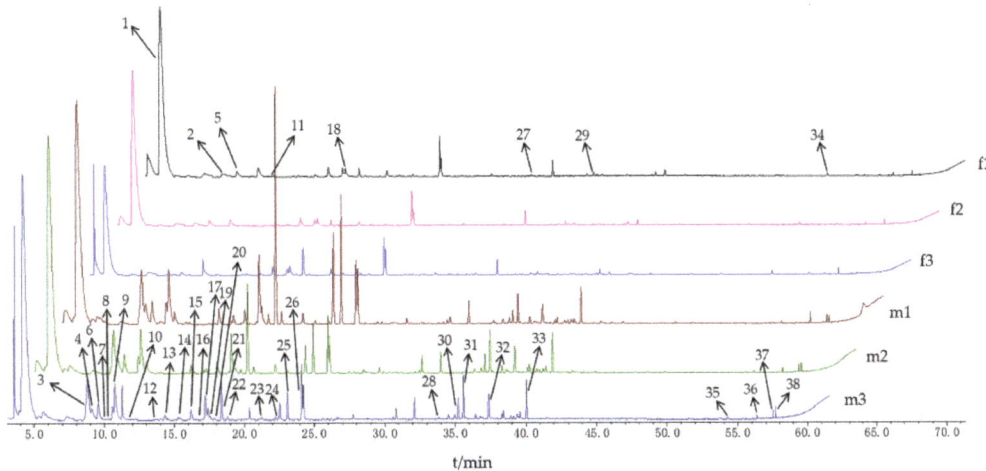

Figure 2. GC-MS total ion chromatogram of volatile components from female and male buds of *H. pedunculosum*.

As a result, in total, 38 volatile compounds were identified from the female and male buds of *H. pedunculosum*, including 18 compounds in female buds and 32 compounds in male buds, covering alkanes, alkenes, alcohols, aldehydes, ketones, and oximes, etc. The female and male buds had 12 common compounds, the female buds had 6 unique compounds, and the male buds had 20 unique compounds.

A compound of the highest content in volatile components from both female and male buds of *H. pedunculosum* was isoamyl alcohol whose content was 83.29%, 90.17%, and 90.19% in female buds, and was 43.23%, 57.90%, and 64.01% in male buds, and the average relative content in female buds was 1.60 times higher than that in male buds. Isoamyl alcohol has the aroma of apple brandy, which has the effect of increasing the aroma and obvious spicy taste, so its higher content may be the main reason for the weaker fragrance of female flowers and the stronger one of male flowers. Among other common compounds, the content of β-ocimene in male buds (5.58%, 5.22%, and 5.69%) was higher than that in female buds (0.56%, 0.79%, and 1.19%) and the average content in female buds was 6.47 times higher than that in male buds. The content of δ-limonene in male buds (2.10%, 2.95%, and 4.20%) was also higher than that in female buds (1.19%, 1.64%, and 1.95%), and the average content in male buds was 1.94 times higher than that in female buds.

The characteristic volatile compounds in female buds were (E)-4-hexenol, isoamyl acetate, 2-ethylhexanol, *n*-pentadecane, *n*-hexadecane, and neophytadiene. The characteristic volatile compounds in male buds had 20 compounds, including 2 oximes, 3 terpenes (β-myrcene, α-pinene, and α-copaene), and 4 phenyl compounds (*p*-xylene, *o*-xylene, benzaldehyde, and benzyl alcohol).

Table 1. Volatile compounds of GC-MS analysis from female and male buds of H. pedunculosum.

No.	Rt/min	Compound	F	CAS	MW	Fragment (m/z)	Female 1	Female 2	Female 3	Male 1	Male 2	Male 3
1	4.044	Isoamyl alcohol	$C_5H_{12}O$	123-51-3	88	70/55/42	90.19	83.29	90.17	43.23	57.90	64.01
2	8.555	(E)-4-Hexenol	$C_6H_{12}O$	928-92-7	100	82/67/55//41	1.24	0.64	1.92	-	-	-
3	8.675	(Z)-3-Methylbutanal oxime	$C_5H_{11}NO$	5780-40-5	101	86/59/41	-	-	-	7.15	6.71	6.66
4	8.960	p-Xylene	C_8H_{10}	106-42-3	106	91/76	-	-	-	2.57	2.11	1.83
5	9.513	3-Methyl-1-butanol acetate	$C_7H_{14}O_2$	123-92-2	130	70/55/43	1.66	1.31	1.69	-	-	-
6	9.438	(E)-3-Methylbutanal oxime	$C_5H_{11}NO$	5775-74-6	101	86/59/41	-	-	-	2.11	2.13	1.94
7	10.000	o-Xylene	C_8H_{10}	95-47-6	106	91/76	-	-	-	0.3	0.15	0.11
8	10.393	2,2-Dimethyl-1-butanol	$C_6H_{14}O$	1185-33-7	102	71/43/27	-	-	-	1.38	1.30	1.13
9	10.578	1-Nitropentane	$C_5H_{11}NO_2$	628-05-7	117	71/43/29	-	-	-	5.35	5.39	4.82
10	11.791	2-Methyl-5-(1-methylethyl)-bicyclo[3.1.0]hex-2-ene	$C_{10}H_{16}$	2867-05-2	136	93/77	-	-	-	0.13	0.10	0.07
11	12.055	1R-α-Pinene	$C_{10}H_{16}$	7785-70-8	136	121/93/77	0.17	0.29	0.15	0.12	0.11	0.11
12	13.451	Benzaldehyde	C_7H_6O	100-52-7	106	77/51	-	-	-	0.10	0.08	0.07
13	14.158	Sabinene	$C_{10}H_{16}$	3387-41-5	136	93/77/41	0.17	0.33	0.14	1.02	0.53	0.45
14	15.227	β-Myrcene	$C_{10}H_{16}$	123-35-3	136	93/69/41/27	-	-	-	0.61	0.36	0.25
15	16.423	(+)-4-Carene	$C_{10}H_{16}$	29050-33-7	136	121/93/79	-	-	-	0.07	0.06	0.05
16	16.846	p-Cymene	$C_{10}H_{14}$	535-77-3	134	119/91	-	-	-	0.17	0.13	0.11
17	17.044	TMδ-Limonene	$C_{10}H_{16}$	5989-27-5	136	121/107/93/68	1.64	1.95	1.19	2.20	2.95	2.10
18	17.258	2-Ethyl-1-hexanol	$C_8H_{18}O$	104-76-7	130	70/57/41/29	1.16	2.86	1.61	-	-	-
19	17.351	Benzyl alcohol	C_7H_8O	100-51-6	108	91/79/77/51	-	-	-	1.17	0.81	0.86
20	17.695	α-Pinene	$C_{10}H_{16}$	80-56-8	136	93/77/39/27	-	-	-	0.50	0.21	0.10
21	18.209	β-Ocimene	$C_{10}H_{16}$	13877-91-3	136	93/79/53/39	1.19	0.56	0.79	5.69	5.22	5.58
22	18.638	γ-Terpinene	$C_{10}H_{16}$	99-85-4	136	121/93/77/43	0.19	0.33	0.15	0.65	0.39	0.26
23	21.073	Nonanal	$C_9H_{18}O$	124-19-6	142	98/70/57/41	0.15	0.39	0.15	0.15	0.14	0.14
24	22.362	2-Methyl-1-undecene	$C_{12}H_{24}$	18516-37-5	168	69/56/41	-	-	-	1.36	1.70	1.08
25	22.934	4-Oxoisophorone	$C_9H_{12}O_2$	1125-21-9	152	96/68/39	-	-	-	5.57	5.79	5.62
26	24.110	2,2,6-Trimethyl-1,4-cyclohexanedione	$C_9H_{14}O_2$	20547-99-3	154	139/69/56/42	-	-	-	2.68	1.97	1.88
27	30.453	n-Pentadecane	$C_{15}H_{32}$	629-62-9	212	85/71/57/43	0.19	0.27	0.19	-	-	-
28	33.704	α-Copaene	$C_{10}H_{16}$	3856-25-5	136	161/119/105/91	-	-	-	0.09	0.10	0.11
29	34.834	n-Hexadecane	$C_{14}H_{30}$	629-73-2	198	85/57/43	0.35	0.51	0.22	-	-	-

Table 1. Cont.

No.	Rt/min	Compound	F	CAS	MW	Fragment (m/z)	Female			Male		
							1	2	3	1	2	3
30	35.068	(+)-7-epi-Sesquithujene	$C_{15}H_{24}$	159407-35-9	204	119/93/77/69/41	0.13	0.13	0.13	0.50	0.88	0.97
31	35.435	trans-α-Bergamotene	$C_{15}H_{24}$	13474-59-4	204	119/107/93/69/41	0.25	0.13	0.17	1.11	1.94	2.17
32	37.199	(1S,5S)-4-Methylene-1-((R)-6-methylhept-5-en-2-yl)bicyclo[3.1.0]hexane	$C_{15}H_{24}$	58319-04-3	204	93/69/41	-	-	-	1.03	1.71	1.85
33	39.901	(1R,5R)-4-Methylene-1-((R)-6-methylhept-5-en-2-yl)bicyclo[3.1.0]hexane	$C_{15}H_{24}$	58319-04-3	204	93/69/41	0.64	0.45	0.57	1.39	1.78	1.99
34	51.476	Neophytadiene	$C_{20}H_{38}$	504-96-1	278	123/95/82/68/55/41	0.27	0.76	0.35	-	-	-
35	54.139	(E,E,E)-3,7,11,15-Tetramethylhexadeca-1,3,6,10,14-pentaene	$C_{20}H_{32}$	77898-97-6	272	93/81/69/55/41	-	-	-	0.10	0.10	0.10
36	56.232	(E,E)-7,11,15-Trimethyl-3-methylene-1,6,10,14-hexadecatetrene	$C_{20}H_{32}$	70901-63-2	272	93/81/69/41	0.21	0.67	0.17	0.24	0.14	0.13
37	57.391	(Z)-9-Tricosene	$C_{23}H_{46}$	27519-02-4	322	97/83/55	-	-	-	0.14	0.14	0.18
38	57.569	n-Eicosane	$C_{20}H_{42}$	112-95-8	282	85/71/57/43	0.21	0.13	0.23	0.12	0.18	0.26

F: formula; MW: molecular weight; -: not be detected.

2.2. Multivariate Statistical Analysis

In order to further explore the difference of volatile compounds between female and male buds, multivariate statistical analysis such as principal component analysis (PCA) and orthogonal partial least squares discrimination analysis (OPLS-DA) were carried out. The PCA was performed on the female and male buds and the result was shown in Figure 3A. The contribution rates of principal component 1 (PC1) and principal component 2 (PC2) were 88.20% and 9.54%, respectively, and the total contribution rate was 97.74%. R^2X and Q^2 were 0.977 and 0.917, respectively, where R^2X represented the fitting degree and Q^2 represented the prediction degree of the model. Additionally, it was indicated that the model was stable and reliable from these data. The classification results of the female and male buds were ideal, the female buds were distributed on the right side of the Y axis, and the male buds were distributed on the left side of the Y axis, which suggested that the volatile compounds of different sexes were significantly different. The Loadings analysis was shown in Figure 3B. The volatile compounds with the largest contribution rate to distinguishing the female and male buds were screened with the coordinate axis 0.2 or −0.2 as the limit value. It was found that the five compounds had the largest contribution rate, i.e., isoamyl alcohol (1), 4-oxoisophorone (25), β-ocimene (21), (Z)-3-methylbutyraldehyde oxime (3), and 1-nitropentane (9).

As an unsupervised analysis method, PCA cannot ignore within-group and eliminate irrelevant random errors. Therefore, supervised OPLS-DA was used to further determine the difference of volatile compounds between female and male buds. However, overfitting was easy to occur while expanding the differences between groups, so it was necessary to arrange experiments with the help of external model validation methods to prove the validity of the model. Different random Q^2 values were obtained by changing the arrangement order of the categorical variable y randomly multiple times ($n = 200$). In this experiment, the greater the slope of the regression line, the smaller the difference between R^2 and Q^2, indicating that there were more data with which to explain the model; the predictive ability of the model was better. The test result was shown in Figure 4A. Among them, the R^2 and Q^2 values generated by any random arrangement on the left end were smaller than those on the right, the slope of the regression line was large, and the lower regression line intersected the negative half-axis of the Y-axis, indicating that the model was effective, stable, and predictable, and could continue to screen the difference compounds. From the OPLS-DA scatter plot of female and male buds (R^2X: 0.977, R^2Y: 0.998, and Q^2: 0.994) (Figure 4B), it could be seen that the two types of samples were clearly distinguished. According to the corresponding VIP value (Figure 4C), these variables whose VIP value was greater than 1 and whose confidence interval did not contain 0 were screened, and it was determined that there were nine differential volatile compounds between the female and male buds (Table 2). Among them were section A for isoamyl alcohol of higher levels in female buds and section B for (Z)-3-methylbutyraldehyde oxime, 1-nitropentane, β-ocimene, 4-oxoisophorone, 2-methyl-1-undecene, 2,2,6-trimethyl-1,4-cyclohexanedione, p-xylene, and (E)-3-methylbutyraldehyde oxime of higher levels in male buds.

Table 2. Nine differential volatile compounds between female and male buds of *H. pedunculosum*.

Section	No.	Compound	VIP
A	1	Isoamyl alcohol	4.00
B	3	(Z)-3-Methylbutanal oxime	1.83
B	9	1-Nitropentane	1.59
B	21	β-Ocimene	1.48
B	25	4-Oxoisophorone	1.26
B	24	2-Methyl-1-undecene	1.07
B	26	2,2,6-Trimethyl-1,4-cyclohexanedione	1.03
B	4	p-Xylene	1.03
B	6	(E)-3-Methylbutanal oxime	1.00

A: higher in females than in males; B: higher in males than in females.

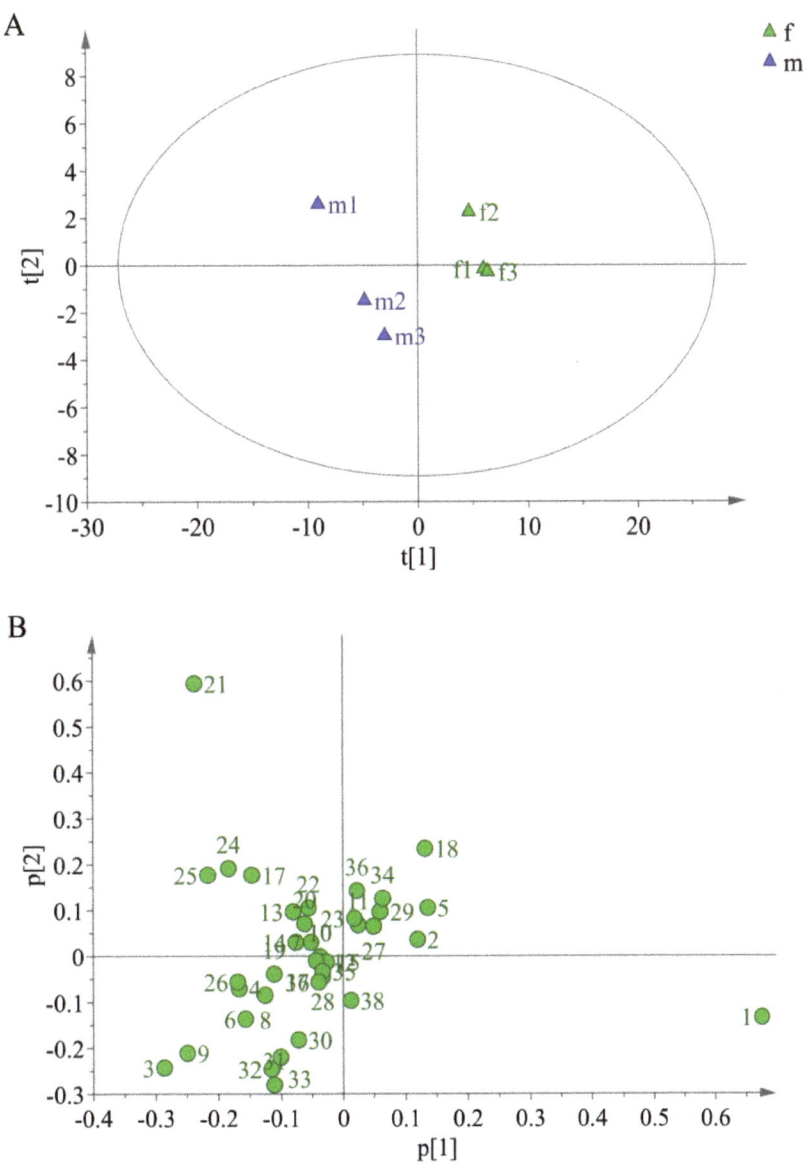

Figure 3. PCA scores (**A**) and Loadings (**B**) for female and male buds of *H. pedunculosum*. The green numbers in figure B were consistent with the No. in Table 1.

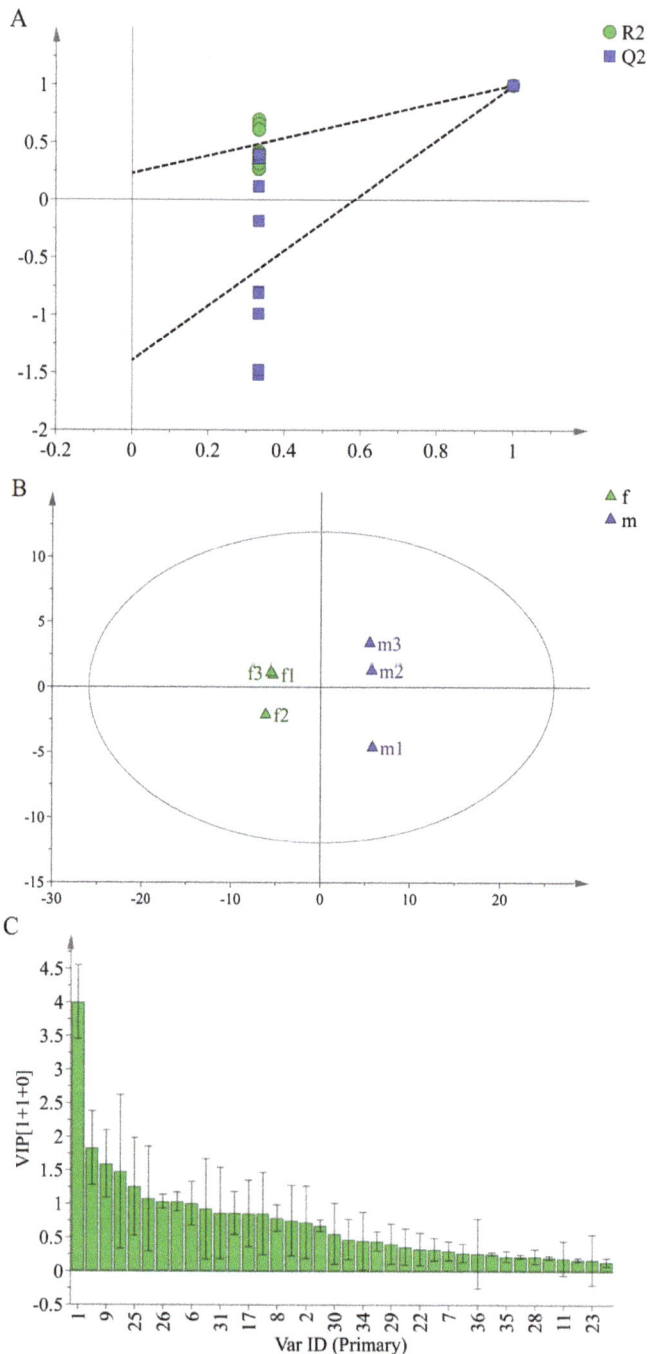

Figure 4. OPLS-DA arranges the model verification diagram (**A**), scores (**B**), and VIP scores (**C**) between female and male buds of *H. pedunculosum*. The numbers of Var ID in Figure C were consistent with the No. in Table 1.

3. Discussion

There were a lot of differences in the morphological structure between female and male flowers and buds of the dioecious plant *H. pedunculosum*, but the difference in volatile compounds has not been reported in the literature. Buds are the state when flowers are about to bloom; their chemical components have not been lost and polluted by foreign substances, thus the research on chemical components of buds has a more realistic effect. Therefore, there were many studies which took buds as research objects, such as *Eugenia caryophyllus* (Myrtaceae) [18–20], *Capparis spinosa* (Capparaceae) [21], *Magnolia biondii* and *M. sirindhorniae* (Magnoliaceae) [22,23], *Populus tomentosa* (Salicaceae) [24], and *Pyrus pyrifolia* (Rosaceae) [25]. In this study, it was found that there was significant difference in the volatile compounds between female and male buds of *H. pedunculosum*.

Isoamyl alcohol is a fragrance ingredient used in cosmetics, detergents, fine fragrances, household cleaners, shampoos, toilet soaps, and other toiletries [26]. Because of its advantages of high energy density, low hygroscopicity, and compatibility with the current infrastructure, isoamyl alcohol has attracted considerable attention as one of the biofuels [27]. As a result, it was showed that the content (up to 90.19%) of isoamyl alcohol was the highest, which provided a new perspective for improving the economic value of *H. pedunculosum*.

Some unique compounds in female or male buds made this study more meaningful. Two kinds of oximes and phenyl compounds were only present in male buds. For (Z)-3-methylbutyraldehyde oxime and (E)-3-methylbutyraldehyde oxime, the reports on plant volatile compounds were extremely rare: only the night-flowering plants of *Gaura drummondii* (Onagraceae) and *Trichosanthes kirilowii* (Cucurbitaceae) [28,29]. For phenyl compounds, benzaldehyde was only detected in male flowers and the content of benzyl alcohol in male flowers was significantly higher than that in female flowers of *T. kirilowii* [29]. Similarly, 4-oxoisophorone was detected only in male buds in this study, which was one of nine differential compounds filtered by multivariate statistical analysis. It was shown that 4-oxoisophorone may be a key compound for honeybees' perceptions of flower odor, which was likely to be implicated in bee foraging behavior [30].

(E)-4-Hexenol, isoamyl acetate, and 2-ethylhexanol were only detected in female buds. It was found that the volatile compounds such as isoamyl alcohol and isoamyl acetate were more attractive to pollinators in the field behavior measurements [31]. High contents of alcohol and ester compounds were beneficial to better pollination of the female flowers of *H. pedunculosum*, which could promote the yield of fruits and improve the economic value of female plants.

In addition, seven terpenes were detected in female or male buds, including β-myrcene, α-pinene, α-copaene, sabinene, β-ocimene, δ-limonene, and γ-terpinene. It was shown that terpenes had the function of attracting pollinating insects, which caused the plant to attract a variety of flower visiting insects, and could promote the natural pollination of flowers [32].

There were few reports on the difference in chemical composition of dioecious plants. For example, the contents of 2-cyclohexenone and methyl benzoate in female buds were significantly higher than those in male buds, and the content of ethyl benzoate in male buds was significantly higher than that in female buds of *P. tomentosa* [23]. The content of linalool in female flowers was significantly higher than that in male flowers, whereas the content of benzyl alcohol in male flowers was significantly higher than that in female flowers of *T. kirilowii* [29]. The comparative study on the volatile compounds of the buds revealed the characteristic difference between female and male buds of *H. pedunculosum*, pointed out the preliminary relationship between the volatile compounds and plant physiology, and enriched the basic research of dioecious plants.

Based on the differences of the volatile compounds from female and male buds of *H. pedunculosum*, the synthesis mechanism of these differential compounds and their specific effects on plant growth should be studied in the future. In addition, the differences in non-volatile compounds and biological activities between female and male plants are also worthy of further study.

4. Materials and Methods

4.1. Apparatus

The SPME fiber 50 mm carboxen/polydimethylsiloxane/divinylbenzene (CAR/PDMS/DVB) was purchased from Supelco (Bellefonte, PA, USA). An AOC-5000 auto injector solid-phase microextraction device and a SHIMADZU QP 2010 gas chromatography–mass spectrometer were purchased from Shimadzu (Tokyo, Japan). An ME204/02 electronic balance was purchased from Mettler Toledo Instruments Co., Ltd. in Shanghai in China.

4.2. Sample Collection

Because *H. pedunculosum* was a night-flowering plant, the buds of both female and male plants were collected at 11:00 pm just before they bloomed. The collection date was 31 August 2021, and the collection site was located in the Shangri-La Alpine Botanical Garden, Diqing Tibetan Autonomous Prefecture, Yunnan Province (27°54'17.07" N, 99°38'15.02" E, and 3269 m altitude). The female and male buds were obtained by pinching off the flower stalk with tweezers, which were identified as fresh female and male buds of *Herpetospermum pedunculosum* (Ser.) C. B. Clarke (Family Cucurbitaceae) by Prof. Zhimao Chao (Institute of Chinese Materia Medica, China Academy of Chinese Medical Sciences) according to the description in Flora of China (Editorial Board of Flora of China, 1984). The voucher specimens (HPF1-3 and HPM1-3) were deposited at 1022 laboratory of Institute of Chinese Materia Medica, China Academy of Chinese Medical Sciences, Beijing, China.

4.3. Sample Preparation

A fresh bud sample was placed in a 15 mL headspace vial. A SPME fiber 50 mm CAR/PDMS/DVB was extended through the needle and exposed into the headspace vial to adsorb volatile compounds at 40 °C for 30 min and then immediately injected into the gas chromatography injection port at 250 °C for 3 min to desorb volatile compounds.

4.4. Chromatographic Condition

The volatile compounds from the buds were analyzed by GC-MS method. A GC-MS QP 2010 HP-5MS (SHIMADZU, Tokyo, Japan) was used coupled to UI elastic quartz capillary column (0.25 μm × 0.25 mm × 30 m). The splitless injection mode was used. The carrier gas was high purity helium which was used at a constant flow rate of 1 mL/min. The temperature of injection port was 250 °C. The heating program was as follows: the initial temperature was 40 °C and held for 5 min; the temperature was increased to 190 °C at a rate of 3 °C/min; then increased to 300 °C at a rate of 20 °C/min and held for 1 min.

4.5. MS Condition

The ion source temperature was 200 °C. The interface temperature was 230 °C. The solvent delay time was 2 min. The scanning range was *m/z* 45–500. The mass spectra would be obtained at 70 eV with electron ionization (EI) mode.

4.6. Data Processing

The mass spectral identification of volatile compounds was obtained by comparing with the National Institute of Standards and Technology (NIST) 14. The qualitative analysis of mass spectral data was verified by comparing the retention time and mass spectra. The quantitative analysis of volatile compounds was determined by peak area normalization of total ion chromatography.

4.7. Statistical Analysis

All multivariate analysis and calculations were performed on SIMCA-P software (version 14.1, Umetrics, Malmö, Sweden). The data were imported into the software and scaled by Pareto scaling method to reduce the relative importance of large values and to keep the data structure partially intact. Then, the data were submitted to PCA and OPLS-DA analysis. The differential compounds between female and male buds were filtered and used for the subsequent in-depth analysis.

5. Conclusions

In this study, HS-SPME-GC-MS technology coupled to multivariate statistics analysis was used to explore some difference in volatile compounds of *H. pedunculosum*. It was found that there was a significant difference between female and male buds. In total, 38 volatile compounds were identified by GC-MS, and 18 and 20 volatile compounds were detected in female and male buds, respectively. Among them, 6 unique compounds were detected, such as (*E*)-4-hexenol and isoamyl acetate in female buds, 20 unique compounds were detected such as (*Z*)-3-methylbutyraldehyde oxime and benzyl alcohol in male buds. In total, 9 differential volatile compounds of female and male buds were screened out, including isoamyl alcohol, (*Z*)-3-methylbutanal oxime, and 1-nitropentane based on multivariate statistical analysis. This is the first time that the differences in volatile compounds of the dioecious plant *H. pedunculosum* with HS-SPME-GC-MS have been reported, which reveals the essential differences in female and male buds.

Author Contributions: Data curation, Z.L. (Zhenying Liu), C.W. and X.H.; Formal analysis, Y.F.; Funding acquisition, Z.C.; Investigation, Z.L. (Zhenying Liu), P.S. and H.W.; Methodology, Z.L. (Zhenying Liu), Y.F. and X.H.; Resources, Z.C.; Software, B.X. and Z.L. (Zhuojun Li); Supervision, Z.C.; Validation, B.X. and Z.L. (Zhuojun Li); Visualization, P.S. and H.W.; Writing—original draft, Z.L. (Zhenying Liu); Writing—review and editing, Z.C. All authors have read and agreed to the published version of the manuscript.

Funding: Thanks for the support of China Agriculture Research System of the Ministry of Finance and the Ministry of Agriculture and Rural Areas (CARS-21) and Zhimao Chao Expert Workstation of Yunnan Province (202005AF150055).

Institutional Review Board Statement: Not applicable.

Informed Consent Statement: Not applicable.

Data Availability Statement: All the relevant data have been provided in the manuscript. The authors will provide additional details if required.

Conflicts of Interest: The authors confirm that there are no conflict of interest associated with this article.

Sample Availability: Not available.

References

1. Ma, L.; Yang, L.; Zhao, J.; Wei, J.; Kong, X.; Wang, C.; Zhang, X.; Yang, Y.; Hu, X. Comparative proteomic analysis reveals the role of hydrogen sulfide in the adaptation of the alpine plant *Lamiophlomis rotata* to altitude gradient in the northern Tibetan plateau. *Planta* **2015**, *241*, 887–906. [CrossRef] [PubMed]
2. Xu, B.; Liu, S.; Fan, X.D.; Deng, L.Q.; Ma, W.H.; Chen, M. Two new coumarin glycosides from *Herpetospermum caudigerum*. *J. Asian Nat. Prod. Res.* **2015**, *17*, 738–743. [CrossRef] [PubMed]
3. Yang, L.; Suo, Y.; Zhou, C.; Li, Y.; Wang, H. Study on trace elements in seeds of *Herpetospermum penduculosum* (Ser) Baill. *Trace Elem. Sci.* **2003**, *10*, 45–47.
4. Zhao, X.; Suo, Y.; Wang, L.; You, J.; Ding, C. Analysis of carbohydrates in a Tibetan medicine using new labeling reagent, 1-(2-naphthyl)-3-methyl-5-pyrazolone, by HPLC with DAD detection and ESI-MS identification. *J. Liq. Chromatogr. Relat. Technol.* **2008**, *31*, 2375–2400. [CrossRef]
5. Li, Q.; Li, H.J.; Xu, T.; Du, H.; Huan, G.C.; Fan, G.; Zhang, Y. Natural medicines used in the traditional Tibetan medical system for the treatment of liver diseases. *Front. Pharmacol.* **2018**, *9*, 1–16. [CrossRef] [PubMed]
6. Li, G.; Wang, X.Y.; Suo, Y.R.; Wang, H.L. Protective effect of seed oil of *Herpetospermum pedunculosum* against carbon tetrachloride-induced liver injury in rats. *Saudi Med. J.* **2014**, *35*, 981–987.
7. Hang, L.Y.; Shen, B.D.; Shen, C.Y.; Yang, K.; Yuan, H.L. Advances in modern research on anti-liver disease of Tibetan medicine Semen Hertospermi. *Chin. Trad. Her. Drug.* **2020**, *51*, 549–556.

8. Huang, D.; Ma, Y.X.; Wei, L.; Sun, Y.; Zeng, Q.H.; Lan, X.Z.; Liao, Z.H.; Chen, M. One new coumarin from seeds of *Herpetospermum pedunculosum*. *China J. Chin. Mater Med.* **2021**, *46*, 2514–2518. [CrossRef]
9. Liu, J.; Chen, X.; Zhang, Y.; Zhang, M. Chemical constituents of the ethyl acetate extract from Semen Herpetospermi. *Pharm Clin. Chin. Mater Med.* **2010**, *1*, 15–18.
10. Wang, C.; Wang, X.; Tseringand, T.; Song, Y.; Zhu, R. The plastid genome of *Herpetospermum pedunculosum* (Cucurbitaceae), an endangered traditional Tibetan medicinal herbs. *Mitochondrial DNA B Resour.* **2020**, *5*, 495–497. [CrossRef]
11. Xie, K. Molecular Sex Identification in *Herpetospermum pedunculosum* (Ser.) C.B. Clarke via RAPD and SCAR Markers. Master's Thesis, Chengdu University, Chengdu, China, 2021.
12. Feng, X.; Dengba, D.J.; Kong, S.X.; Li, H.K.; Shuya, K.; Zhong, G.J. Forecasting the seed yield of Tibetan herb medicine *Herpetospermum pedunculosum* Baill and analyzing for its influencing factors based on linear regression analysis model. *Mod. Chin. Med.* **2020**, *22*, 409–411+426. [CrossRef]
13. Zhao, Y.; Xu, F.; Liu, J.; Guan, F.; Quan, H.; Meng, F. The adaptation strategies of H*erpetospermum pedunculosum* (Ser.) Baill at altitude gradient of the Tibetan plateau by physiological and metabolomic methods. *BMC Genom.* **2019**, *20*, 451. [CrossRef] [PubMed]
14. Feng, Z.Y.; Ma, Y.X.; Wang, H.; Chen, M. Studies on chemical constituents of stems of *Herpetospermum pedunculosum*. *China J. Chin. Mater. Med.* **2020**, *45*, 2571–2577. [CrossRef]
15. Liu, R.; Tang, D.; Liu, Q. Experimental study on the extraction process of water-soluble polysaccharide from Tibetan medicine *Herpetospermum pedunculosum* flower. *J. Southwest Nat. Univ. (Nat. Sci. Ed.)* **2019**, *45*, 57–578. [CrossRef]
16. Zhang, Y.X.; Li, C.Y.; Liu, C.; Zhang, Y.; De, L. Preliminary study of quality standards of Tibetan medicine *Herpetospermum pedunculosum* flower. *J. Southwest Nat. Univ. (Nat. Sci. Ed.)* **2015**, *41*, 432–435. [CrossRef]
17. Chen, Z.Q.; Zhou, Z.L.; Wang, L.L.; Meng, L.H.; Duan, Y.W. Development of microsatellite markers for a dioecious *Herpetospermum pedunculosum* (Cucurbitaceae). *Evol. Bioinform. Online* **2020**, *16*, 1176934320908261. [CrossRef]
18. Chaieb, K.; Hajlaoui, H.; Zmantar, T.; Kahla-Nakbi, A.B.; Rouabhia, M.; Mahdouani, K.; Bakhrouf, A. The chemical composition and biological activity of clove essential oil, *Eugenia caryophyllata* (*Syzigium aromaticum* L. Myrtaceae): A short review. *Phytother. Res.* **2007**, *21*, 501–506. [CrossRef]
19. Hemalatha, R.; Nivetha, P.; Mohanapriya, C.; Sharmila, G.; Muthukumaran, C.; Gopinath, M. Phytochemical composition, GC-MS analysis, in vitro antioxidant and antibacterial potential of clove flower bud (*Eugenia caryophyllus*) methanolic extract. *J. Food Sci Technol.* **2016**, *53*, 1189–1198. [CrossRef]
20. Bitterling, H.; Schäfer, U.; Krammer, G.; Meier, L.; Brückner, S.I.; Hartmann, B.; Ongouta, J.; Carle, R.; Steingass, C.B. Investigations into the natural occurrence of 1-phenylethyl acetate (styrallyl acetate). *J. Agric. Food Chem.* **2020**, *68*, 8613–8620. [CrossRef]
21. Sonmezdag, A.S.; Kelebek, H.; Selli, S. Characterization of aroma-active compounds, phenolics, and antioxidant properties in fresh and fermented capers (*Capparis spinosa*) by GC-MS-olfactometry and LC-DAD-ESI-MS/MS. *J. Food Sci.* **2019**, *84*, 2449–2457. [CrossRef]
22. Hu, M.; Bai, M.; Ye, W.; Wang, Y.; Wu, H. Variations in volatile oil yield and composition of "Xin-yi" (*Magnolia biondii* Pamp. Flower Buds) at different growth stages. *J. Oleo Sci.* **2018**, *67*, 779–787. [CrossRef] [PubMed]
23. Ghosh, D.; Chaudhary, N.; Uma Kumari, K.; Singh, J.; Tripathi, P.; Meena, A.; Luqman, S.; Yadav, A.; Chanotiya, C.S.; Pandey, G.; et al. Diversity of essential oil-secretory cells and oil composition in flowers and buds of magnolia sirindhorniae and its biological activities. *Chem. Biodivers.* **2021**, *18*, e2000750. [CrossRef]
24. Xu, L.; Liu, H.P.; Ma, Y.C.; Wu, C.; Li, R.Q.; Chao, Z.M. Comparative study of volatile components from male and female flower buds of *Populus tomentosa* by HS-SPME-GC-MS[J]. *Nat. Prod. Res.* **2019**, *33*, 2105–2108. [CrossRef] [PubMed]
25. Horikoshi, H.M.; Sekozawa, Y.; Kobayashi, M.; Saito, K.; Kusano, M.; Sugaya, S. Metabolomics analysis of 'Housui' Japanese pear flower buds during endodormancy reveals metabolic suppression by thermal fluctuation. *Plant Physiol. Biochem.* **2018**, *126*, 134–141. [CrossRef] [PubMed]
26. McGinty, D.; Lapczynski, A.; Scognamiglio, J.; Letizia, C.S.; Api, A.M. Fragrance materials review on isoamyl alcohol. *Food Chem. Toxicol.* **2010**, *48*, S102–S109. [CrossRef] [PubMed]
27. Tu, R.; Lv, T.; Sun, L.; He, R.; Wang, Q. Development of a simple colorimetric assay for determination of the isoamyl alcohol-producing strain. *Appl. Biochem. Biotechnol.* **2020**, *192*, 632–642. [CrossRef]
28. Shaver, T.N.; Lingren, P.D.; Marshall, H.F. Nighttime variation in volatile content of bud of the night blooming plant *Gaura drummondii*. *J. Chem. Ecol.* **1997**, *23*, 2673–2682. [CrossRef]
29. Sun, W.; Chao, Z.M.; Wang, C.; Wu, X.Y.; Tan, Z.G. Research on the difference of volatile compounds in female and male buds of *Trichosanthes kirilowii* Maxim based on HS-SPME-GC-MS. *China J. Chin. Mater. Med.* **2012**, *37*, 1570–1574.
30. Twidle, A.M.; Mas, F.; Harper, A.R.; Horner, R.M.; Welsh, T.J.; Suckling, D.M. Kiwifruit flower odor perception and recognition by honey bees, *Apis mellifera*. *J. Agric. Food Chem.* **2015**, *63*, 5597–5602. [CrossRef]
31. Tang, Y.C.; Zhou, C.L.; Chen, X.M.; Zheng, H. Visual and olfactory responses during butterfly foraging. *J. Insect. Behav.* **2013**, *26*, 387–401. [CrossRef]
32. Zhu, X.Z.; Lu, Q.B.; Hu, X.H.; Deng, T.; Duan, Y.B.; Fang, Z.M.; Huang, S.X. Leaf volatile components and foraging insects of dioecious *Siraitia grosvenorii*: Sexual differences and its ecological effects. *Guihaia* **2020**, *40*, 1259–1268.

Article

Identification of Biomarker Volatile Organic Compounds Released by Three Stored-Grain Insect Pests in Wheat

Lijun Cai [1,2,3], Sarina Macfadyen [3], Baozhen Hua [2], Haochuan Zhang [4], Wei Xu [4,*] and Yonglin Ren [4,*]

[1] State Key Laboratory of Ecological Pest Control for Fujian/Taiwan Crops, Institute of Applied Ecology, Fujian Agriculture and Forestry University, Fuzhou 350002, China; cai-lijun@live.cn
[2] Key Laboratory of Plant Protection Resources and Pest Management, Ministry of Education, Northwest A&F University, Xianyang 712100, China; huabz@nwafu.edu.cn
[3] Agriculture and Food, Commonwealth Scientific and Industrial Research Organisation, Acton, ACT 2601, Australia; sarina.macfadyen@csiro.au
[4] College of Science, Health, Engineering and Education, Murdoch University, Murdoch, WA 6150, Australia; 32985118@student.murdoch.edu.au
* Correspondence: w.xu@murdoch.edu.au (W.X.); y.ren@murdoch.edu.au (Y.R.)

Abstract: Monitoring and early detection of stored-grain insect infestation is essential to implement timely and effective pest management decisions to protect stored grains. We report a reliable analytical procedure based on headspace solid-phase microextraction coupled with gas chromatography–mass spectrometry (HS-SPME-GC-MS) to assess stored-grain infestation through the detection of volatile compounds emitted by insects. Four different fibre coatings were assessed; 85 μm CAR/PDMS had optimal efficiency in the extraction of analytes from wheat. The headspace profiles of volatile compounds produced by *Tribolium castaneum* (Herbst), *Rhyzopertha dominica* (Fabricius), and *Sitophilus granarius* (Linnaeus), either alone or with wheat, were compared with those of non-infested wheat grains. Qualitative analysis of chromatograms showed the presence of different volatile compound profiles in wheat with pest infestation compared with the wheat controls. Wheat-specific and insect-specific volatile compounds were identified, including the aggregation pheromones, dominicalure-1 and dominicalure-2, from *R. dominica*, and benzoquinones homologs from *T. castaneum*. For the first time, the presence of 3-hydroxy-2-butanone was reported from *S. granarius*, which might function as an alarm pheromone. These identified candidate biomarker compounds can be utilized in insect surveillance and monitoring in stored grain to safeguard our grain products in future.

Keywords: volatile organic compounds (VOCs); headspace solid-phase microextraction (HS-SPME); SPME-GC-MS; stored-product insect pests; *Tribolium castaneum*; *Rhyzopertha dominica*; *Sitophilus granarius*

1. Introduction

Maintenance of quality parameters in stored grain over extended periods is of critical importance to global food security [1,2]. Annually, around one-third of food production, about 1.3 billion tonnes worth, approximately, US $1 trillion, is lost after harvest operations to infestation by pests and microorganisms [3]. Stored grains can be damaged by multiple insect pest species including *Tribolium castaneum* (Herbst), *Rhyzopertha dominica* (Fabricius), and *Sitophilus granarius* (Linnaeus), three of the most damaging stored-grain insect pests around the world. *R. dominica* and *S. granarius* are notorious primary pests of stored products that live in whole grain kernels, while *T. castaneum* is commonly considered as a secondary pest, appearing in stored products after the primary infestation. These species not only cause direct damage to stored grains through feeding, but also deteriorate grain quality [4]. Reliable and simple methods of early detection of these pests in stored grains are critical to their control throughout the supply chain, with implications for storage and the multi-billion-dollar international trade in grains.

To detect stored-product insects in grains, current techniques include near-infrared spectroscopy [5], digital imaging [6–9], and aroma sensing (techniques such as electronic nose, or "eNose") [9–13]. However, these methods are either labour intensive, susceptible to environmental influences, inefficient in detecting immature stages of insects and internal infestation, or not sensitive enough [14]. Novel methods of sample collection, insect detection, and data analysis should be developed to report the real-time quality of stored grains, allowing fast and appropriate management decisions. A potential detection approach is to investigate the air samples from a grain mass to detect specific volatile compounds (VOCs) released by damaging insects, which can be biomarkers for monitoring grain infestation.

Since its original use for the isolation and identification of volatile chemicals in the food industry, headspace solid-phase micro-extraction (HS-SPME), coupled with a gas chromatograph–flame ionisation detection (GC-FID) or gas chromatograph–mass spectrometry (GC-MS) technique, has been widely used to examine pheromones and other volatile secretions of coleopteran insects [15–19]. For example, this technique has been utilized to identify the aggregation pheromones, dominicalure 1 and dominicalure 2, of *R. dominica*, and how pest infestation influences the volatile profiles of grains, as well as volatile compounds of wheat alone [4]. The compound 4,8-dimethyldecanal (DMD), which is a well-known male aggregation pheromone in Tenebrionidae, and other volatile metabolites, such as benzoquinones in *Tribolium* spp., are potential indicators of infestation by these insects [20,21]. Conversely, little attention has been paid to investigating other volatile chemicals of *S. granarius*. A well-known exception is the male-produced aggregation pheromone identified as 1-ethylpropyl (2R, 3S)-2-methyl-3-hydroxypentanoate [22,23], commonly known as 2R, 3S-sitophilate. In addition, the HS-SPME technique has also been reported in detecting musty–earthy off-odours in wheat [24].

The aim of this study was to identify candidate VOCs released by pest infestation in stored grains using the HS-SPME-GC-MS technique. The volatile compound profiles obtained from non-infested grains were compared with those of grains infested with *R. dominica*, *T. castaneum*, or *S. granarius*. The volatile profiles of the three pest species alone were also examined. This study verifies changes in the relative responses of the target compounds among treatments and identified specific volatiles that can be used as potential indicators of pest infestations in stored grains.

2. Results

2.1. SPME Fibre Selection

Air only, as a blank check, was tested for each fibre as well, but they did not demonstrate any meaningful peaks, or the peaks were too tiny when compared with the signals released by the treatments such as wheat and insects. The chromatograms obtained from the four fibres tested are shown in Figure 1. The CAR/PDMS fibre showed the best extraction efficiency of the analytes from wheat because it detected the highest number of peaks and most peaks exhibited sharply. For example, we compared the identified compound peaks that were higher than 250 mV and were detected by each fibre. CAR/PMDS detected a total of 14 peaks while the other fibres detected less than two peaks. When the retention time (RT) was <5 min, CAR/PMDS detected more and larger peaks than the other three fibres. This result accords with the guide for fibre selection provided by the manufacturer for relatively small molecular weight compounds. Therefore, the 85 µm CAR/PDMS fibre was selected for subsequent method optimization experiments.

2.2. Selection of Sampling Time

The general volatile patterns of headspace tested at different time points are quite similar (Figure 2). The main peaks identified for wheat only were acetone, methanol, ethanol, 2,3-butanedione, 2-butanol, hexanal, 2-methyl-1-propanol, 1-butanol, 1-penten-3-ol, 3-methyl-1-butanol, 1-pentanol, 3-methyl-2-buten-1-ol, 1-hexanol, and 1-pentadecene, numbered from 1 to 14, respectively (Figure 2a). Other common compounds, such as dimethyl sulphide and ethyl formate, were also detected and identified but in quite low amounts. The peak areas of the main compounds detected are very close among the three time points (at 2–4, 22–24, and 46–48 h) (Table 1), except for acetone (peak 1) (Figure 2b).

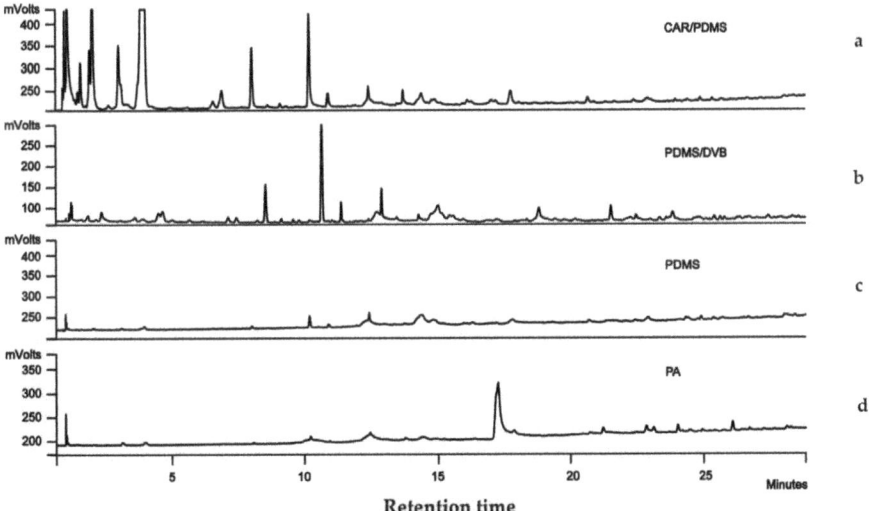

Figure 1. Comparison of different fibres. GC-FID chromatograms of headspace volatiles from wheat extracted using (**a**) CAR/PDMS; (**b**) PDMS/DVB; (**c**) PDMS; and (**d**) PA SPME-type fibres.

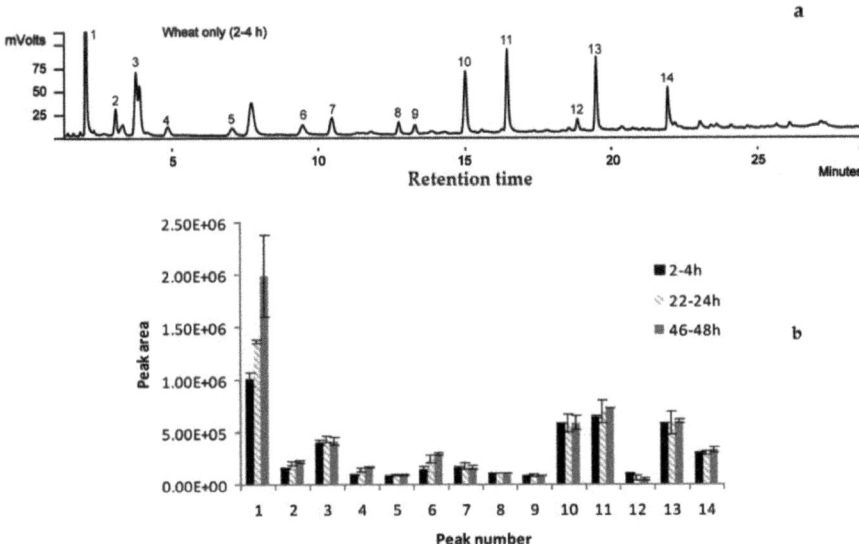

Figure 2. Comparison of mean GC responses of main peaks from healthy wheat collected at 2–4, 22–24, and 46–48 h after being sealed in an airtight container: (**a**) GC chromatograms of headspace volatiles from wheat only and (**b**) the GC peak areas of main peaks from healthy wheat collected at 2–4, 22–24, and 46–48 h after being sealed in an airtight container. Numbered peaks are: 1 = Acetone; 2 = Methanol; 3 = Ethanol; 4 = 2,3-Butanedione; 5 = 2-Butanol; 6 = Hexanal; 7 = 2-Methyl-1-propanol; 8 = 1-Butanol; 9 = 1-Penten-3-ol; 10 = 3-Methyl-1-butanol; 11 = 1-Pentanol; 12 = 3-Methyl-2-buten-1-ol; 13 = 1-Hexanol; and 14 = 1-Pentadecene. Each bar represents the average of three replicates and the error bars indicate standard deviation.

Table 1. The seven treatments used in GC, GC/MS analysis.

Sample	Volatile Collection Time (h)
20 g Homogenized wheat only	2–4, 22–24, 46–48
20 *Tribolium castaneum* (Herbst) adults only	2–4
20 *T. castaneum* adults + 20 g homogenized wheat	2–4
80 *Rhyzopertha dominica* (F.) adults only	22–24
80 *R. dominica* adults + 20 g homogenized wheat	22–24
100 *Sitophilus granarius* (L.) only	46–48
100 *S. granarius* adults + 20 g homogenized wheat	46–48

2.3. Single-Species Experiments

For all three insect species tested, treatments with insects only produced fewer volatile compounds when compared with the wheat-only controls. Each species had its own diagnostic volatile compounds which could be visually distinguished from each other (Figure 3a). *T. castaneum* adults only produced three significant peaks, namely T1, T2, and T3, the most abundant ions of which showed that they might be benzoquinones. The volatile pattern was the same with mixed gender samples or with adult male only or adult female only.

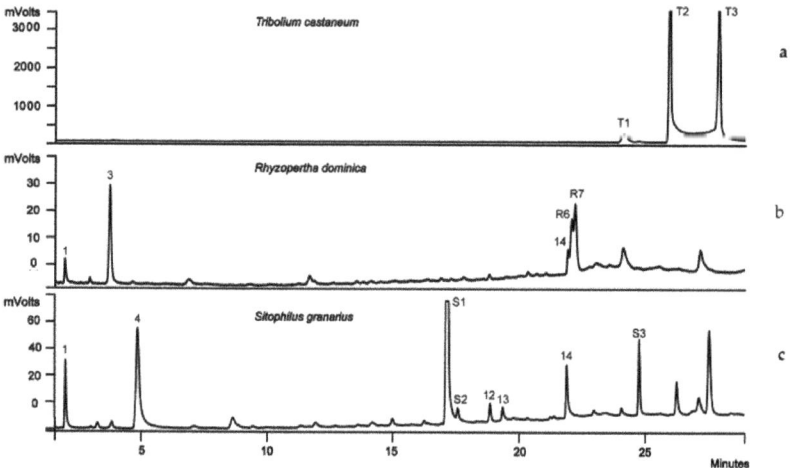

Figure 3. GC chromatograms of headspace volatiles from (**a**) *Tribolium castaneum*, (**b**) *Rhyzopertha dominica*, and (**c**) *Sitophilus granarius*. Peaks T1, T2, T3 are benzoquinones; peak 1 = Acetone; peak 3 = Ethanol; peak 4 = 2,3-Butanedione; peak 12 = 3-Methyl-2-buten-1-ol; peak 13 = 1-Hexanol; peak 14 = 1-Pentadecene; peak R6 = Dominicalure 2; peak R7 = Dominicalure 1; peaks S1–3 = typical peaks of *S. granarius* only; and peak S1 = 3-Hydroxy-2-butanone.

R. dominica adults exhibited two typical detectable peaks, R6 and R7, which were identified as domimicalure 2 (D2) and dominicalure 1(D1), respectively (Figure 3b). Peak 14, identified as 1-pentadecane, might be associated with traces of culture medium. Peak intensities of D1 and D2 were very low and did not increase over time. Besides the characteristic compounds of *R. dominica*, acetone, ethanol and 1-pentadecene (Peaks 1, 3, and 14, Figure 2) were also detected but peak heights were very small.

Detectable and reproducible signals from *S. granarius* adults could only be obtained after being held in a flask for over 40 h. The peaks characterizing *S. granarius* volatile emissions were marked as peaks S1, S2, and S3. Peak S1 was identified as 3-hydroxy-2-butanone (Figure 3c). In addition, several volatile compounds (peaks 1, 4, 12–14) originating from wheat were detected as trace amounts, and probably resulted from the wheat culture medium. These are numbered in Figure 2.

The volatile patterns of wheat only and wheat plus insects, for the three species tested, showed distinct patterns (Table 2). For wheat plus *T. castaneum*, the chromatogram turned out to be a simple combination of that of wheat only and that of *T. castaneum* only, but *T. castaneum* generated relevant high boiling point Benzene, 1-ethoxy-4-isothiocyanato, stearic acid, and one unknown compound (Table 2). Wheat plus *R. dominica* showed the most complex pattern; it not only had all the peaks from both wheat alone and *R. dominica* alone, but also seven new VOCs, such as ethyl acetate, 11-methylpentacosane, palmitic acid, 1-pentadecene, dominicalure 1, apparent homologs of dominicalure 1, and one unknown compound which could only be detected when *R. dominica* adults were added to wheat in the same flask (Peaks R1~R5 and Peaks R8~9 in Table 2). In addition, when *R. dominica* and wheat were combined, R6 (dominicalure 1) and R7 (dominicalure 2) increased dramatically and dwarfed any other components. In the case of wheat plus *S. granarius*, most of the typical VOCs of this species disappeared except VOC of 1-pentadecene, which was 2–3 folds enhanced in peak intensity than in wheat alone or in *S. granarius* alone (Table 2).

Table 2. Volatile organic compounds (VOCs) collected from wheat only, wheat plus *T. castaneum*, wheat plus *R. dominica*, and wheat plus *S. granarius* identified from gas chromatography analysis and Kovats' values calculation.

Compounds	RT	NIST RI	Kovats	Match Quality	GC Response (10^5) ± SD, n = 4			
	(min)		indices	(%)	Wheat	Wheat + *T. castaneum*	Wheat + *R. dominica*	Wheat + *S. granarius*
Acetone	1.18	1116	862	79.5	112.53 ± 6.77	124.36 ± 8.03	17.53 ± 3.35	127.18 ± 9.06
Methanol	2.51	1157	901	77.0	32.42 ± 4.29	43.28 ± 5.51	33.09 ± 4.52	32.57 ± 4.19
Ethanol	3.48	1215	943	79.1	75.52 ± 6.14	5.049 ± 2.75	49.51 ± 5.02	27.55 ± 3.66
2,3-Butanedione	4.43	1248	977	91.8	9.88 ± 2.22	15.49 ± 3.93	65.16 ± 5.15	12.17 ± 3.83
Ethyl acetate	5.38	1291	988	85.0	nd	nd	27.16 ± 3.33	nd
2-Butanol	7.02	1332	996	71.3	8.83 ± 2.04	7.59 ± 2.09	7.65 ± 2.08	6.94 ± 2.11
Hexanal	9.47	1360	1087	77.2	11.27 ± 3.15	19.08 ± 3.31	26.14 ± 3.81	23.51 ± 3.05
2-Methyl-1-propanol	10.49	1381	1125	80.1	23.53 ± 3.27	6.86 ± 2.88	7.15 ± 2.11	7.22 ± 2.02
11-Methylpentacosane	11.47	1418	1150	89.0	nd	nd	102.27 ± 8.04	nd
1-Butanol	12.53	1435	1173	71.9	13.84 ± 4.09	9.55 ± 2.73	9.17 ± 2.50	8.58 ± 2.22
1-Penten-3-ol	13.25	1463	1198	97.1	11.62 ± 3.15	8.11 ± 3.11	9.59 ± 2.72	8.06 ± 3.06
3-Methyl-1-butanol	14.82	1488	1216	94.3	76.77 ± 5.26	58.92 ± 5.38	61.49 ± 4.85	51.64 ± 5.17
1-Pentanol	16.34	1502	1238	84.4	103.52 ± 6.33	95.13 ± 7.09	117.25 ± 8.88	99.64 ± 7.61
Palmitic acid	17.57	1527	1251	96.0	nd	nd	55.72 ± 4.92	nd
3-Methyl-2-buten-1-ol	18.72	1544	1280	91.0	10.08 ± 2.01	4.73 ± 2.10	22.57 ± 3.27	4.08 ± 2.06
1-Hexanol	19.44	1581	1305	75.9	85.83 ± 5.09	78.59 ± 7.77	101.55 ± 8.21	91.21 ± 5.06
Unknown	20.81	-	-	-	nd	nd	31.09 ± 4.48	nd
1-Pentadecene	21.91	1640	1349	92.2	nd	nd	>250	nd
Dominicalure 1	23.58	1685	1377	88.7	nd	nd	26.05 ± 4.17	nd
Apparent homologs of dominicalure 1	24.23	1751	1391	94.0	nd	nd	143.95 ± 8.49	nd
Benzene, 1-ethoxy-4-isothiocyanato-	24.68	1830	1424	95.7	nd	74.93 ± 6.28	nd	nd
Stearic acid	26.11	1892	1451	75.9	nd	68.37 ± 5.92	nd	nd
Unknown	27.84	-	-	-	nd	11.5.08 ± 2.50	nd	nd

RT = retention time. NIST RI = retention indices obtained from National Institute of Standards and Technology database (NIST). SD = standard deviation (n = 4). nd = not detected.

3. Discussion

The HS-SPME technique coupled to GC-MS was utilized to investigate volatiles in infested stored grains for biomarkers of insect infestation. The CAR/PDMS coating fibre was recommended for the extraction of volatiles in grains infested with the tested species *R. dominica*, *T. castaneum*, and *S. granarius*. Previous research showed that the CAR/PDMS fibre had the highest efficiency of absorption of the overall volatile organic compounds from disturbed *T. castaneum* [21,25]. Our results confirmed that the CAR/PDMS fibre was efficient in extracting volatiles emitted by one single individual of *T. castaneum*. Whether using only one single individual or 20, the signals were detectable in the first 2 to 4 h after secretion (data not shown). In this study, only *R. dominica*, *T. castaneum*, and *S. granarius* adults were used to identify VOCs. Insect eggs, larvae, or pupae may produce different biomarker volatile compounds, which needs further investigation in future.

The samples containing wheat only showed the presence of the common volatile compounds, including dimethyl sulphide, acetone, methanol, ethanol, and butanol, which have been reported in previous research on clean wheat grains [26,27]. Nevertheless, fewer constituents were observed in our investigations, which were conducted under controlled laboratory temperatures, as compared with heated or distilled methods [26]. In addition, we found that more simple compounds in the form of alcohols were obtained in whole grains than in polished–graded wheat flours and distillers' grains, in which aldehydes and ketones are more common [15,26]. The increase of acetone levels (shown as increased peak areas) over time might be due to the natural accumulation of respiration products.

Volatile compounds of *T. castaneum* were abundant in peak intensity but were of a limited variety. Secretions of benzoquinones and the male-produced aggregation pheromone 4,8-dimethyldecanal (DMD), by *T. castaneum* have been well studied [20,21,28]. The extractable level and detected amounts vary between sampling methods and assay design. Our data showed that using the CAR/PDMS fibre was an efficient method for detecting benzoquinones. However, further identification is needed to confirm whether they are methyl-1,4-benzoquinone (MBQ) and ethyl-1,4-benzoquinone (EBQ), as reported in previous studies [20,21,29]. Similar to a previous study [21], no DMD was detected in our investigations. It was reported that amounts of DMD varied from 0.7 ng/male in a 4-day collection period with SPME analysis [20], to over 600 ng/day per male by using a Super Q column [30]. However, both two studies sexed and maintained adult *T. castaneum* separately for 30 days. Here, no DMD was detected in our experiment, which utilized unsexed *T. castaneum*. It was shown that high population densities have negative effects on DMD production [31], which were not apparent when food was absent. Nevertheless, DMD was not detected even when *T. castaneum* was combined in a sealed flask with wheat. The presence of primary pest species, such as *R. dominica* and *S. granarius*, in wheat might facilitate the release of aggregation pheromones in secondary pests.

The typical peaks R6 and R7 that characterise the presence of *R. dominica* increased over time when they were added to wheat grains. This finding is consistent with previous research which concluded that *R. dominica* releases aggregation pheromones when they identified food resources. However, it has been reported that the pheromone released by this species was highest in late afternoon and the variance of actual quantities could reach up to 10 times more among individuals [32]. Here, we used a large population density of 80 adults per flask, so differences between individuals were not examined. In our work, we reported only a few compounds compared with around 90 compounds identified previously [4]. In the latter work, *R. dominica* was killed and removed before the headspace collection of samples. A long infestation period of 25 days, metabolites of the insects, and microbial contamination might also have contributed to the larger number of compounds detected. Moreover, heating can increase mass transfer, and this can lead to a greater recovery of higher molecular weight volatile compounds. Our experiments were carried out at room temperature and in a relatively natural storage environment. This is a useful way to avoid artificial reactions of volatile reagents in the headspace, but it is less efficient in completely absorbing all the volatiles, particularly those of low volatility. In another study, with both polar and non-polar columns at 25 °C, 114 compounds were detected in the headspace of three different samples, healthy wheat, *R. dominica*, and wheat, with *R. dominica*, indicating that the use of both polar and non-polar columns is essential to capturing the full range of VOCs produced [33].

S. granarius adults presented the most characteristic peaks compared with the other two pest species tested, besides some components in common with those of wheat. Here, for the first time, the compound 3-hydroxy-2-butanone was found in *S. granarius*. It has been reported as one of the three male-produced sex pheromones in the lobster cockroach, *Nauphoeta cinerea* (Olivier), and functions in agonistic interaction [34]. We suggest that this compound may be an alarming pheromone among *S. granarius* adults when they are under crowded and starving conditions. Further bioassay or behavioural function studies are needed to confirm its bio-function. To our knowledge, this is the first study examining the volatile compounds of grains infested by *S. granarius*. Moreover, 1-pentadecene increased two-fold in the headspace of wheat plus *S. granarius*, and it appeared to be a potentially useful indicator of infestation by *S. granarius*. It would be of interest to determine whether the compounds coming from *S. granarius* adults alone (S2 and S3) are absorbed by wheat or are not released once insects have found a food resource. In a recent study, wheat bran, with a low amount of 1-pentadecene added, was more attractive than wheat bran alone to *T. castaneum*, whereas higher concentrations of 1-pentadecene were repellent [35].

Sensitivity and reproducibility are the critical factors to consider in developing detection techniques for grain infestation. The fact that bulk grains are good sorbents to weak physical signals in the form of electrical, magnetic, nuclear, acoustic, or thermal energy is one of the greatest challenges for detection techniques. Our study showed that detectable signals could be recorded for only a single adult *T. castaneum*. However, further work needs to be conducted to determine the detection limits of this technique for *R. dominica* and *S. granarius*, mainly in non-laboratory conditions.

In summary, a simple and reliable detection method for stored-product pests is urgently needed in the grain industry considering the great annual losses during storage due to pest infestations. The results, here, demonstrated that SPME-GC-MS is an efficient technique and has the potential to identify volatile chemical compounds of different insect species which can be used as early warning indicators of insect presence or infestation. A complete VOC profile of all the main stored-product pests would be necessary for future rapid diagnosis and monitoring to draw practical conclusions.

4. Materials and Methods

4.1. Insect Materials

Tribolium castaneum (TC4), *Rhyzopertha dominica* (RD2), and *Sitophilus granarius* (LG2) were obtained from the CSIRO Entomology Culture Collection, Canberra, Australia [36]. Single-species insect cultures were set up using adults and reared in 2 L glass culture jars under dark laboratory conditions at 30 °C and 60% relative air humidity. Two grams of *T. castaneum* were cultured on 734 g of whole organic wheat flour and 66 g of brewer's yeast at a ratio of 12:1 (*w/w*). *R. dominica* (1.5 g) were cultured on 721 g of organic whole wheat grains with one cup of wheat flour on top. *S. granarius* (2.8 g) were cultured on 800 g of organic whole wheat grain. The adult insects were cultured for 4–5 weeks until adults of the next generation emerged.

4.2. SPME Fibre Selection

Four SPME fibres with different coatings were tested and compared with the extraction efficiency of volatile compounds: 85 μm Carboxen/Polydimethylsiloxane (CAR/PDMS) (Sigma-Aldrich, Sydney, Australia, Cat. 57334-U), 65 μm Polydimethylsiloxane/Divinylbenzene (PDMS/DVB) (Sigma-Aldrich, Sydney, Australia, 57326-U), 85 μm polyacrylate (PA) (Sigma-Aldrich, Sydney, Australia, 57304), and 100 μm Polydimethylsiloxane (PDMS) (Sigma-Aldrich, Sydney, Australia, 57300-U). All fibres were conditioned prior to use in accordance with the manufacturer's recommendations. The wheat variety 'Rosella' was used for testing the fibres. Grain of this variety was stored at 10 °C until used. Two hundred grams of wheat grain was added to a 250 mL Erlenmeyer flask (Bibby Sterilin, Staffordshire, UK, Cat. No. FE 250) and sealed with a Teflon-faced septum lid. There were three replicates for each treatment and the control. Sample collection was performed manually using a SPME fibre holder. The coated fibre was exposed to the upper portion of the headspace of wheat grains, just below the septum, for 3 h without agitation. The compounds were then desorbed from the fibre in the GC injector at 250 °C for 5 min under splitless mode.

4.3. Selection of Sampling Time

To standardize sampling time according to volatile emissions for the different insects used (treatments with insects) and monitor possible changes in volatile compound profiles in 250 mL sealed flasks over 48 h (treatment with wheat only), the fibres were suspended in the flask above the insects or wheat, or wheat plus insects, and out of reach of the insects to extract the volatiles for 3 h. According to our pre-experiments, the collection time used in this research was selected as the earliest timepoint that each species can give out a consistent and stable release (Table 1). For 20 g of homogenized wheat only, headspace samples were collected at 2–4, 22–24, and 46–48 h, respectively, as controls. For 20 *T. castaneum* only and 20 *T. castaneum* plus 20 g homogenized wheat, headspace samples were collected at 2–4 h for investigation. For 20 *R. dominica* only and 20 *R. dominica* plus 20 g homogenized wheat, headspace samples were collected at 22–24 h for investigation. For 20 *S. granarius* only and 20 *S. granarius* plus 20 g homogenized wheat, headspace samples were collected at 46–48 h for investigation (Table 1).

4.4. Single-Species Experiments

Single-species experiments were performed for the three pest species using 250 mL Erlenmeyer flasks sealed with a Teflon-faced septum lid. The experimental design consisted of a wheat control without insects, and treatments with either insects only or with wheat plus insects (Table 1). Mixed-age and mixed-gender batches of insects were used in all runs. According to our pre-experiments, the amount of the insects used in this research was the limit that can provide a consistent and stable VOCs release. The test insects were transferred on wet filter paper for half an hour allowing them to crawl, cleaning the insect body, and the insects were then cleaned further by transferring them to filter paper. Thirty minutes were allowed for the insects to settle before the flasks were sealed. There were three replicates for each treatment and control. The wheat grains used in all the other experiments were harvested and stored at 10 °C until used. Samples were then transferred into a 25 °C cabinet overnight. The volatile collection method referred to is described in the above "fibre selection". Periodic blank flasks were tested as procedural controls to confirm the stability and purity of the system.

4.5. GC and GC-MS Analysis

A Varian 3400CX GC (Varian Instruments, Sunnyvale, CA, USA) equipped with a split–splitless injector, a ZB-WAXplus column (30 m × 0.32 mm i.d. × 0.25 μm film thickness), and a Flame Ionization Detector (FID) was used to analyse the volatile profiles extracted by SPME. The gas chromatograph oven was operated under the following temperature program conditions: 35 °C for 8 min, increasing to 120 °C at a rate of 5 °C/min, and held at 120 °C for 10 min. FID temperature was set at 250 °C. Nitrogen was used as the carrier gas, injected at 250 °C in a constant flow of 1.1 mL min^{-1}.

Analysis and identification of analytes was carried out on a Shimazu gas chromatograph–mass spectrometer (GC-MS-QP2010 Plus). A Stabilwax® Restek column (30 m × 0.25 mm i.d. × 0.25 μm film thickness) was used with helium as the carrier gas at a constant speed of 30 cm/s. Oven temperature programme conditions included being held at 35 °C for 8 min, increased from 35 to 120 °C at 5 °C/min where it was held for 10 min, then increased from 120 to 245 °C at 15 °C/min where it was held for 10 min. Total run time was 53.3 min. Electron ionization was at 70 eV. Qualitative analysis of the samples was carried out by scanning the mass range between 40 and 300 amu. Peaks were identified by mass spectra comparisons against the standards of the National Institute of Standards and Technology (NIST) spectral library database and by comparison with the retention times of known authentic standards (Sigma-Aldrich, Pty, Ltd., MD, USA).

4.6. Data Analysis

The GC analysis data, including the retention time and the peak area, were collected and integrated into the chromatography software, Agilent Chem Station, and then exported to Microsoft Excel and SPSS 20.0 for statistical analysis. The chromatogram pattern features, including detected peak retention times and peak areas, were analysed and compared to verify the repeatability of replicates from the same treatment. The variance between peak areas was analysed using an ANOVA (single factor) with post hoc Tukey's test by using SPSS 20.0 to compare volatile emissions between different treatments, including: wheat only, wheat plus *T. castaneum*, wheat plus *R. dominica*, and wheat plus *S. granarius*.

Author Contributions: Conceptualization, L.C. and Y.R.; methodology, L.C. and Y.R.; formal analysis, S.M. and H.Z.; resources, W.X. writing—original draft preparation, L.C.; writing—review and editing, W.X. and B.H.; supervision, Y.R. and S.M.; project administration, Y.R. and S.M.; funding acquisition, Y.R. All authors have read and agreed to the published version of the manuscript.

Funding: We thank the financial support from the Special Funds from the Civil Services Bureau of Fujian Province for Early Career Researcher (132120030). Wei Xu is funded by an Australian Research Council Discovery Early Career Researcher Award (DECRA) (DE160100382).

Institutional Review Board Statement: Not applicable.

Informed Consent Statement: Not applicable.

Data Availability Statement: The datasets during and/or analysed during the current study are available from the corresponding author on reasonable request.

Acknowledgments: We are indebted to Amalia Berna, Rosangela A. Devilla, Daphne Mahon, Benjamin Padovan, Julie Cassells, James Darby (Division of Entomology, CSIRO Black Mountain Laboratories), and Hui Cheng (Stored Grain Research Laboratory, Murdoch University, Australia), for their assistance during our investigation. Thanks also go to Martin Parkes for his kind comments on the manuscript. The lead author's research in Australia is supported by the CSIRO Ministry of Education of the China Joint-Supervising Ph.D. Student Project. This research was financially supported by the Grains Research Development Corporation (GRDC).

Conflicts of Interest: The authors declare no conflict of interest.

Sample Availability: Samples of the compounds, including acetone, methanol, ethanol, 2,3-butanedione, 2-butanol, hexanal, 2-methyl-1-propanol, 1-butanol, 1-penten-3-ol, 3-methyl-1-butanol, 1-pentanol, 3-methyl-2-buten-1-ol, 1-hexanol, and 1-pentadecene are available from the authors.

References

1. Orriss, G.D.; Whitehead, A.J. Hazard analysis and critical control point (HACCP) as a part of an overall quality assurance system in international food trade. *Food Control* **2000**, *11*, 345–351. [CrossRef]
2. Whitehead, A.J. Elements of an effective national food control-system. *Food Control* **1995**, *6*, 247–251. [CrossRef]
3. Kumar, D.; Kalita, P. Reducing postharvest losses during storage of grain crops to strengthen food security in developing countries. *Foods* **2017**, *6*, 8. [CrossRef] [PubMed]
4. Seitz, L.M.; Ram, M.S. Metabolites of lesser grain borer in grains. *J. Agric. Food Chem.* **2004**, *52*, 898–908. [CrossRef] [PubMed]
5. Cassells, J.A.; Reuss, R.; Osborne, B.G.; Wesley, I.J. Near infrared spectroscopic studies of changes in stored grain. *J. Near Infrared Spec.* **2007**, *15*, 161–167. [CrossRef]
6. Brosnan, T.; Sun, D.W. Inspection and grading of agricultural and food products by computer vision systems: A review. *Comput. Electron. Agric.* **2002**, *36*, 193–213. [CrossRef]
7. Chen, Y.R.; Chao, K.L.; Kim, M.S. Machine vision technology for agricultural applications. *Comput. Electron. Agric.* **2002**, *36*, 173–191. [CrossRef]
8. Luo, X.; Jayas, D.S.; Symons, S.J. Identification of damaged kernels in wheat using a colour machine vision system. *J. Cereal Sci.* **1999**, *30*, 49–59. [CrossRef]
9. Olsson, J.; Borjesson, T.; Lundstedt, T.; Schnurer, J. Detection and quantification of ochratoxin A and deoxynivalenol in barley grains by GC-MS and electronic nose. *Int. J. Food Microbiol.* **2002**, *72*, 203–214. [CrossRef]
10. Evans, P.; Persaud, K.C.; McNeish, A.S.; Sneath, R.W.; Hobson, N.; Magan, N. Evaluation of a radial basis function neural network for the determination of wheat quality from electronic nose data. *Sens. Actuators B-Chem.* **2000**, *69*, 348–358. [CrossRef]
11. Keshri, G.; Magan, N. Detection and differentiation between mycotoxigenic and non-mycotoxigenic strains of two *Fusarium* spp. using volatile production profiles and hydrolytic enzymes. *J. Appl. Microbiol.* **2000**, *89*, 825–833. [CrossRef] [PubMed]
12. Keshri, G.; Voysey, P.; Magan, N. Early detection of spoilage moulds in bread using volatile production patterns and quantitative enzyme assays. *J. Appl. Microbiol.* **2002**, *92*, 165–172. [CrossRef] [PubMed]
13. Ridgway, C.; Chambers, J.; Portero-Larragueta, E.; Prosser, O. Detection of mite infestation in wheat by electronic nose with transient flow sampling. *J. Sci. Food Agric.* **1999**, *79*, 2067–2074. [CrossRef]
14. Neethirajan, S.; Karunakaran, C.; Jayas, D.S.; White, N.D.G. Detection techniques for stored-product insects in grain. *Food Control* **2007**, *18*, 157–162. [CrossRef]
15. Biswas, S.; Staff, C. Analysis of headspace compounds of distillers grains using SPME in conjunction with GC/MS and TGA. *J. Cereal Sci.* **2001**, *33*, 223–229. [CrossRef]
16. Ginzel, M.D.; Hanks, L.M. Role of host plant volatiles in mate location for three species of longhorned beetles. *J. Chem. Ecol.* **2005**, *31*, 213–217. [CrossRef]

17. Lacey, E.S.; Ginzel, M.D.; Millar, J.G.; Hanks, L.M. Male-produced aggregation pheromone of the Cerambycid beetle *Neoclytus acuminatus acuminatus*. *J. Chem. Ecol.* **2004**, *30*, 1493–1507. [CrossRef]
18. Pelusio, F.; Nilsson, T.; Montanarella, L.; Tilio, R.; Larsen, B.; Facchetti, S.; Madsen, J.O. Headspace solid-phase microextraction analysis of volatile organic sulfur-compounds in black-and-white truffle aroma. *J. Agric. Food Chem.* **1995**, *43*, 2138–2143. [CrossRef]
19. Rochat, D.; Ramirez-Lucas, P.; Malosse, C.; Aldana, R.; Kakul, T.; Morin, J.P. Role of solid-phase microextraction in the identification of highly volatile pheromones of two *Rhinoceros* beetles *Scapanes australis* and *Strategus aloeus* (Coleoptera, Scarabaeidae, Dynastinae). *J. Chromatogr. A* **2000**, *885*, 433–444. [CrossRef]
20. Arnaud, L.; Lognay, G.; Verscheure, M.; Leenaers, L.; Gaspar, C.; Haubruge, E. Is dimethyldecanal a common aggregation pheromone of *Tribolium* flour beetles? *J. Chem. Ecol.* **2002**, *28*, 523–532. [CrossRef]
21. Villaverde, M.L.; Juarez, M.P.; Mijailovsky, S. Detection of Tribolium castaneum (Herbst) volatile defensive secretions by solid phase micro extraction-capillary gas chromatography (SPME-CGC). *J. Stored Prod. Res.* **2007**, *43*, 540–545. [CrossRef]
22. Phillips, J.K.; Chong, J.M.; Andersen, J.F.; Burkholder, W.E. Determination of the enantiomeric composition of (R-Star, S-Star)-1-ethylpropyl 2-methyl-3-hydroxypentanoate, the male-produced aggregation pheromone of *Sitophilus granarius*. *Entomol Exp. Appl.* **1989**, *51*, 149–153. [CrossRef]
23. Phillips, J.K.; Miller, S.P.F.; Andersen, J.F.; Fales, H.M.; Burkholder, W.E. The chemical-identification of the granary weevil aggregation pheromone. *Tetrahedron Lett.* **1987**, *28*, 6145–6146. [CrossRef]
24. Jelen, H.H.; Majcher, M.; Zawirska-Wojtasiak, R.; Wiewiorowska, M.; Wasowicz, E. Determination of geosmin, 2-methylisoborneol, and a musty-earthy odor in wheat grain by SPME-GC-MS, profiling volatiles, and sensory analysis. *J. Agric. Food Chem.* **2003**, *51*, 7079–7085. [CrossRef]
25. Laopongsit, W.; Srzednicki, G.; Craske, J. Preliminary study of solid phase micro-extraction (SPME) as a method for detecting insect infestation in wheat grain. *J. Stored Prod. Res.* **2014**, *59*, 88–95. [CrossRef]
26. Maeda, T.; Kim, J.H.; Ubukata, Y.; Morita, N. Analysis of volatile compounds in Polished-graded wheat flours using headspace sorptive extraction. *Eur. Food Res. Technol.* **2008**, *227*, 1233–1241.
27. Mcwilliams, M.; Mackey, A.C. Wheat flavor components. *J. Food Sci.* **1969**, *34*, 493–496. [CrossRef]
28. Yezerski, A.; Gilmor, T.P.; Stevens, L. Genetic analysis of benzoquinone production in *Tribolium confusum*. *J. Chem. Ecol.* **2004**, *30*, 1035–1044. [CrossRef]
29. Alnajim, I.; Agarwal, M.; Liu, T.; Ren, Y.L. A novel method for the analysis of volatile organic compounds (VOCs) from red glour beetle *Tribolium castaneum* (H.) using headspace-SPME technology. *Curr. Anal. Chem.* **2020**, *16*, 404–412. [CrossRef]
30. Qazi, M.C.B.; Boake, C.R.B.; Lewis, S.M. The femoral setiferous glands of *Tribolium castaneum* males and production of the pheromone 4,8-dimethyldecanal. *Entomol. Exp. Appl.* **1998**, *89*, 313–317. [CrossRef]
31. Suzuki, T.; Sugawara, R. Isolation of an aggregation pheromone from the flour beetles, *Tribolium castaneum* and *Tribolium confusum* (Coleoptera, Tenebrionidae). *Appl. Entomol. Zool.* **1979**, *14*, 228–230. [CrossRef]
32. Bashir, T.; Birkinshaw, L.A.; Farman, D.; Hall, D.R.; Hodges, R.J. Pheromone release by *Rhyzopertha dominica* (F) (Coleoptera: Bostrichidae) in the laboratory: Daily rhythm, inter-male variation and association with body weight and/or boring activity. *J. Stored Prod. Res.* **2002**, *39*, 159–169. [CrossRef]
33. Niu, Y.; Hua, L.; Hardy, G.; Agarwal, M.; Ren, Y. Analysis of volatiles from stored wheat and *Rhyzopertha dominica* (F.) with solid phase microextraction-gas chromatography mass spectrometry. *J. Sci. Food Agric.* **2016**, *96*, 1697–1703. [CrossRef] [PubMed]
34. Kou, R.; Chen, S.C.; Chen, Y.R.; Ho, H.Y. 3-Hydroxy-2-butanone and the first encounter fight in the male lobster cockroach, *Nauphoeta cinerea*. *Naturwissenschaften* **2006**, *93*, 286–291. [CrossRef]
35. Dukic, N.; Andric, G.; Glinwood, R.; Ninkovic, V.; Andjelkovic, B.; Radonjic, A. The effect of 1-pentadecene on *Tribolium castaneum* behaviour: Repellent or attractant? *Pest. Manag. Sci.* **2021**, *77*, 4034–4039. [CrossRef]
36. Stevenson, B.J.; Cai, L.; Faucher, C.; Michie, M.; Berna, A.; Ren, Y.; Anderson, A.; Chyb, S.; Xu, W. Walking responses of *Tribolium castaneum* (Coleoptera: Tenebrionidae) to its aggregation pheromone and odors of wheat infestations. *J. Econ. Entomol.* **2017**, *110*, 1351–1358. [CrossRef]

Article

Response of Aphid Parasitoids to Volatile Organic Compounds from Undamaged and Infested *Brassica oleracea* with *Myzus persicae*

Qasim Ahmed [1], Manjree Agarwal [2], Ruaa Alobaidi [3], Haochuan Zhang [2] and Yonglin Ren [2,*]

[1] Department of Plant Protection, College of Agricultural Engineering Sciences, Al-Jadriya Campus, University of Baghdad, Baghdad 10071, Iraq; qasim.h@coagri.uobaghdad.edu.iq
[2] Department of Agricultural Sciences, College of Science, Health, Engineering and Education, Murdoch University, South Street, Murdoch, WA 6150, Australia; m.agarwal@murdoch.edu.au (M.A.); 32985118@student.murdoch.edu.au (H.Z.)
[3] Department of Clinical Laboratory Sciences, College of Pharmcy, Al-Qadisyia Campus, Al-Mustansiriyah University, Baghdad 10052, Iraq; ruaa_abdulsattar@uomustansiriyah.edu.iq
* Correspondence: y.ren@murdoch.edu.au

Abstract: Headspace solid microextraction (HS-SPME) and GC-MS were used to investigate volatile organic compounds (VOCs) from cabbage plants infested and uninfested with green peach aphid *Myzus persicae*. The HS-SPME combined with GC-MS analysis of the volatiles described the differences between the infested and uninfested cabbage. Overall, 28 compounds were detected in infested and uninfested cabbage. Some VOCs released from infested cabbage were greater than uninfested plants and increased the quantity of the composition from infested plants. According to the peak area from the GC-MS analysis, the VOCs from infested cabbage consisted of propane, 2-methoxy, alpha- and beta pinene, myrcene, 1-hexanone, 5-methyl-1-phenyl-, limonene, decane, gamma-terpinen and heptane, 2,4,4-trimethyl. All these volatiles were higher in the infested cabbage compared with their peak area in the uninfested cabbage. The results of the study using a Y-shape olfactometer revealed that the VOCs produced by infested cabbage attracted *Myzus persicae* substantially more than uninfested plants or clean air. The percentage of aphid choice was 80% in favor of infested cabbage; 7% were attracted to the clean air choice and uninfested plants. A total of aphids 7% were attracted to clean air. Comparing between infested and uninfested cabbage plants, the aphid was attracted to 63% of the infested cabbage, versus 57% of the uninfested cabbage. The preferences of *Aphidus colemani* and *Aphelinus abdominalis* to the infested or uninfested plants with *M. persicae* and compared with clean air indicated that parasitoids could discriminate the infested cabbage. Both parasitoids significantly responded to the plant odor and were attracted to 86.6% of the infested cabbage plants.

Keywords: green peach aphids; VOCs; parasitoids; *Aphidus colemani*; *Aphelinus abdominalis*; cabbage

1. Introduction

Myzus persicae (Hemiptera: Aphididae) has a universal distribution, including Australia, and is considered a serious pest that has caused damage to hundreds of agricultural crops in more than 66 families [1,2]. The aphid mainly exists in young plant tissues, causing reduced leaf size, delayed growth of the plant and reduced yield [3]. *M. persicae* is considered a common pest insect of cruciferous crops, and sucks plant sap, leading to yellowing and curling of plant leaves. Additionally, the excretion of honeydew by aphids affects plant photosynthesis and encourages fungal growth [4]. Cabbage plants are commonly attacked by different species of aphids, such as turnip aphid *Lipaphis erysimi*, cabbage aphid *Brevicoryne brassicae* and green peach aphid *M. persicae*, which economically damage these crops [5].

Chemical insecticides play a significant role in controlling insects on crop plants. Insecticides have been extensively used in horticultural systems; however, they can cause the appearance of secondary pests instead of primary pests, pesticide resistance, contamination of environment and affect non-target organisms [6,7]. Therefore, it is necessary to find alternative methods for pest management. In biological control, aphid parasitoids from families such as Braconidae and Aphelinidae are important and can cause a high percentage of mortality on aphids [8,9]. Natural enemies of aphids can reduce the rate of population increase, and the use of wasp parasitoids in biological control of aphids has been successful [10].

Plants VOCs play an important role in plant–insect interactions by influencing insect communication and plant defense [11]. When sucking insect pests such as the green peach aphid feed on the plant, one response from the plant is to release odors in the form of VOCs. The VOCs have an important role in plant–insect interactions because they can be used by parasitoids to locate their host [12].

Cabbage plants attacked by aphids may emit volatile compounds that attract parasitoid wasps or predators [13,14]. Previous studies have concluded that natural enemies can identify the VOCs released from the infested plants; the response of parasitoids and predators were confirmed and this provided an explanation how natural enemies were attracted by the host plant using the olfactory scale [15,16].

Aphidius colemani (Hymenoptera, Braconidae) and *Aphelinus abdominalis* (Hymenoptera, Aphelinidae) are endoparasitoids of many species of aphids and both attack *M. perisecae* [17]. The VOCs released from infested *Brassica* plants by aphids can bring with lure parasitoids, which showed the family of Brassicaceae possess chemical defense [18]. When the aphids feed on the plant leaves, the plant produces blends of volatiles as a response to the infestation by aphids, releases volatile compounds in different quantities and qualities from damaged Brassica plants, and these differences in the VOCs can attract other pests and natural enemies [17]. *A. colemani* and *A. abdominalis* are parasitic wasps specific to green peach aphids, whose females use VOC signals to detect and locate aphids feeding on host plants and lay their eggs into aphids [19,20]. Additionally, honeydew excreted by aphids on plants could lead to the release of semiochemicals or VOCs attracting and guiding parasitoids to the aphid [21,22].

In Y-tube olfactometer tests, Reed [8] reported no attraction of the parasitoid *Diaeretiella rapae* to the cabbage leaves. However, the choice of wasps to infest cabbage plants by *B. Brassicae* was more significant than other plants infested by different species of aphid, such as Russian wheat aphid *Diuraphis noxia*. These results indicate that the cabbage plant VOCs are more important than other plants in attracting the parasitoid to the aphid location [8]. The heavy population of *M. persicae* on the plant can accumulate wasps, while the uninfested plant sees few parasitoids come to the plant because wasps fail to locate the uninfested plant [23,24].

The identification of VOCs can be a signal for aphids and their parasitoids' receivers, and it is necessary to develop methods to analyze VOCs as diagnostic indicators that involve aphid management. Therefore, this study aims to determine the VOCs released from *M. persicae*-infested and uninfested cabbage plants to elucidate the responses of *M. persicae* and their parasitoids (*A. colemani* and *A. abdominalis*) to aphid-infested and uninfested cabbage plants in the Y-tube olfactometer. Understanding the treatments influencing the attraction of the parasitoids may provide fundamental data for controlling green peach aphids and generating new methods for aphid biological control.

2. Results

2.1. VOCs Released from M. persicae Infested and Uninfested Plant

Analysis of the volatiles of cabbage induced by *M. persicae* for the infested and uninfested plant treatments shows significant differences. Several compounds were present in all samples that were trapped by SPME and identified by GC-MS. Plants damaged by *M. persicae* can change in plant odor emission, and the volatiles of samples were significantly higher than uninfested plants. The volume and the variety of VOCs released from infested cabbage were greater than the uninfested plant in some compounds, and the qualitative differences in the composition of the odor from these treated plants consisted of propane, 2-methoxy that was released from uninfested cabbage, which was greater than the VOCs released from infested cabbage, with an average peak area in the uninfested plant of 23.10 compared with the peak area in the infested plant of 7.84. Meanwhile, alpha- and beta pinene were much higher in the infested than uninfested plants (Table 1). There was a significantly larger quantity of (E)-3-hexen-1-ol (p-value 0.223), beta-pinene (p-value 0.930) and decane (p-value 0.020) released from the infested plant but not detected in uninfested cabbage plants. Moreover, the peak area for the following volatile compounds, which were detected from infested cabbage, were higher in the infested cabbage compared with their peak area in the uninfested cabbage: myrcene, 1-hexanone, 5-methyl-1-phenyl-, limonene, decane, gamma-terpinen and heptane, 2,4,4-trimethyl. However, some of the volatile compounds from uninfested cabbage were released in a high amount based on peak area detected by GC-MS as compared with the infested plant. These compounds were eucalyptol, cyclohexasiloxane, 3,4-dihydroxyphenylglycol, 1,5-pentanediamine, octamethyl and decamethyl. VOCs lead to odor differences between aphid infested plants and uninfested plants. Figure 1 shows the heat map that graphically displays results by hierarchical clustering of the volatile compounds from the infested and uninfested cabbage. This work was conducted to find the closeness of individual compounds released from both samples (uninfested and infested plants with M. persicae). Distances between samples and assays were calculated for hierarchical clustering based on Pearson's Correlation Distance. Each volatile compound has a peak area detected by GC-MS, presented by the color scale that illustrates the differences between the replicates of the infested and uninfested cabbage. The heat map indicated that the detected compounds and the difference between uninfested and infested cabbage plant with the scale of color and each color corresponds to one detected VOC. The value of the compound is represented by red, orange and dark blue for the maximum (2), average (0) and minimum (−2) (Figure 1). In addition, principal component analysis (PCA) was performed and the PCA score plot (Figure 2) shows the separation of the two samples (uninfested and infested plants with M. persicae) into two different groups based on their profile of volatile organic compound using the significant difference ($p < 0.05$), relationship between the VOCs within infested and uninfested as shown in Figure 2.

Table 1. Volatile compounds detected in the headspace of infested and uninfested cabbage with M. persicae by using solid phase microextraction (SPME).

No	Compound Name	RT [1]	Uninfested Plant Area ± SD [2]	Infested Plant Area ± SD	LSD [4]	p-Value
1	Propane, 2-methoxy	3.12	23.10 ± 3.13	7.84 ± 2.70	11.45	0.020 *
2	n-Hexane	3.28	15.38 ± 4.21	8.40 ± 3.83	15.8	0.199
3	Benzene	3.61	72.20 ± 1.55	601.75 ± 28.09	78	0.305
4	3-Hexen-1-ol, (E)	6.38	ND [3]	28.83 ± 1.51	4.197	0.223
5	4,6-Heptadiyn-3-one	9.33	90.28 ± 2.26	601.75 ± 28.09	78.2	1.211
6	Toluene	11.02	12.50 ± 3.48	1.65 ± 0.31	9.7	0.653
7	Oxime-, methoxy-phenyl	12.43	757.69 ± 322.83	680.68 ± 300.96	1223.9	0.200
8	2-Pentenal, (E)-	12.49	16.86 ± 0.82	23.36 ± 0.76	3.105	0.136
9	Alpha-Pinene	13.32	24.44 ± 4.96	131.41 ± 16.53	47.87	0.003 *
10	Sabinene	13.47	72.54 ± 34.72	137.59 ± 37.07	140.8	0.377
11	Myrcene	15.22	20.15 ± 7.96	68.45 ± 30.99	88.7	0.046 *
12	beta-Pinene	16.25	ND	55.75 ± 17.03	47.24	0.930
13	1-Hexanone, 5-methyl-1-phenyl	16.81	21.05 ± 3.78	35.38 ± 7.44	23.14	0.004 *
14	p-Cymene	17.28	422.85 ± 144.03	564.67 ± 82.08	459.7	0.339
15	3-Hexen-1-ol, acetate, (Z)	17.48	394.93 ± 152.39	245.99 ± 62.11	456.3	0.277

Table 1. Cont.

No	Compound Name	RT [1]	Uninfested Plant Area ± SD [2]	Infested Plant Area ± SD	LSD [4]	p-Value
16	Eucalyptol	19.97	129.50 ± 5.22	96.14 ± 34.98	98.1	0.036 *
17	Limonene	20.38	14.66 ± 1.92	247.26 ± 84.09	233.2	0.003 *
18	Decane	23.57	ND	39.31 ± 5.50	15.25	0.020 *
19	gamma-Terpinen	24.81	9.03 ± 1.70	56.55 ± 3.68	11.23	0.007 *
20	Heptane, 2,4,4-trimethyl	26.24	3.75 ± 1.44	91.50 ± 45.46	126.1	0.001 *
21	Cyclopentasiloxane, decamethyl	27.84	1.95 ± 0.23	314.91 ± 12.00	33.29	0.212
22	1-Undecyne	30.22	2.68 ± 0.52	110.55 ± 13.59	37.72	0.036 *
23	Heptane, 2,5,5-trimethyl	30.82	2.17 ± 0.43	33.82 ± 4.85	13.5	0.630
24	Cyclohexasiloxane	34.24	123.62 ± 53.60	1.16 ± 0.17	148.6	0.301
25	3,4-Dihydroxyphenylglycol	37.29	20.15 ± 7.96	1.72 ± 0.41	22.09	0.286
26	1,5-Pentanediamine	40.10	249.45 ± 12.70	10.33 ± 0.64	35.27	0.127
27	octamethyl	42.66	565.00 ± 22.07	7.89 ± 2.42	61.6	0.129
28	decamethyl	41.43	113.05 ± 55.42	ND	153.7	0.401

[1] RT indicated to the retention time of compounds. [2] SD referred to the standard deviation of peak area calculated from three replicates. [3] ND referred to not detected. [4] LDS referred to Least Significant Difference at 0.05 level. * indicated to the significant different 5%.

2.2. Effect of VOCs on Attractive Parasitoid

Results of the laboratory experiments using Y-tube olfactometer bioassays showed the response of the aphids *M. persicae* (n = 30 for each replicate) and their parasitoids *A. colemani* and *A. abdominalis* (n = 15 for each replicate and each parasitoid) to the uninfested and infested cabbage plants by 30 individual aphids and 15 individuals per replicate of parasitoid.

These results indicated that green peach aphids in cabbage were significantly (Chi-square (χ^2) = 18.61, df = 1 and p < 0.0005) more attracted to the VOCs released from infested plant (80%) rather than clean air (7%). Results showed that *M. persicae* were significant different in the preference for cabbage plants, with more attraction to the uninfested plants than clean air. The percentage of attracted aphids was 75.56% versus 3% (χ^2 = 20.16, df = 1 and p < 0.0005). While the results indicated that the aphids were significantly more attracted to the infested cabbage compared with the uninfested plant, the percentage of aphid numbers attracted towards infested cabbage plants was 63%, versus 26.67% attracted to uninfested cabbage plants (χ^2 = 4.48, df = 1 and p < 0.034) (Figure 3).

For the parasitoid experiments, the attraction of parasitoids *A. colemani* and *A. abdominalis* to volatiles released by plants, where they were given a choice between uninfested and infested plants, was analyzed. Both *A. colemani* and *A. abdominalis* were significantly more attracted to volatiles from plants infested with green peach aphids compared with clean air (Figure 4). The frequency of parasitoid attraction was 93.33% and 100% towards the infested cabbage plant versus 7% and 20%, respectively, towards the clean air for both parasitoids *A. colemani* and *A. abdominalis* (χ^2 = 11.26, df = 1 and p = 0.001 for *A. colemani* and χ^2 = 4.57, df = 1 and p = 0.033 for *A. abdominalis*). The statistical analysis showed that both parasitoids were significantly attracted to the infested plant. However, there was no difference between attracted wasps for the odors released from an uninfested plant and clean air, and there were no responses for both parasitoids *A. colemani* and *A. abdominalis* to the healthy plant odor versus clean air (both parasitoids showed no significant response to the treatment). By percentage, 4.44% of *A. colemani* wasp and 7% of *A. abdominalis* were attracted to volatiles released from uninfested plants, versus 7% for both parasitoids headed for clean air treatment, while the percentage of no responses of parasitoids was 88.86% and 86.66% for *A. colemani* and *A. abdominalis*, respectively (χ^2 = 19.20, df = 2 and p = 0.001 for *A. colemani* and χ^2 = 19.20, df = 2 and p = 0.001 for *A. abdominalis*). When given a choice between uninfested and infested cabbage plants, *A. colemani* and *A. abdominalis* parasitoids were significantly more attracted to volatiles released from infested plant rather than attracted towards uninfested cabbage plants. By percentage, 86.67% of the *A. colemani* and 100% of the *A. abdominalis* responded to infested cabbage compared to 9% of the *A. colemani* and 0% of the *A. abdominalis* being attracted to uninfested plants ($\chi^2 = 10.28$, df = 1 and $p = 0.001$ for *A. colemani* and $\chi^2 = 12.25$, df = 1 and $p = 0.0005$ for *A. abdominalis*).

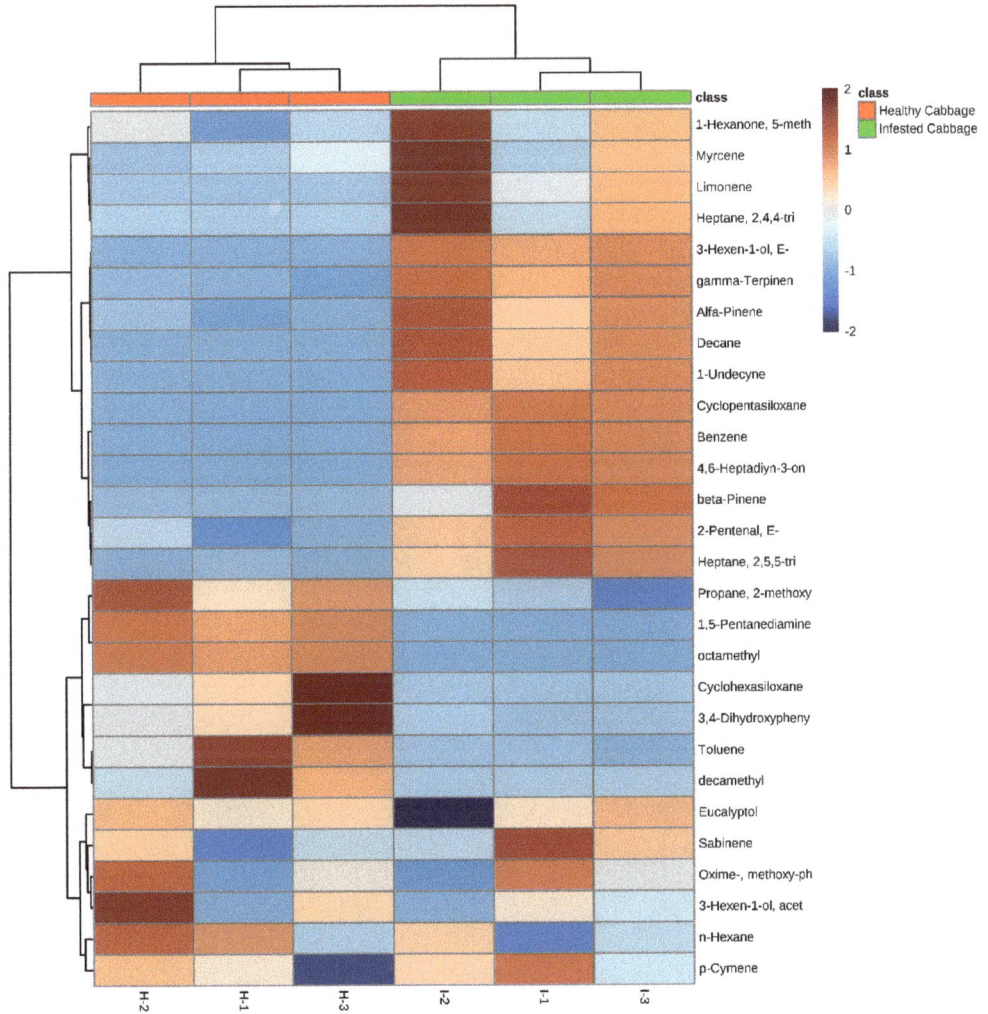

Figure 1. Clustering result is shown as a heat map of volatile compounds released from uninfested and infested cabbage with green peach aphid *M. persicae*. Each volatile compound's peak area detected by GC-MS is shown by colors.

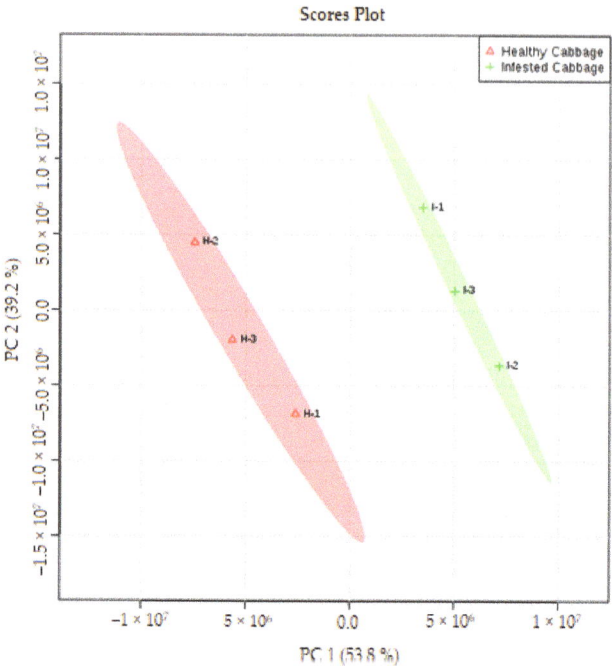

Figure 2. Principal component analysis (PCA) scatter plots reveals between the volatile compounds detected in uninfested and infested cabbage with *M. persicae*. PCA was applied to VOCs from three replicates uninfested and three replicates infested cabbage plants. Red and green circles show results of K-means clustering with k = 2 clusters.

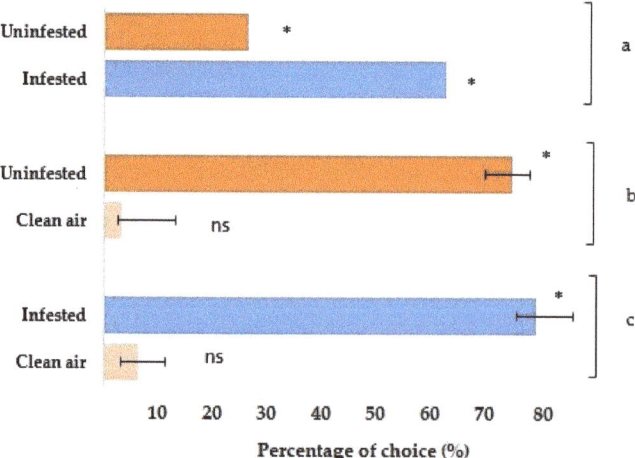

Figure 3. Olfactory response of green peach aphid *M. persicae* in Y-tube olfactometer experiments to volatiles released from infested and uninfested cabbage (**a**) uninfested versus infested plants (**b**) uninfested versus clean air (**c**) infested versus clean air. All treatments presented with standard deviation (SD) bar. Asterisks (*) indicates significant difference $p < 0.05$ (Chi-square test).

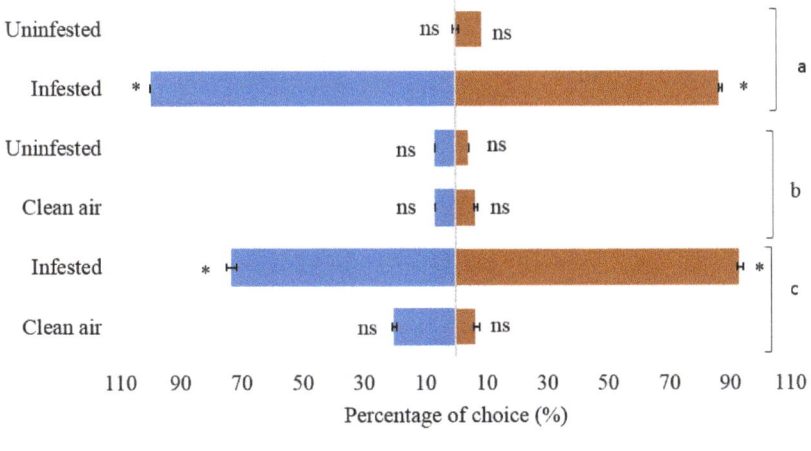

Figure 4. Olfactory response of two parasitoids *Aphidius colemani* and *Aphelinus abdominalis* in Y-tube olfactometer experiments to volatiles released from infested and uninfested cabbage *B. oleracea* (**a**) uninfested versus infested plants (**b**) uninfested versus clean air (**c**) infested versus clean air. All treatments presented with standard deviation (SD) bar. * indicates significant difference $p < 0.05$ (Chi-square test).

3. Discussion

The VOCs that released from infested cabbage plants by *M. persicae* showed many compounds comparing with uninfested plants and reported by previous studies [25–27]. In the current study, volatile compound profiles of uninfested and infested cabbage plants with *M. persicae* were compared to show the differences between treated plants and used as identification tools for the infestation. Taveira et al. [27] reported that a comparison of volatile compounds identified from uninfested and aphid-infested plants from several *Brassica* plants. The damage of cruciferous plants caused by aphids can emit many volatile compounds such as glucosinolate metabolites, phenolics and terpenoids [28,29]. However, our results showed the *M. persicae* preferred damaged *Brassica* plants because the infested plant released different VOCs, such as alpha- and beta pinene, €-3-hexen-1-ol, myrcene, 1-hexanone, 5-methyl-1-phenyl, limonene, decane, gamma-terpinen and heptane, 2,4,4-trimethyl. This finding is consistent with [17], who reported that alpha- and beta pinene and limonene could increase in *Brassica* plants infested by aphids. Some VOCs disappeared from uninfested plants, such as 3-hexen-1-ol-(E) and beta-pinene [17,30]. The increase in (E)-3-hexen-1-ol, beta-pinene and decane in infested plants could be expected because these compounds are well known as green leaf volatiles and are involved in the attraction of natural enemies such as parasitoids and predators [22,31]. The VOCs can be released by an intact and uninfested Brassicaceous plant in large amounts [32]. These compounds were found in the headspace of infested cabbage plants and can be involved in attracting beneficial insects as a response to the aphid infestation [22,31,33]. Thus, the selection of SPME in the extraction of volatile compounds from uninfested and infested cabbage plants with *M. persicae* was based on the peak areas of all compounds identified in the treatments.

The results of Y-tube olfactometer bioassays confirmed the results of aphids *M. persicae* and the parasitoids *A. colemani* and *A. abdominalis* were influenced and attracted to volatiles produced by Cruciferous plants. These wasps significantly preferred, and were attracted to, volatiles from aphid-infested plants over uninfested plants. The use of the Y-tube olfactometer to test the response of aphid *M. persicae* to the host plant, *B. oleracea* var. *capitata*, indicated that *M. persicae* was influenced by the volatiles released from *B. oleracea* var. *capitata* and were significantly attracted to both uninfested and infested plants when compared with the clean air choice.

Aphids can find their host visually and chemically, by chemical, color, size and the shape of the host, and this may be a useful guide to attracting aphids. This result confirms past studies [20,34,35] that show aphids find their host plants by plant odor as well as visual cues. Moreover, the attraction of aphids to the plant volatiles using olfactometer has been reported in experiments testing plant odor against aphids and their host-finding ability [34–36]. Our results showed that aphids tended to be attracted to both damaged and undamaged plants. Our observation is that plant compounds can explain the variance in attraction by aphids and also that plant volatile compounds can increase in response to feeding [37,38]. The population of natural enemies can be increased when adding organic fertilizer [39]. Based on VOCs from cabbage, *M. persicae* was attracted to seven different cabbage varieties in diverse ways. Additionally, the wingless *M. persicae* was considerably attracted to Qingan 80 cabbage cultivar in Y-tube olfactometer bioassays as compared to Yuanbao cabbage cultivar [40,41].

The results from the olfactometer studies demonstrated that parasitoids respond to the plant volatiles and that *A. colemani* and *A. abdominalis* respond to the odor released from infested plants. Both tested parasitoids are significantly responsive to plant volatiles when compared with a clean air treatment. This finding is consistent with [42]. The preference of *A. colemani* and *A. abdominalis* showed no response of parasitoid attraction to clean air and uninfested cabbage, while a statistically significant non-response was noted in the parasitoids. van Emden et al. [43] explain that the attraction of parasitoids can be significantly higher to the infested plant and attack aphids feeding on the same plant as the origin of the mummy offered. The parasitoids *A. colemani* and *A. abdominalis* showed their responses to the infested *B. oleracea*, preferring aphid-induced volatiles. Both parasitoids have significant responses to infested plants with aphids. The results are consistent with [44] who showed parasitoid *A. colemani* could be attracted to volatiles released from *Brassica juncea* and preferred plants damaged by green peach aphids rather than plants damaged by *M. persicae* and *Plutella xylostella* caterpillars.

4. Materials and Methods

4.1. Experimental Plants

Cabbage (*Brassica oleracea* L. var. *capitata*) seeds were sown in a 90 mm square pot filled with potting soil mixture (Richgro Regular Potting Mix, NSW, Australia) and grown under greenhouse conditions at 23–25 °C, 60–70% relative humidity and L16: D8 light cycle. Plants were grown in a glasshouse to the 7–9 leaves stage and used for all experiments. Green peach aphid was reared on cabbage in cages made from plastic and covered by anti-insect white mesh with external dimensions of 40 cm × 40 cm × 40 cm.

4.2. Insect Culture

Myzus persicae for experiments were obtained from the Department of Primary Industries and Regional Development, Entomology Branch (Western Australia) and maintained on potted cabbage seedlings in a greenhouse that were placed into large cages (210 cm × 90 cm) covered by anti-aphid mesh and provided with a control light system set at L16: 8D photoperiod, at the glasshouse temperature 23–25 °C, located at Murdoch University (Western Australia).

Aphidus colemani (Hymenoptera, Braconidae) and *Aphelinus abdominalis* (Hymenoptera, Aphelinidae) were commercially obtained from Biological Services (South Australia) as mummies and maintained on potted cabbage plants infested with *M. persicae* as hosts. Mummies of wasps were removed from the plant leaves on the 12th day for the *A. colemani* and 15th day for the *A. abdominalis* of the parasitism, and placed in open 9 cm Petri dishes inside a small cage of 40 cm × 40 cm × 40 cm, in greenhouse conditions (23–25 °C, 60–70% RH, 16:8 L:D) until emergence. Then, the parasitoids were allowed to mate in the cage for one day with provided 50% honey solution for feeding. After that, the parasitoid was held individually in glass vials (one wasp per vial), a small piece of cotton attached to the vial cap for the drop of 50% honey solution to feed the parasitoid until tested. Female wasps were used for the Y-shape olfactometer choice test [9].

4.3. Volatiles Collection and GCMS Analysis using HS-SPME
4.3.1. VOCs Extraction with HS-SPME

The analysis of volatiles was focused on cabbage for infested and uninfested plants with the green peach aphid. Cabbages were placed individually into 4 L glass jars, and one plant in each jar was analyzed. For each glass jar, a 5 mm port was drilled into the side, into which a septa (20633 Thermogreen® LB-2 Septa, plug) was placed and used for the collection of infested and uninfested plant VOCs. Aluminum foil of 100 m × 44 cm (Vital Packaging Company) was used to carefully cover and wrap the surface of the top of the plant pot, and the glass jar placed upside down on the plant. The reason for selecting glass jars is that it is easy to capture the VOCs emitted and also easy to wash, clean and oven-dry them at 100 °C for a minimum of 30 min to sterilize. VOCs were extracted from samples, which were infested and uninfested cabbage plants with M. persicae. For extracting VOCs from samples, headspace technique analyses were used with three replicates in all experiments, for profiling and characterization of VOCs from both plants. The identification of VOCs was conducted with the SPME fiber by extracting the compound from the headspace of treatments. Three phase fibers 50/30 μm divinylbenzene/carboxen/polydimethyl siloxane (PDMS/CAR/DVB; Sigma-Aldrich, Australia, catalogue number 57347-U) coating was selected for volatiles released from infested and uninfested plants. The SPME fiber is commonly used and this three phase fiber was selected because it was being used for the analysis of a wide range of analysts. The fibers were first conditioned at the range of operating temperature recommended by the manufacturer, before analyses were conducted. For optimizing various conditions, the sealing time was optimized to 2.30 h under laboratory temperature 25 ± 1 °C, and the SPME fiber was exposed to the headspace of the samples by inserting the SPME into the jar through the septum for two hours to extract the VOCs, which characterized the optimum extraction time. The desorption time of SPME fiber was 5 min in the GC injection port. The SPME was used because it is a fast, simple and modern tool for GC-MS analysis.

4.3.2. Samples Analysis with GC-MS

The analysis of VOCs obtained by HS-SPME was performed on a gas chromatography mass spectrometer (GC Agilent GCMS 7820A) equipped with MS detector 5977E (Agilent Technologies, USA) and a DB-35ms column (30 m × 250 μm × 0.25 μm) (Santa Clara, CA, USA). The fiber was desorbed in the splitless injector 270 °C of GCMS with other operation conditions. The initial temperature of the column was 50 °C and held for 2 min, then increased to 250 °C at 5 °C min^{-1} and held for 5 min at 250 °C. Helium gas (He) was used as a carrier and supplied by (BOC Gas, Sydney, Australia) and the flow rate of the column was 1:1 mL/min, while the splitless was 20 mL/min at 1.5 min and the total GC-MS run time was 45 min. The calibration of the SPME fiber was performed by injecting the n-alkanes standard C7–C30.

HS-SPME/GC-MS analysis of the VOCs were identified by using AMDIS software version 2.72 and the US National Institute of Standards and Technology (NIST) 2014 MS database. The VOCs were confirmed by comparing GC retention time data with those of authentic standards or from the published literature [44].

4.4. Evaluation of Olfactory Responses of M. persicae and Its Parasitoids

A glass Y-tube olfactometer was used to determine the responses of M. persicae and its two species parasitoids, A. colemani and A. abdominalis, to each of the following pairs of plant treatments. For the aphid responses, the test was (1) infested (cabbage plants infested with M. persicae) versus clean (filter) air; (2) non-infested versus clean air; and (3) infested versus non-infested plants (Figure 5). For the test of parasitoid wasps, A. colemani and

A. abdominalis, (1) infested plant versus clean air; (2) non-infested plant versus clean air; and (3) infested versus non-infested plants. Bioassays were used to compare their olfactory responses to VOCs released from uninfested plants versus clean air or infested plants with *M. persicae* versus uninfested plants. The infested cabbage plants that were used in this study contained aphids.

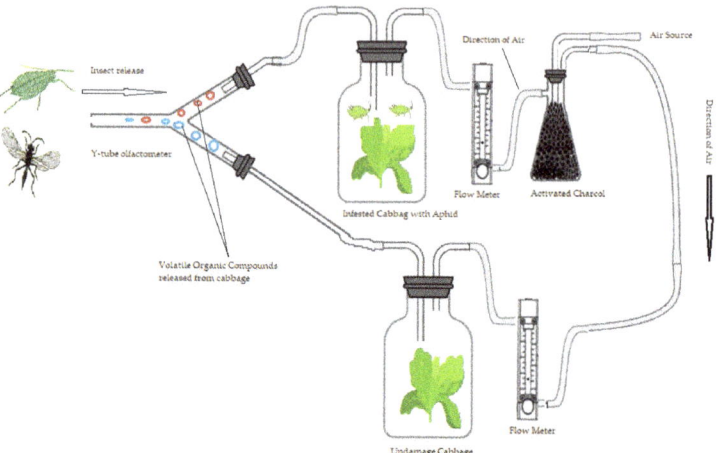

Figure 5. The diagram of the olfactometer, including the glass Y-tube where the aphid *Myzus persicae* and the parasitoid *Aphidus colemani* and *Aphelinus abdominalis* were released individually and exposed to two plant VOCs, blends from uninfested and plants infested with *M. persicae* as shown by the blue and red small circles.

Volatile preference experiments were made using a glass Y-tube olfactometer as previously described [45], with a 7 cm arm length and 2 cm internal diameter, ground glass fitting for the air that passed 200 mL/min through each arm, controlled by air flow meter (SCFH AIR, Dwyer Instruments, Michigan City, IN 46360, USA) (Figure 5). Each arm tube was connected to a glass chamber (2 L desiccator). Couples of blend VOCs (released from different plant treatments) were presented in a sealed glass chamber (2 L each) at the end of either arm. The compressed air was filtered by using activated charcoal passed through two glass chambers, before the treatment plant could be introduced, and then the air passed through the olfactometer. After assembly, the olfactometer was left to stabilize for 15 min prior to use [46].

The Y-tube olfactometer work was carried out under the same conditions as the glasshouse conditions. The area surrounding the olfactometer (below and around) was covered by white paper and white light was placed directly over the olfactometer. For the bioassay, a single aphid or single parasitoid was introduced into the main arm of olfactometer and pushed 1–2 cm inside the main arm. Each aphid or wasp was given up to 3 min in the olfactometer to respond. Once an individual moved beyond 2 cm and into one of the Y-tube arms, it was considered to have made a choice for the conforming plant treatment in that arm. Non-responders that did not make a choice in 5 min were discarded and excluded from the statistical analysis (non-responsive parasitoids counted in statistical analysis in the experiment of comparison of clean air with the uninfested plant).

Three replicates and 30 adults of wingless aphid *M. persicae* were assayed for each replicate, and each aphid was tested only one time. Every 10 aphids were assayed, the volatile treatment resources were removed, and all glass vessels cleaned with ethanol, then washed with water and oven dried at 100 °C for a minimum of 30 min. For the comparison, three replicates were carried out on different days using new aphids and fresh infested and non-infested plants. All plant resources were the same age and same size.

The same procedure above was carried out for the parasitoid *A. colemani* and *A. abdominalis*. Additionally, three replicates were used for the parasitoids with 15 wasps for each replicate and wasps were used only once. Throughout the experiments, after all 15 wasps were assayed for each replicate, the apparatus was cleaned with water and ethanol, then dried and heated in the oven at 100 °C for more than 30 min. Statistical significance between wasp responses to pairwise combinations of plant treatments was determined using Chi-square tests at the 5% level.

4.5. Statistical Analysis

To identify the differences in the emission of volatile compounds between uninfested and infested cabbage by green peach aphids, all peak area analyses were performed with MetaboAnalyst software for the *p*-value, principal component analysis (PCA and PLS-DA) and the hierarchical clustering heat map [47]. The differences in the results were compared by using the least significant difference test ($p \leq 0.05$) for determining the means between infested and uninfested plants. The peak area was divided by 100,000 for every single compound that obtained from GC-MS and subjected to analysis of variance (ANOVA) using Genstat software version 10 (VSNI International Limited, UK) and the least significant difference (LSD) was used at 5% probability level. The data of the Y-tube olfactometer bioassays were analyzed for preference (aphid *M. persicae* and their parasitoids *A. colemani* and *A. abdominalis* choice between two treatments tested) using the Chi-square goodness of fit test by using SPSS software version 24.0.

5. Conclusions

The HS-SPME with GC-MS analysis for the volatiles described the differences between the infested and uninfested cabbage plants and their role in attracting natural enemies of aphids. Collection of volatiles from cabbage occurred by using HS-SPME to detect volatiles compounds between uninfested and plants infested with *M. persicae* and examined the attraction of natural enemies. A total of 28 VOCs were identified in cabbage plant treatments, by using HS-SPME combined with GC-MS. The parasitoids *A. colemani* and *A. abdominalis* laid eggs within the body of *M. persicae* and immature stages completed development inside the hosts, eventually killing them by feeding the wasp larva inside the aphids; the parasitoid pupates inside the aphid mummy and they emerges as an adult. To detect and locate hosts, it is believed that *A. colemani* and *A. abdominalis*, as with many parasitoids, rely on odors released from infested plants as a response to aphids feeding. The results indicated that the preferences of *A. colemani* and *A. abdominalis* to infested plants with *M. persicae* compared with uninfested plants and clean air by using an olfactometer. The results showed that parasitoids can discriminate the infested cabbage and significantly respond to the plant odor. Thus, we believe that aphid parasitoids can find damaged plants and then detect aphids on the plant-by-plant odor. It is likely that the natural enemies' search for aphid infestation may start before landing on the uninfested plant, because parasitoids will first find a damaged plant and then begin searching for aphids. For this reason, many aphid parasitoids efficiently search for damaged plants where aphids will be present, as explained by [20].

Author Contributions: Conceptualization, Q.A., M.A., R.A., H.Z. and Y.R.; data curation, Q.A.; investigation, Q.A.; methodology, Q.A, H.Z. and R.A.; project administration, Q.A. and Y.R.; resources, Q.A. and Y.R.; supervision, M.A. and Y.R.; validation, Q.A., M.A. and Y.R.; writing—original draft: Q.A.; writing—review and editing: Q.A., M.A. and Y.R. All authors have read and agreed to the published version of the manuscript.

Funding: This research received no external funding.

Institutional Review Board Statement: Not applicable.

Informed Consent Statement: Not applicable.

Data Availability Statement: All data are contained within the article.

Acknowledgments: We would like to thank the Iraqi government and the University of Baghdad for supporting to the first author and sponsoring a Ph.D. scholarship. We also appreciate the support of Murdoch University Postharvest Biosecurity and Food Safety Laboratory, and thanks to the technician team in the laboratory.

Conflicts of Interest: The authors declare no conflict of interest.

References

1. Kim, J.H.; Jander, G. *Myzus persicae* (green peach aphid) feeding on *Arabidopsis induces* the formation of a deterrent indole glucosinolate. *Plant J.* **2007**, *49*, 1008–1019. [CrossRef]
2. Valenzuela, I.; Hoffmann, A.A. Effects of aphid feeding and associated virus injury on grain crops in Australia. *Austral Entomol.* **2015**, *54*, 292–305. [CrossRef]
3. Yoon, C.; Seo, D.K.; Yang, J.O.; Kang, S.H.; Kim, G.H. Attraction of the predator, *Harmonia axyridis* (Coleoptera: Coccinellidae), to its prey, *Myzus persicae* (Hemiptera: Aphididae), feeding on Chinese cabbage. *J. Asia-Pac. Entomol.* **2010**, *13*, 255–260. [CrossRef]
4. Amarawardana, L.; Bandara, P.; Kumar, V.; Pettersson, J.; Ninkovic, V.; Glinwood, R. Olfactory response of *Myzus persicae* (Homoptera: Aphididae) to volatiles from leek and chive: Potential for intercropping with sweet pepper. *Acta Agric. Scand. B Soil Plant Sci.* **2007**, *57*, 87–91. [CrossRef]
5. Liu, T.X.; Sparks, A.N. Aphids on cruciferous crops: Identification and management. *Texas Agric. Life Exten.* **2001**, *B-6109*, 1–11.
6. Anstead, J.A.; Williamson, M.S.; Denholm, I. Evidence for multiple origins of identical insecticide resistance mutations in the aphid *Myzus persicae*. *Insect Biochem. Mol. Biol.* **2005**, *35*, 249–256. [CrossRef]
7. Nauen, R.; Denholm, I. Resistance of insect pests to neonicotinoid insecticides: Current status and future prospects. *Arch. Insect Biochem. Physiol.* **2005**, *58*, 200–215. [CrossRef]
8. Reed, H.; Tan, S.; Haapanen, K.; Killmon, M.; Reed, D.; Elliott, N. Olfactory responses of the parasitoid *Diaeretiella rapae* (Hymenoptera: Aphidiidae) to odor of plants, aphids, and plant-aphid complexes. *J. Chem. Ecol.* **1995**, *21*, 407–418. [CrossRef]
9. Takemoto, H.; Takabayashi, J. Parasitic wasps *Aphidius ervi* are more attracted to a blend of host-induced plant volatiles than to the independent compounds. *J. Chem. Ecol.* **2015**, *41*, 801–807. [CrossRef]
10. Goh, H.G.; Kim, J.H.; Han, M.W. Application of *Aphidius colemani* Viereck for control of the aphid in greenhouse. *J. Asia-Pac. Entomol.* **2001**, *4*, 171–174. [CrossRef]
11. Guerrieri, D.M. Aphid-plant interactions. *J. Plant Interact.* **2008**, *3*, 223–232. [CrossRef]
12. De Farias, A.M.; Hopper, K.R. Responses of female *Aphelinus asychis* (Hymenoptera: Aphelinidae) and *Aphidius matricariae* (Hymenoptera: Aphidiidae) to host and plant-host odors. *Environ. Entomol.* **1997**, *26*, 989–994. [CrossRef]
13. Shiojiri, K.; Takabayashi, J.; Yano, S.; Takafuji, A. Flight response of parasitoids toward plant-herbivore complexes: A comparative study of two parasitoid-herbivore systems on cabbage plants. *Appl. Entomol. Zool.* **2000**, *35*, 87–92. [CrossRef]
14. Vuorinen, T.; Nerg, A.M.; Ibrahim, M.; Reddy, G.; Holopainen, J.K. Emission of *Plutella xylostella* induced compounds from cabbages grown at elevated CO2 and orientation behavior of the natural enemies. *Plant Physiol.* **2004**, *135*, 1984–1992. [CrossRef]
15. Steinberg, S.; Dicke, M.; Vet, L.; Wanningen, R. Response of the braconid parasitoid *Cotesia* (= *Apanteles*) *glomerata* to volatile infochemicals: Effects of bioassay set-up, parasitoid age and experience and barometric flux. *Entomol. Exp. Appl.* **1992**, *63*, 163–175. [CrossRef]
16. Vet, L.E.; Dicke, M. Ecology of info chemical use by natural enemies in a tritrophic context. *Annu. Rev. Entomol.* **1992**, *37*, 141–172. [CrossRef]
17. Najar-Rodriguez, A.J.; Friedli, M.; Klaiber, J.; Dorn, S. Aphid-deprivation from Brassica plants results in increased isothiocyanate release and parasitoid attraction. *Chemoecology* **2015**, *25*, 303–311. [CrossRef]
18. Girling, R.; Hassall, M. Behavioural responses of the seven-spot ladybird *Coccinella septempunctata* to plant headspace chemicals collected from four crop *Brassicas* and *Arabidopsis thaliana*, infested with *Myzus persicae*. *Agric. For. Entomol.* **2008**, *10*, 297–306. [CrossRef]
19. Godfray, H.C.J. *Parasitoids: Behavioral and Evolutionary Ecology*; Princeton University Press: Princeton, NJ, USA, 1994.
20. Hatano, E.; Kunert, G.; Michaud, J.; Weisser, W.W. Chemical cues mediating aphid location by natural enemies. *Eur. J. Entomol.* **2008**, *105*, 797. [CrossRef]
21. Hågvar, E.; Hofsvang, T. Effect of honeydew and hosts on plant colonization by the aphid parasitoid *Ephedrus cerasicola*. *Entomophaga* **1989**, *34*, 495–501. [CrossRef]
22. Leroy, P.D.; Sabri, A.; Heuskin, S.; Thonart, P.; Lognay, G.; Verheggen, F.J.; Francis, F.; Brostaux, Y.; Felton, G.W.; Haubruge, E. Microorganisms from aphid honeydew attract and enhance the efficacy of natural enemies. *Nat. Commun.* **2011**, *2*, 348. [CrossRef] [PubMed]
23. de Rijk, M.; Dicke, M.; Poelman, E.H. Foraging behaviour by parasitoids in multiherbivore communities. *Anim. Behav.* **2013**, *85*, 1517–1528. [CrossRef]
24. Hagvar, E.; Hofsvang, T. Aphid parasitoids (Hymenoptera, Aphidiidae): Biology, host selection and use in biological control. *Biocontrol. News Inf.* **1991**, *12*, 13–42.

25. Bruinsma, M.; Posthumus, M.A.; Mumm, R.; Mueller, M.J.; van Loon, J.J.; Dicke, M. Jasmonic acid-induced volatiles of *Brassica oleracea* attract parasitoids: Effects of time and dose, and comparison with induction by herbivores. *J. Exp. Botany* **2009**, *60*, 2575–2587. [CrossRef]
26. Taveira, M.; Fernandes, F.; de Pinho, P.G.; Andrade, P.B.; Pereira, J.A.; Valentão, P. Evolution of *Brassica rapa* var. *rapa* L. volatile composition by HS-SPME and GC/IT-MS. *Microchem. J.* **2009**, *93*, 140–146. [CrossRef]
27. Mathur, V.; Tytgat, T.O.; Hordijk, C.A.; Harhangi, H.R.; Jansen, J.J.; Reddy, A.S.; Harvey, J.A.; Vet, L.E.; Dam, N.M. An ecogenomic analysis of herbivore-induced plant volatiles in *Brassica juncea*. *Mol. Ecol.* **2013**, *22*, 6179–6196. [CrossRef] [PubMed]
28. de Vos, M.; Jander, G. Volatile communication in plant–aphid interactions. *Curr. Opin. Plant Biol.* **2010**, *13*, 366–371. [CrossRef] [PubMed]
29. Hien, T.T.; Heuskin, S.; Delaplace, P.; Francis, F.; Lognay, G. VOC emissions and protein expression mediated by the interactions between herbivorous insects and *Arabidopsis* plant. A review. *Biotechnol. Agron. Soc. Environ.* **2014**, *18*, 455–464.
30. Winde, I.; Wittstock, U. Insect herbivore counteradaptations to the plant glucosinolate–myrosinase system. *Phytochemistry* **2011**, *72*, 1566–1575. [CrossRef] [PubMed]
31. Li, Y.; Weldegergis, B.T.; Chamontri, S.; Dicke, M.; Gols, R. Does aphid infestation interfere with indirect plant defense against lepidopteran caterpillars in wild cabbage? *J. Chem. Ecol.* **2017**, *43*, 493–505. [CrossRef]
32. Mumm, R.; Posthumus, M.A.; Dicke, M. Significance of terpenoids in induced indirect plant defence against herbivorous arthropods. *Plant Cell Environ.* **2008**, *31*, 575–585. [CrossRef] [PubMed]
33. Pinto-Zevallos, D.M.; Bezerra, R.H.; Souza, S.R.; Ambrogi, B.G. Species-and density-dependent induction of volatile organic compounds by three mite species in cassava and their role in the attraction of a natural enemy. *Exp. Appl. Acarol.* **2018**, *74*, 261–274. [CrossRef] [PubMed]
34. Hori, M. Role of host plant odors in the host finding behaviors of aphids. *Appl. Entomol. Zool.* **1999**, *34*, 293–298. [CrossRef]
35. Döring, T.F. How aphids find their host plants, and how they do not. *Annal. Appl. Biol.* **2014**, *165*, 3–26. [CrossRef]
36. Verheggen, F.J.; Haubruge, E.; De Moraes, C.M.; Mescher, M.C. Aphid responses to volatile cues from turnip plants (*Brassica rapa*) infested with phloem-feeding and chewing herbivores. *Arthropod Plant Interact.* **2013**, *7*, 567–577. [CrossRef]
37. Züst, T.; Agrawal, A.A. Mechanisms and evolution of plant resistance to aphids. *Nat. Plants* **2016**, *2*, 15206. [CrossRef]
38. Hopkins, D.P.; Cameron, D.D.; Butlin, R.K. The chemical signatures underlying host plant discrimination by aphids. *Sci. Rep.* **2017**, *7*, 8498. [CrossRef]
39. Khidr, S.K. Effects of organic fertilizers and wheat varieties on infestation by, corn leaf aphid, *Rhopalosiphum maidis* and wheat thrips, *Haplothrips tritici* and their predators. *Iraqi J. Agric. Sci.* **2018**, *49*, 93–104.
40. Ahmed, Q.; Agarwal, M.; Alsabte, A.; Aljuboory, A.B.; Ren, Y. Evaluation of Volatile Organic Compounds from Broccoli Plants Infested with Myzus persicae and Parasitoids Aphidius colemani Attraction. In Proceedings of the IOP Conference Series: Earth and Environmental Science, Baghdad, Iraq, 1 May 2021; IOP Publishing: Bristol, UK, 2021; Volume 761, p. 012029.
41. Ahmed, N.; Darshanee, H.L.; Khan, I.A.; Zhang, Z.F.; Liu, T.X. Host selection behavior of the green peach aphid, Myzus persicae, in response to volatile organic compounds and nitrogen contents of cabbage cultivars. *Front. Plant Sci.* **2019**, *12*, 10–79.
42. Kalule, T.; Wright, D. The influence of cultivar and cultivar-aphid odours on the olfactory response of the parasitoid *Aphidius colemani*. *J. Appl. Entomol.* **2004**, *128*, 120–125. [CrossRef]
43. van Emden, H.F.; Storeck, A.P.; Douloumpaka, S.; Eleftherianos, I.; Poppy, G.M.; Powell, W. Plant chemistry and aphid parasitoids (Hymenoptera: Braconidae): Imprinting and memory. *Euro. J. Entomol.* **2008**, *105*, 477. [CrossRef]
44. da Silva, S.E.; França, J.F.; Pareja, M. Olfactory response of four aphidophagous insects to aphid-and caterpillar-induced plant volatiles. *Arthropod Plant Interact.* **2016**, *10*, 331–340. [CrossRef]
45. Saad, K.A.; Roff, M.M.; Hallett, R.H.; Idris, A. Aphid-induced defences in chilli affect preferences of the whitefly, *Bemisia tabaci* (Hemiptera: Aleyrodidae). *Sci. Rep.* **2015**, *5*, 13697. [CrossRef] [PubMed]
46. Pope, T.; Girling, R.; Staley, J.; Trigodet, B.; Wright, D.; Leather, S.; Van Emden, H.; Poppy, G. Effects of organic and conventional fertilizer treatments on host selection by the aphid parasitoid *Diaeretiella rapae*. *J. Appl. Entomol.* **2012**, *136*, 445–455. [CrossRef]
47. Chong, J.; Soufan, O.; Li, C.; Caraus, I.; Li, S.; Bourque, G.; Wishart, D.S.; Xia, J. Metabo Analyst 4.0: Towards more transparent and integrative metabolomics analysis. *Nucl. Acids Res.* **2018**, *46*, W486–W494. [CrossRef]

Article

Headspace Solid-Phase Microextraction/Gas Chromatography–Mass Spectrometry for the Determination of 2-Nonenal and Its Application to Body Odor Analysis

Keita Saito *, Yoshiyuki Tokorodani, Chihiro Sakamoto and Hiroyuki Kataoka

School of Pharmacy, Shujitsu University, Nishigawara, Okayama 703-8516, Japan; sh1908055@outlook.jp (Y.T.); sh1909027@outlook.jp (C.S.); hkataoka@shujitsu.ac.jp (H.K.)
* Correspondence: ksaito@shujitsu.ac.jp

Abstract: The odors and emanations released from the human body can provide important information about the health status of individuals and the presence or absence of diseases. Since these components often emanate from the body surface in very small quantities, a simple sampling and sensitive analytical method is required. In this study, we developed a non-invasive analytical method for the measurement of the body odor component 2-nonenal by headspace solid-phase microextraction coupled with gas chromatography–mass spectrometry by selective ion monitoring. Using a StableFlex PDMS/DVB fiber, 2-nonenal was efficiently extracted and enriched by fiber exposition at 50 °C for 45 min and was separated within 10 min using a DB–1 capillary column. Body odor sample was easily collected by gauze wiping. The limit of detection of 2-nonenal collected in gauze was 22 pg (S/N = 3), and the linearity was obtained in the range of 1–50 ng with a correlation coefficient of 0.991. The method successfully analyzed 2-nonenal in skin emissions and secretions and was applied to the analysis of body odor changes in various lifestyles, including the use of cosmetics, food intake, cigarette smoking, and stress load.

Keywords: body odor; 2-nonenal; wiping method; pasting method; solid-phase microextraction (HS–SPME); gas chromatography–mass spectrometry (GC–MS)

1. Introduction

Body odor may be an indicator of stress and lifestyle diseases, including psychosomatic disorders, and may be an obstacle to a healthy social life [1–4]. As people are reluctant to indicate that others have body odor and the judgment criteria are ambiguous, the development of an objective quantitative method to evaluate body odor is desired. Body odor components include fatty acids, spices, and sulfurous odors [3,4]. Secretions by sweat glands and sebaceous glands are derived from volatile components decomposed by skin bacteria as well as components external to the body [3,4]. Among these compounds are unsaturated aldehydes, such as 2-nonenal, which is produced by the oxidation and decomposition of 9-hexadecenoic acid secreted by the sebaceous glands [1] (Figure S1). 2-Nonenal is also known as the odor contained in foods such as fresh juice and beer, and several analytical methods have been reported [5–15]. To date, however, sampling and analytical methods required to quantitatively analyze body odor compounds have not been established, so the actual state and dynamics of their occurrence have not been fully determined [3,4,16]. As "odors" generated from the body surface are complex and quantitatively small, their components may be difficult to detect with an analytical instrument. Highly sensitive analytical methods with efficient sampling and pretreatment are therefore needed to separate, identify, and quantify odor compounds. Although stir bar sorptive extraction [10], single-drop microextraction [11], and headspace solid-phase microextraction (HS–SPME) [12] methods have been reported for the efficient preconcentration of 2-nonenal in foods, they have not yet been used for body odor analysis. A conventional sampling method of skin emissions and secretions consists of placing the palm of the hand in a PVF bag and collecting the generated gas [1], but this method has drawbacks, including a possible lack of airtightness and questions about quantitative sampling. Body odor compounds have also been analyzed by gas chromatography (GC) after headspace sampling or passive flux sampling [1,17,18], but these methods have problems with cost and storage. The SPME, using a fused-silica fiber externally coated with an appropriate stationary phase, is an effective sample preparation technique for integrating several operations such as sample collection, extraction, analyte enrichment, and isolation from sample matrices and is easily coupled with GC [4,19–21]. It is fast, solvent-free, and cost-effective and can improve the detection limits.

2-Nonenal is produced by reactions between fatty acids and lipid peroxides and is involved in the intake of fats and foods rich in fatty acids and in the peroxidation of lipids [1]. Thus, this production and intake of this compound are associated with the generation of active oxygen species [2]. Drinking, smoking, stress, and other lifestyle factors may therefore affect 2-nonenal secretion and alter body odor. Therefore, the effects of internal and external factors such as dietary intake, cosmetic use, smoking, and stress on body odor changes were analyzed using the method developed in this study.

In this study, we developed an efficient sampling method for the collection of body odor and an HS–SPME coupled with GC–mass spectrometry (GC–MS) for the selective and sensitive analysis of 2-nonenal in skin emissions and secretions. The method was applied to the body odor analysis in various lifestyles.

2. Results and Discussion

2.1. Detection of 2-Nonenal by Gas Chromatography–Mass Spectrometry

Full scan mass spectrum was measured at an m/z range from 30 to 150 amu in order to determine the selected monitoring ion (SIM) for 2-nonenal (Mw: 140 daltons). Figure 1a shows a typical total ion chromatogram of 2-nonenal fragments obtained by 1 μL direct injection analysis of standard 2-nonenal (1 μg/mL solution). 2-Nonenal was eluted as a single and symmetrical peak within 10 min using a DB–1 column. As shown in Figure 1b, no molecular ions peak was detected, but several fragment ion peaks produced by cleavage of the 2-nonenal structure were observed on the spectra. Among these fragment ions, $m/z = 55$ [CH=CHCHO], $m/z = 83$ [CH$_2$CH$_2$CH=CHCHO], and $m/z = 111$ [CH$_2$CH$_2$CH$_2$CH$_2$CH=CHCHO]) were selected for SIM mode detection.

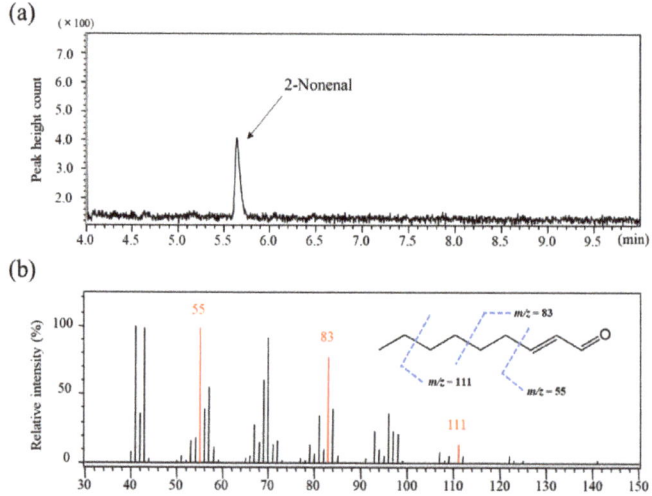

Figure 1. Chromatogram (**a**) and mass spectrum (**b**) obtained from standard 2-nonenal. The dotted lines in the structure represent the fragments of 2-nonenal.

2.2. Optimization of Headspace Solid-Phase Microextraction and Desorption

In the SPME method, fiber coatings with high distribution coefficients must be used to increase the extraction efficiency of the compounds. To select an SPME fiber, the four different commercially available fibers (PDMS/DVB, DVB/CAR/PDMS, CAR/PDMS, and polyacrylate) were tested for extraction of 100 ng of the 2-nonenal standard solution by HS–SPME. Of these fibers, the 65 µM StableFlex PDMS/DVB fiber extracted 2-nonenal most efficiently (Figure 2A), and the fiber was used in subsequent experiments. To optimize HS–SPME extraction conditions, 100 ng of 2-nonenal standard solution in a 40 mL vial was exposed to 65 µM StableFlex PDMS/DVB fiber at various temperature and times. These experiments showed that 2-nonenal could be efficiently extracted and concentrated by exposing the fiber to the headspace at 50 °C (Figure 2B) for 45 min (Figure 2C). Although the extraction efficiency was highest at 30 °C, the extraction temperature set at 50 °C was less affected by the outside air temperature and could be kept. On the other hand, the time of the fiber exposition in the GC sample vaporization chamber at 230 °C was tested to optimize the desorption condition. The 2-nonenal extracted on the fiber was easily desorbed within 5 min by heating in the GC sample vaporization chamber, and carryover was not observed because the fiber was washed during exposition.

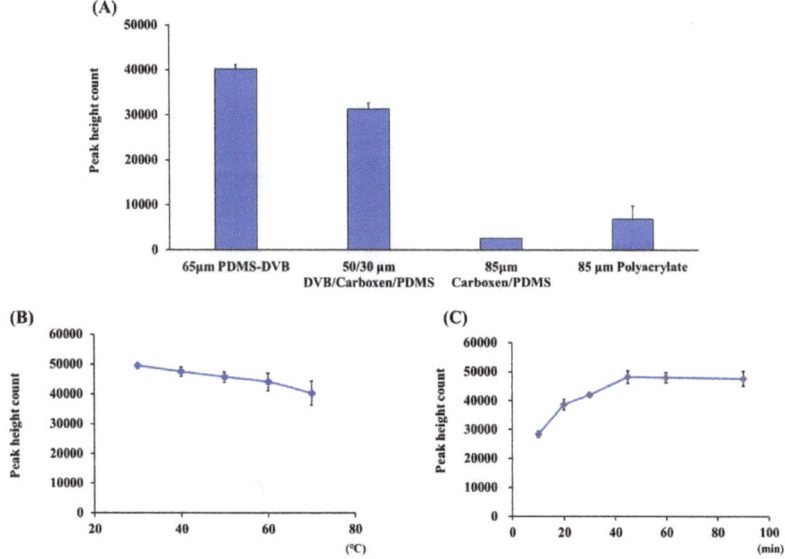

Figure 2. Effects of (**A**) fiber coatings, (**B**) temperature, and (**C**) time on the HS–SPME of 2-nonenal. For HS–SPME, 100 ng of 2-nonenal in gauze was extracted by (**A**) fiber exposition at 50 °C for 45 min, (**B**) for 45 min with PDMS-DVB, and (**C**) at 50 °C with PDMS-DVB.

The absolute amount of 2-nonenal extracted by the HS–SPME method was calculated by comparing peak height count with direct injection onto the GC column. From the sample of 20 ng, 8.0 ng (40%) of 2-nonenal was extracted onto the PDMS/DVB fiber by HS–SPME under optimized conditions. Although the extraction yield was low, the reproducibility was good (relative standard deviation: RSD = 4.7%, n = 3). Figure 3 shows the chromatograms obtained by direct injection and HS–SPME techniques. The sensitivity of the SPME method showed a 40-fold increase in sensitivity compared with when a solution of the same concentration was directly injected (1 µL).

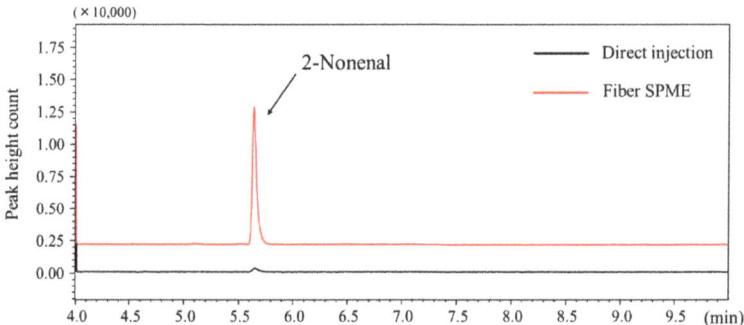

Figure 3. Typical total ion chromatogram of 2-nonenal obtained by (a) direct injection (200 ng/mL solution, 1 µL) and HS–SPME (200 ng/mL solution, 100 µL in 40-mL vial).

2.3. Sampling of Skin Emissions and Secretions

Four sampling methods (see the Materials and Methods section) for the collection of skin emissions and secretions from the palm were examined. As shown in Table 1, the wiping method (method C), in which the palm surface is wiped with gauze, was found to be the most efficient in collecting 2-nonenal. On the other hand, 2-nonenal was collected by the solution method (method A) using water in a cup and by the pasting method (method D) using gauze taped to the palm but not by the headspace method (method B) using HS–SPME directly in a suction cup. These results indicate that 2-nonenal released from the skin surface is difficult to collect from the headspace at body temperature and that active wiping is more effective than passive sampling, even when the sampling material is in direct contact with the skin. Therefore, the wiping method was used as the sampling method for the collection of skin emissions and secretions in this study. The recovery of 2-nonenal from the gauze soaked with 5 ng of 2-nonenal was about 60% compared to that of 2-nonenal added directly to the vial without gauze.

Table 1. Comparison of skin emission sampling methods.

Classification	Sampling Method	2-Nonenal Content (ng) Mean ± SD (n = 3)
A	Glass cup sampling	0.47 ± 0.51
B	Direct SPME sampling	N.D.[1]
C	Wiping method	0.90 ± 0.13
D	Pasting method	0.18 ± 0.23

[1] N.D.: not detectable.

To examine the stability of 2-nonenal collected on gauze by the wiping method, 0.1 g of gauze soaked with 100 ng of 2-nonenal was stored in vials at various times and temperatures, and the amount of 2-nonenal remaining on the gauze was determined by HS–SPME/GC–MS. As shown in Table 2, when stored at room temperature, there was almost no decrease in 2-nonenal up to 6 h, but it was found to decrease to 60% at 24 h. However, if the gauze was kept refrigerated at 4 °C, there was almost no decrease even after 24 h. Therefore, the gauze containing 2-nonenal collected by the wiping method should be kept refrigerated until analysis.

Table 2. Comparison of storage conditions after sampling by the wiping method.

Storage Condition [1]	Storage Time (h)	Peak Height Count Mean ± SD (n = 3)
Room temperature	0	25,694 ± 1486
	1	25,025 ± 2830
	2	24,537 ± 1454
	3	25,968 ± 2162
	6	24,208 ± 3058
	24	15,250 ± 1600
4 °C (refrigerator)	24	24,594 ± 3679

[1] The gauze soaked with 50 ng of 2-nonenal was stored in vials.

2.4. Analytical Method Validation by Wiping Method

To validate a proposed method based on the wiping sample collection, standard 2-nonenal placed in a petri dish was wiped with 0.1 g of dry gauze and analyzed by HS–SPME/GC–MS. Linearity was validated by triplicate analyses for 2-nonenal at 0, 1, 2, 5, 10, 20, and 50 ng. The calibration curve was linear with a correlation coefficient of 0.991, and the relative standard deviations (RSDs) of peak height counts at each point ranged from 1.8 to 9.6 % (n = 3). The LOD (S/N = 3) and LOQ (S/N = 10) of 2-nonenal in gauze were 22 pg and 74 pg, respectively. Intra-day and inter-day precisions expressed as RSD (%) were found to be 9.6 and 3.6%, respectively. These results show that the method is highly sensitive and reproducible for the determination of 2-nonenal by the wiping method.

2.5. Analyses of 2-Nonenal in the Body Odor Collected from Several Body Parts

Body odor samples were collected by the wiping method from several body parts 6 h after washing with soap and then analyzed by HS–SPME/GC–MS. Figure 4 shows the 2-nonenal content in body odor samples collected from the arms, axillae, back, chest, forehead, palm, and behind the ears of three subjects. Relatively high amounts of 2-nonenal were detected in the samples from the foreheads, arms, and ears, although this varied between individuals. To determine the amount of 2-nonenal secreted at different times of the day, body odor samples were collected from the forehead every 3 h by the wiping method and analyzed by HS–SPME/GC–MS. As shown in Figure 5, the secretion of 2-nonenal was highest between 13:00 and 16:00 in the afternoon (average contents of 2-nonenal from three people). The chromatogram of the 2-nonenal sampled from the forehead is shown in Figure 6. It was found that 2-nonenal was detected with clear separation and was not affected by the interfering peak.

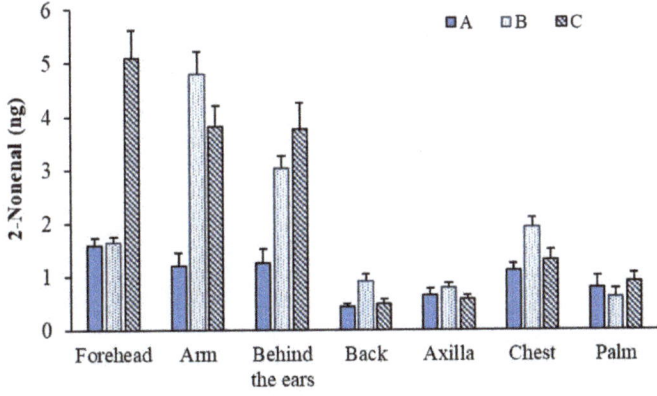

Figure 4. Comparison of 2-nonenal contents in body odor samples collected from several body parts (A: female in their 20 s, B: male in their 20 s, C: male in their 30 s).

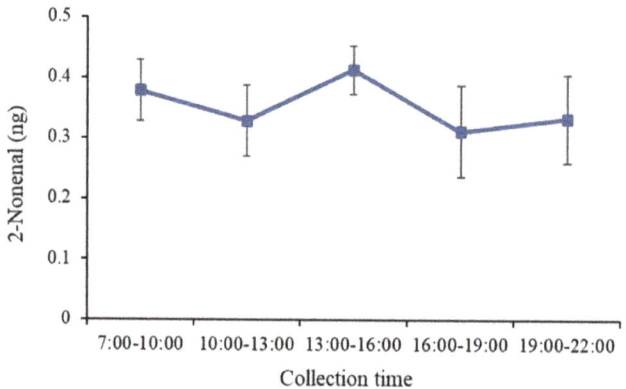

Figure 5. Circadian rhythm of 2-nonenal emission from skin (forehead) (average contents of 2-nonenal from three people).

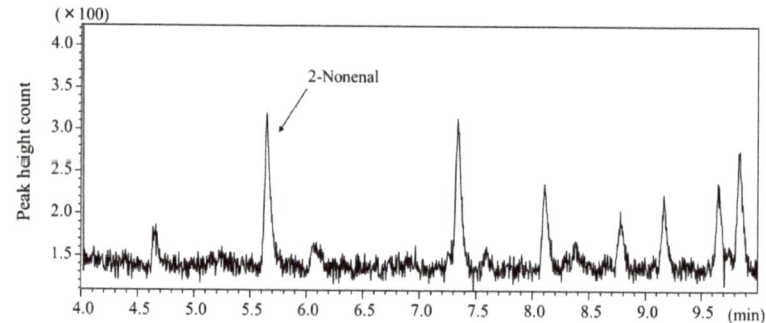

Figure 6. Total ion chromatogram of skin emission sample (forehead, male in their 30 s).

2.6. Application to the Analysis of Body Odor Changes in Various Lifestyles

Table 3 shows the results of HS–SPME/GC–MS analysis of the samples collected from the forehead by the wiping method after a certain period of time after each treatment. 2-Nonenal secretion was significantly increased on the high-fat diet compared with the low-fat diet. 2-Nonenal secretion was also slightly increased in lotion use and smoking compared with non-use and non-smoking, but the difference was not statistically significant. In contrast, mental stress caused by long meetings significantly increased 2-nonenal secretion. This may be because stress-induced sweating increased the secretion of fatty acids and lipid peroxides from the sebaceous glands, increasing oxidative degradation of these compounds to 2-nonenal by indigenous bacteria on the body surface [1].

Table 3. 2-Nonenal contents detected in body odor samples in various life styles.

Treatment	Sampling Time after Treatment	2-Nonenal Content (ng) Mean ± SD (n = 3)	T-Test p Value
Low fat diet	12	0.49 ± 0.09	$p < 0.05$
High fat diet	12	0.70 ± 0.09	
Without using cosmetics	6	0.23 ± 0.04	$p > 0.05$
Using cosmetics	6	0.26 ± 0.08	
Before smoking	2	0.27 ± 0.08	$p > 0.05$
After smoking	2	0.37 ± 0.09	
Before meeting	6	0.23 ± 0.04	$p < 0.05$
After meeting	6	0.35 ± 0.09	

3. Materials and Methods

3.1. Reagents and Materials

2-Nonenal was purchased from Nacalai Tesque (Kyoto, Japan). Methanol and distilled water (LC–MS grade) were purchased from Kanto Chemical (Tokyo, Japan). A stock solution of 1 mg mL^{-1} of 2-nonenal was prepared by dissolving in methanol (LC–MS grade), tightly sealed, and stored at 4 °C. This stock solution was diluted with distilled water (LC–MS grade) to the desired concentration prior to use. All other chemicals were of analytical grade.

SPME assemblies with a replaceable and reusable extraction fiber coated with StableFlexTM polydimethylsiloxane/divinylbenzene (PDMS/DVB, 65 µM), StableFlexTM carboxene/PDMS/DVB (DVB/CAR/PDMS, 50/30 µM), (CAR/ PDMS, 85 µM), StableFlexTM CAR/PDMS (85 µM), and polyacrylate (85 µM) were purchased from Supelco (Supelco Japan, Tokyo). These fibers were conditioned in a GC injection port at adequate temperature prior to use. One fiber can be used repeatedly at least more than one hundred times.

3.2. Gas Chromatography–Mass Spectrometry

GC–MS analysis was performed in the scan and SIM modes on a Shimadzu QP–2010 (Kyoto, Japan)gas chromatograph–mass spectrometer in conjunction with a GCMS solution Ver.2 workstation. GC separation was performed using a fused-silica capillary column of cross-linked DB−1 (J&W, Folsom, CA, USA: 60 m × 0.25 mm i.d., 1.0 µM film thickness) under the following operating conditions: injection and detective temperatures, 230 °C; column temperature, held at a temperature of 190 °C for 2 min and increased to 230 °C at a rate of 5 °C min^{-1}; inlet helium carrier gas flow rate, 1 mL min^{-1} maintained by an electronic pressure controller; and split ratio, 10:1. The electron impact (EI)–MS conditions were as follows: ion-source temperature, 230 °C; ionizing voltage, 70 eV. The full scan mass spectra were obtained at an m/z range from 40 to 150 amu. Selected ion monitoring (SIM) mode detection for 2-nonenal was selected at m/z = 55, 83, 111. The peak height count was measured to construct calibration curve and to determine concentrations of 2-nonenal in samples.

3.3. Headspace Solid-Phase Microextraction

The sample was placed in a 40-mL screw-cap vial with a PTFE septum and heated on a hot plate at 50 °C. The SPME needle was passed through the septum of the vial, and the fiber was exposed in the headspace (HS) above sample for 45 min to adsorb the compounds vaporized by heating. After extraction, the fiber was retracted into the needle, the needle was withdrawn and introduced directly into the GC–MS sample vaporization chamber to desorb the extracted compound by heating. Then, the fiber was retracted into the needle, the needle was removed from the sample vaporization chamber and used for the HS–SPME of the next sample. An outline of the procedure for extraction by HS–SPME and desorption for GC–MS analysis is shown in Figure S2.

3.4. Sampling of Skin Emissions and Secretions

Four sampling methods for the collection of skin emissions and secretions were compared. Figure 7 shows the sampling methods from the palm of the hand. Sampling was carried out 3 h after washing the skin surface with soap. In method A (solution method), 2 mL of distilled water were added to a glass cup, and the cup was inverted so that the water touched the skin on the palm of one hand for 5 min. The aqueous solution was subjected to the above HS–SPME method. In method B (headspace method), a suction cup is placed on the palm, the SPME fiber needle is inserted through the septum into the cup under reduced pressure, and then, the fiber is exposed for 5 min to extract the skin emissions. In method C (wiping method), a 2 cm square area of the palm was wiped for 1 min with 0.1 g of dry gauze as a collecting material of skin emissions and secretions, which was then placed in a vial and subjected to the above HS–SPME method. In method

D (pasting method), 0.1 g of dry gauze was taped to the palm and extracted for 3 h. These collecting materials were then subjected to the HS–SPME as in method C.

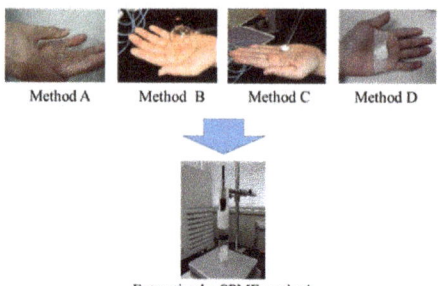

Figure 7. Sampling methods of skin emissions and secretions.

3.5. Analysis of Body Odor Samples

The experimental protocol was approved by the Research Ethics and Safety Committee of Shujitsu University, and body odor samples were provided from one female and two male subjects in their 20 s and 30 s by informed consent. The samples were collected from the arm, axilla, back, chest, forehead, palm, and behind the ears by method C described in the above section. 2-Nonenal in the body odor collected on the gauze was subsequently analyzed by HS–SPME/GC–MS. To analyze changes in body odor due to the various life styles, body odor samples were collected from the forehead by method C before and after treatments and analyzed by HS–SPME/GC–MS as described above. As a concrete examination, we compared before and after food intake (eating 200 g of vegetables for the low fat diet or 200 g of beef for the high fat diet, sampling 12 h later), using cosmetics (applying 1 mL lotion to face, sampling 6 h later), smoking (smoking one cigarette, sampling 2 h later), and a 3 h meeting as a stress load (sampling 6 h later).

4. Conclusions

The method developed in this study can easily collect 2-nonenal, a component of human body odor, from the skin surface by gauze wiping and can be selectively and sensitively analyzed by HS–SPME/GC–MS. Gauze wiping is a rapid and non-invasive sampling method for skin emissions and secretions and has better storage stability than other tested methods. As shown in Table 4, the proposed method is more sensitive than previously reported methods, even when converted into the skin surface area and the amount of recovery per hour. We believe that this method is a useful tool for 2-nonenal analysis in body odor. Although this study focused on 2-nonenal, body odor is a complex mixture of compounds, and therefore, it is necessary to analyze not only 2-nonenal but also a wide range of related compounds in the future. The method developed in this study may be applicable to their comprehensive analysis.

Table 4. Comparison of analytical methods for 2-nonenal and other aldehydes in body odor.

Compounds	Analytical Method	Sampling and Preconcentration	Detection Amount	Reference
2-Nonenal	GC–MS	Tedlar bag and TENAX-TA column	15.1 ± 20.4 ng·cm^{-3} (1)	[1]
Five aldehydes and acetone	HPLC-UV	Passive flux sampler (trapping filter: DNPH impregnated filter)	17 ng·cm^{-2}·h^{-1} (2)	[17]
2-Nonenal and diacetyl	GC–MS	Passive flux sampler (trapping media: Monotrap, DCC18)	0.020 to 5.8 ng·cm^{-2}·h^{-1} (3)	[18]
2-Nonenal	GC–MS	Fiber SPME (Wiping method)	2.4 pg·cm^{-2}·h^{-1} (4)	This study

(1) Sampling from the back of the shirt worn for 3 days; (2) Sampling from the surface of human skin, sampling time 1 h, aldehydes (LOD); (3) Sampling from the nape of the neck, sampling time 7 h; (4) LOD.

Supplementary Materials: The following are available online. Figure S1: Presumed formation mechanism of 2-nonenal by lipid peroxidation. Data from Haze et al., 2001 [1], Figure S2: Procedure for extraction by headspace fiber SPME and desorption for GC–MS analysis. Data from Kataoka and Saito, 2011 [20].

Author Contributions: Conceptualization, K.S. and H.K.; methodology, K.S. and H.K.; software, K.S. and H.K.; validation, K.S. and H.K.; formal analysis, Y.T., C.S., and K.S.; investigation, K.S. and H.K.; resources, K.S. and H.K.; data curation, Y.T., C.S., and K.S.; writing—original draft preparation, K.S. and H.K.; writing—review and editing, K.S.; visualization, K.S. and H.K.; supervision, K.S.; project administration, K.S. All authors have read and agreed to the published version of the manuscript.

Funding: This research received no external funding.

Institutional Review Board Statement: The study was conducted according to the guidelines of the Declaration of Helsinki and was approved by the ethics committee of Shujitsu University.

Informed Consent Statement: Informed consent was obtained from all subjects involved in the study.

Data Availability Statement: The data presented in this study are available from the corresponding author upon request.

Conflicts of Interest: The authors declare no conflict of interest.

Sample Availability: Samples of the compounds are not available from the authors.

References

1. Haze, S.; Gozu, Y.; Nakamura, S.; Kohno, Y.; Sawano., K.; Ohta, H.; Yamazaki, K. 2-Nonenal newly found in human body odor tends to increase with aging. *J. Investig. Dermatol.* **2001**, *116*, 520–524. [CrossRef] [PubMed]
2. Ishino, K.; Wakita, C.; Shibata, T.; Toyokuni, S.; Machida, S.; Matsuda, S.; Matsuda, T.; Uchida, K. Lipid peroxidation generates body odor component trans-2-nonenal covalently bound to protein in vivo. *J. Biol. Chem.* **2010**, *285*, 15302–15313. [CrossRef] [PubMed]
3. Pandey, S.K.; Kim, K.H. Human body-odor components and their determination. *Trends Anal. Chem.* **2011**, *30*, 784–796. [CrossRef]
4. Kataoka, H.; Saito, K.; Kato, H.; Masuda, K. Non-invasive analysis of volatile biomarkers in human emanations for health and early disease diagnosis. *Bioanalysis* **2013**, *5*, 1443–1459. [CrossRef] [PubMed]
5. Santos, J.R.; Carneiro, J.R.; Guido, L.F.; Almeida, P.J.; Rodrigues, J.A.; Barros, A.A. Determination of E-2-nonenal by high-performance liquid chromatography with UV detection assay for the evaluation of beer ageing. *J. Chromatogr. A* **2003**, *985*, 395–402. [CrossRef]
6. Guido, L.F.; Carneiro, J.R.; Santos, J.R.; Almeida, P.J.; Rodrigues, J.A.; Barros, A.A. Simultaneous determination of E-2-nonenal and beta-damascenone in beer by reversed-phase liquid chromatography with UV detection. *J. Chromatogr. A* **2004**, *1032*, 17–22. [CrossRef] [PubMed]
7. Kuroda, H.; Kojima, H.; Kaneda, H.; Takashio, M. Characterization of 9-fatty acid hydroperoxide lyase-like activity in germinating barley seeds that transforms 9(S)-hydroperoxy-10(E),12(Z)-octadecadienoic acid into 2(E)-nonenal. *Biosci. Biotechnol. Biochem.* **2005**, *69*, 1661–1668. [CrossRef]
8. Chen, Q.; Song, J.; Bi, J.; Meng, X.; Wu, X. Characterization of volatile profile from ten different varieties of Chinese jujubes by HS-SPME/GC-MS coupled with E-nose. *Food Res. Int.* **2018**, *105*, 605–615. [CrossRef]
9. Liu, Y.; He, C.; Song, H. Comparison of fresh watermelon juice aroma characteristics of five varieties based on gas chromatography-olfactometry-mass spectrometry. *Food Res. Int.* **2018**, *107*, 119–129. [CrossRef]
10. Ochiai, N.; Sasamoto, K.; Daishima, S.; Heiden, A.C.; Hoffmann, A. Determination of stale-flavor carbonyl compounds in beer by stir bar sorptive extraction with in-situ derivatization and thermal desorption-gas chromatography-mass spectrometry. *J. Chromatogr. A* **2003**, *986*, 101–110. [CrossRef]
11. Ligor, T.; Buszewski, B. Single-drop microextraction and gas chromatography-mass spectrometry for the determination of volatile aldehydes in fresh cucumbers. *Anal. Bioanal. Chem.* **2008**, *391*, 2283–2289. [CrossRef] [PubMed]
12. Olivero, S.J.P.; Trujillo, J.P.P. A new method for the determination of carbonyl compounds in wines by headspace solid-phase microextraction coupled to gas chromatography-ion trap mass spectrometry. *J. Agric. Food Chem.* **2010**, *58*, 12976–12985. [CrossRef] [PubMed]
13. Carrillo, G.; Bravo, A.; Zufall, C. Application of factorial designs to study factors involved in the determination of aldehydes present in beer by on-fiber derivatization in combination with gas chromatography and mass spectrometry. *J. Agric. Food Chem.* **2011**, *59*, 4403–4411. [CrossRef] [PubMed]
14. Bi, S.; Sun, S.; Lao, F.; Liao, X.; Wu, J. Gas chromatography-mass spectrometry combined with multivariate data analysis as a tool for differentiating between processed orange juice samples on the basis of their volatile markers. *Food Chem.* **2020**, *311*, 125913. [CrossRef] [PubMed]

15. Tsuzuki, S. Higher Straight-Chain Aliphatic Aldehydes: Importance as Odor-Active Volatiles in Human Foods and Issues for Future Research. *J. Agric. Food Chem.* **2019**, *67*, 4720–4725. [CrossRef] [PubMed]
16. Jiang, R.; Cudjoe, E.; Bojko, B.; Abaffy, T.; Pawliszyn, J. A non-invasive method for in vivo skin volatile compounds sampling. *Anal. Chim. Acta* **2013**, *804*, 113–119. [CrossRef] [PubMed]
17. Sekine, Y.; Toyooka, S.; Watts, S.F. Determination of acetaldehyde and acetone emanating from human skin using a passive flux sampler–HPLC system. *J. Chromatogr. B* **2007**, *859*, 201–207. [CrossRef] [PubMed]
18. Kimura, K.; Sekine, Y.; Furukawa, S.; Takahashi, M.; Oikawa, D. Measurement of 2-nonenal and diacetyl emanating from human skin surface employing passive flux sampler-GCMS system. *J. Chromatogr. B* **2016**, *1028*, 181–185. [CrossRef] [PubMed]
19. Kataoka, H. Recent developments and applications of microextraction techniques in drug analysis. *Anal. Bioanal. Chem.* **2010**, *396*, 339–364. [CrossRef]
20. Kataoka, H.; Saito, K. Recent advances in SPME techniques in biomedical analysis. *J. Pharm. Biomed. Anal.* **2011**, *54*, 926–950. [CrossRef]
21. Kataoka, H.; Ishizaki, A.; Saito, K. Recent progress in solid-phase microextraction and its pharmaceutical and biomedical applications. *Anal. Methods* **2016**, *8*, 5773–5788. [CrossRef]

molecules

Article

In Vitro and In Vivo Human Body Odor Analysis Method Using GO:PANI/ZNRs/ZIF−8 Adsorbent Followed by GC/MS

Sehyun Kim and Sunyoung Bae *

Department of Chemistry, Seoul Women's University, 621 Hwarang-ro, Nowon-gu, Seoul 01797, Korea; shgiwork@naver.com
* Correspondence: sbae@swu.ac.kr; Tel.: +82-2-970-5652

Abstract: Among various volatile organic compounds (VOCs) emitted from human skin, *trans*-2-nonenal, benzothiazole, hexyl salicylate, α-hexyl cinnamaldehyde, and isopropyl palmitate are key indicators associated with the degrees of aging. In our study, extraction and determination methods of human body odor are newly developed using headspace-in needle microextraction (HS-INME). The adsorbent was synthesized with graphene oxide:polyaniline/zinc nanorods/zeolitic imidazolate framework-8 (GO:PANI/ZNRs/ZIF−8). Then, a wire coated with the adsorbent was placed into the adsorption kit to be directly exposed to human skin as in vivo sampling and inserted into the needle so that it was able to be desorbed at the GC injector. The adsorption kit was made in-house with a 3D printer. For the in vitro method, the wire coated with the adsorbent was inserted into the needle and exposed to the headspace of the vial. When a cotton T-shirt containing body odor was transferred to a vial, the headspace of the vial was saturated with body odor VOCs. After volatile organic compounds were adsorbed in the dynamic mode, the needle was transferred to the injector for analysis of the volatile organic compounds by gas chromatography/mass spectrometry (GC/MS). The conditions of adsorbent fabrication and extraction for body odor compounds were optimized by response surface methodology (RSM). In conclusion, it was able to synthesize GO:PANI/ZNRs/ZIF−8 at the optimal condition and applicable to both in vivo and in vitro methods for body odor VOCs analysis.

Keywords: human body odor; volatile organic compounds; GO:PANI/ZNRs/ZIF−8; response surface methodology; needle-based adsorbent; headspace-in needle microextraction

1. Introduction

Human body odor is a mixture of compounds generated by skin bacteria resulting in volatile organic compounds (VOCs) such as alcohol, aldehyde, acid, amine, hydrocarbons, and so on [1–6]. The primary body odor, which is distinguishable from others, is very unique in individuals that vary with age, gender, ethnic, body parts, and so on [1,5]. It can be important in areas of medical, security, safety, and cosmetic applications. On the basis of the literature, body odor VOCs carry age-related information and can be analyzed by various methods [1,6–9]. Previous studies have shown that *trans*-2-nonenal, benzothiazole, and isopropyl palmitate are commonly found in the elderly [1,7,8], while hexyl salicylate and α-hexyl cinnamaldehyde occur more frequently in young people [7]. In particular, *trans*-2-nonenal is known to be secreted more as ω7 unsaturated fatty acids are oxidized due to the increase of ω7 unsaturated fatty acids and lipid peroxides as skin aging progresses [1]. In addition, reactive oxygen species (ROS) generated by ultraviolet rays may also react with unsaturated fatty acids to affect the generation of body odor [9].

Organic solvents, gauze, solid phase microextraction (SPME) fiber, and stir bar have been used to extract human body odor [6,10–12]. The SPME method has been commonly used to avoid using organic solvents. There are two methods for analyzing body odor with SPME: in vivo and in vitro [13]. In both methods, target compounds are absorbed onto SPME fiber and then thermally desorbed and analyzed by gas chromatography/mass spectrometry (GC/MS). In the in vitro method, a cotton swab or T-shirt that has absorbed body odor is transferred to a SPME vial and extracted with SPME fiber. The in vivo method is a method to directly extract body odor by installing the SPME device on the skin [13]. However, it is quite inconvenient because of a glass funnel that fasten the SPME device on the skin. To address these limitations, a new method of body odor analysis has been developed by in-needle microextraction (INME) using graphene oxide:polyaniline/zinc nanorods/zeolitic imidazolate framework-8 (GO:PANI/ZNRs/ZIF−8) adsorbent. The in-needle microextraction (INME) technique has been used for extraction of volatile organic compounds in various matrix due to easy fabrication and cost effectiveness [14–20]. The adsorbent has been synthesized with various polarity based on polyaniline modified by additives such as multiwall carbon nanotube and ionic liquid [14,19]. The adsorbent containing analytes was desorbed by gas chromatograph/mass spectrometer (GC/MS) for qualitative and quantitative analyses.

INME has an adsorbent inside the needle and is used for solid dynamic microextraction. INME uses the needle coated with the adsorbent inside or the wire coated with adsorbent being inserted inside of the needle for extraction and analysis without using solvents followed by chromatographic analysis [14–20]. Especially, dynamic headspace-INME (HS-INME) has been applied to the sample by exposing the INME needle with a pump to the headspace of the vial saturated with VOCs. Polymeric adsorbents have been synthesized by sol-gel reaction [15,17,19] and/or electrochemical deposition methods [14,16,18]. It has demonstrated that INME could be customized due to easy fabrication of adsorbent and economically feasible to extract target compounds without using solvents.

The objective of this study is to fabricate GO:PANI:ZNRs:ZIF−8 adsorbent and utilize this for direct extraction of the body odor from human skin in vivo and for indirect extraction from the media adsorbed with body odor in vitro, followed by GC/MS. The mixing ratio of ZNRs and ZIF−8 of the adsorbent was varied to optimize the adsorption by modifying polarity [21]. Porous GO:PANI can be easily coated onto stainless-steel wire from the electrochemical method, while ZNRs exhibit good affinity with polar compounds and ZIF−8 with polar compounds [21,22]. Characterization of the adsorbent was measured by Fourier-transform infrared spectroscopy (FTIR), Thermogravimetric Analyzer (TGA), X-ray diffractometer (XRD), Brunauer–Emmett–Teller (BET) surface area method, and field emission-scanning electron microscopy (FE-SEM). Fabrication and extraction conditions were optimized by the Box-Behnken design (BBD) combined with response surface methodology (RSM), and the developed method was validated including limit of detection (LOD), limit of quantification (LOQ), recovery, and reproducibility.

2. Materials and Methods

2.1. Chemicals and Instrumentation

Five target compounds (Supplementary Materials Table S1), *trans*-2-nonenal (97%), benzothiazole (96%), hexyl salicylate (99%), α-hexyl cinnamaldehyde (85%), and isopropyl palmitate (90%), were obtained from Sigma-Aldrich (St. Louis, MO, USA). Stock solution of all target compounds was prepared using methanol (HPLC grade, Samchun Pure Chemical Co., Pyeongtaek, Korea) at a concentration of 1000 mg L^{-1}. Stainless-steel wire (0.22 mm O.D., 09 one Science, Goyang, Korea) was used as an adsorbent wire and a working electrode. A Hamilton 91022 needle (metal hub, 22 gauge, 718 μm O.D., 413 μm I.D., bevel tip, 51 mm length) and 1 mL Hamilton 1001 TLL (gas tight syringe, luer lock gas tight syringe barrel, Reno, NV, USA) with plunger (polytetrafluoroethylene, Reno, NV, USA) were used for INME adsorption and GC/MS thermal desorption. Chemicals used in fabrication of adsorbent was shown in the supporting information.

Agilent 6970 gas chromatograph (GC) with a 5973 mass selective detector (MS) was used to analyze target compounds. HP-5MS (30 m × 0.25 mm × 0.25 μm, (5%-phenyl)-methylpolysiloxane) was used as the column and helium with the flow rate of 1 mL min^{-1} as the carrier gas. The injector temperature was 240 °C using split mode (30:1), and the oven program was 60 °C (4 min) to 230 °C (10 min) at 6 °C min^{-1}. The range of 50~350 m/z was scanned in full scan mode.

Adsorbent fixing part and cover of the adsorption kit were 3D printed on a Cubicon Single Plus (3DP-310F) with polylactic acid (PLA)-plus filament.

2.2. Fabrication of Wire Coated with Adsorbent

The adsorbent was prepared in 3 steps: (1) fabrication of GO:PANI layer, (2) fabrication of GO:PANI/ZNRs layer, and (3) fabrication of GO:PANI/ZNRs/ZIF−8 layer, as illustrated in Figure 1.

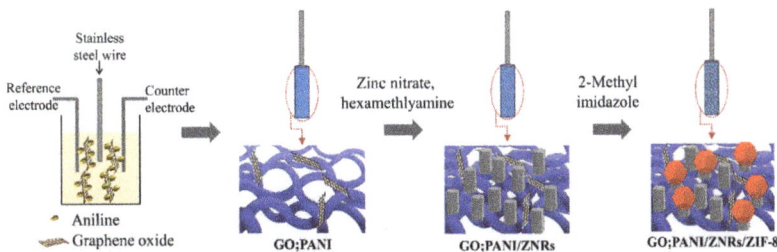

Figure 1. Fabrication process of the adsorbent.

GO:PANI layer was fabricated on stainless-steel wire using cyclic voltammetry in which 0.6–1.0 V (sweeping rate of 50 mV s^{-1} for 30 cycles) was applied to an electrolyte containing GO and aniline in a H$_2$SO$_4$ (0.5 mol L^{-1}) solvent. Three-electrode system consisting of stainless-steel wire (working electrode), Pt wire (counter electrode), and Ag/AgCl electrode (reference electrode) was used. After electrochemical deposition, the coating layer was rinsed several times using distilled water to remove unreacted chemicals. Finally, the stainless-steel wire coated with porous GO:PANI was dried at 80 °C for 0.5 h.

In the GO:PANI/ZNR fabrication step, the GO:PANI coated stainless-steel wire was dipping into an aqueous mixture of zinc nitrate hexahydrate (0.5 mol L^{-1}) and hexamethylenetetramine (0.5 mol L^{-1}) for 30 s and dried in 180 °C oven for 2 min. This step was repeated five times. In this process, ZnO seeds formed by OH$^-$ and Zn^{2+} were deposited on the porous GO:PANI surface. Then, the stainless-steel wire was dipped into the aqueous mixture of zinc nitrate hexahydrate (0.05 mol L^{-1}) and hexamethylenetetramine (0.05 mol L^{-1}) and placed in 95 °C oven for 2 h to hydrothermally grow ZnO nanorods [22].

Finally, the stainless-steel wire coated with GO:PANI/ZNRs was dipped into a 2-MI solution (DMF:H$_2$O = 3:1 (v/v)) and placed in 90 °C oven for 8 h [22]. In this procedure, ZIF−8 was formed by coordination bonding between 2-MI and Zn^{2+} on ZNRs surface. Prior to the extraction process, the adsorbent coated on stainless-steel wire was placed in a 220 °C oven for 1 h to remove impurities from the coated wires. Then, it was kept in a desiccator at room temperature until further use.

2.3. Characterization of GO:PANI/ZNRs/ZIF−8 Adsorbent

The functional groups of the adsorbent generated at each step to confirm the synthesis were measured by the Fourier-transform infrared (FT-IR) spectroscopy (Perkin Elmer Spectrometer 100, Waltham, MA, USA). The crystallinity of synthesized graphene oxide and coating layer of each synthesis step were confirmed using X-ray diffractometer (XRD, Bruker DE/D8 Advance, Bruker, Karlsruhe, Germany). The thermal stability of GO:PANI, GO:PANI/ZNRs, and GO:PANI/ZNRs/ZIF−8 were evaluated by thermogravimetric analyzer (TGA, TG209 F1 Libra, Netzsch, Selb, Germany).

The specific surface area, pore size distribution, and N$_2$ adsorption-desorption isotherm of GO:PANI/ZNRs and GO:PANI/ZNRs/ZIF−8 were observed by the Brunauer–Emmet–Teller (BET, 3flex, Micromeritics, Norcross, GA, USA). The morphology of GO:PANI,

GO:PANI/ZNRs, and GO:PANI/ZNRs/ZIF−8 was characterized using field emission-scanning electron microscope (FE-SEM, JSM-6700F, JEOL Ltd., Tokyo, Japan).

2.4. INME Process

As shown in Figure 2, extraction of the target compounds was performed by the INME method. In the preparation of the INME adsorption device, a GO:PANI/ZNRs/ZIF−8-coated stainless-steel wire was inserted vertically into a Hamilton needle and then connected to a gas tight syringe. After spiking the target compounds into the SPME vial, it was sealed with a mini-nut cap. Then, saturation was performed at 40 °C for 15 min. Next, dynamic adsorption was carried out by penetrating the needle to the headspace of the SPME vial. A homemade reciprocating pump was used at 6 cycles/min to suck the analytes saturated in the headspace inside the needle during this process. Upon completion of adsorption, the analytes were thermally desorbed from the adsorbent by placing it in a GC/MS injector and analyzed by GC/MS.

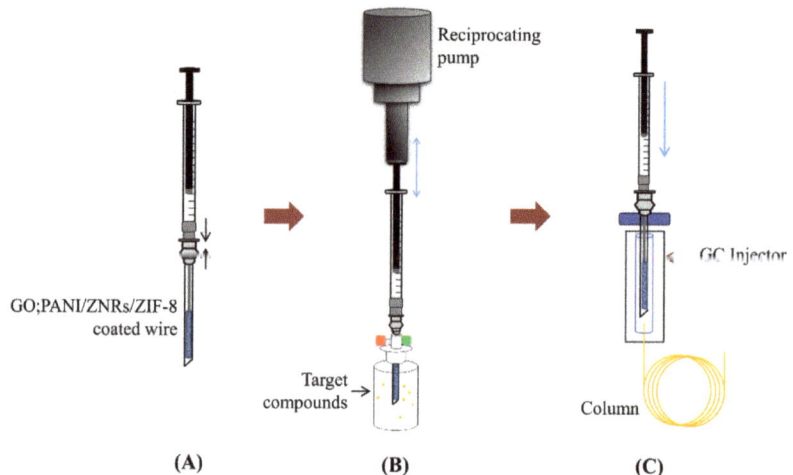

Figure 2. A scheme of INME process. (**A**) A needle inserted with a coated wire is connected to a gas tight syringe. (**B**) The needle was plugged into the mini-nut cap to adsorb target compounds in the headspace using a reciprocating pump. (**C**) The target compounds adsorbed on the adsorbent was thermally desorbed in GC injector and analyze using GC/MS.

2.5. Adsorption Kit Fabrication

The body odor adsorption kit was designed to adsorb VOCs directly from the skin (Figure 3). It consists of adsorbent, adsorbent fixing part, cover, and medical tape. The adsorbent was coated on the middle section of 5.0 cm stainless-steel wire to be placed in the fixing part. An adsorbent fixing part and its cover were manufactured by a 3D printer with PLA plus filament. The adsorbent kit was made as a cylinder with a hollow in the middle and a closed top, which could collect VOCs from the skin and protect the adsorbent from external odors and impacts. In addition, the adsorbent was about 10 mm away from the skin, which could avoid contamination from dead skin cells and sweat. The adsorbent fixing part has a hole in the center of the top to insert the adsorbent. PDMS was applied on the bottom of the adsorbent fixing part to reduce skin irritation due to direct contact on skin. The adsorbent kit was assembled to conduct the experiments; insert the adsorbent into the fixing part, close the cover, put it on the skin, and attach the medical tape to fix it to the skin. After the body odor was collected for 2 h, the wire coated with adsorbent was removed from the kit, inserted into a Hamilton needle, connected to a gas tight syringe, thermally desorbed at GC/MS inlet, and the body odor components were analyzed in GC/MS system. For the feasibility test of the adsorption kit, the target compounds were spiked to the artificial silica skin, and the adsorption kit was assembled to place it and adsorbed for 2 h at 36 °C.

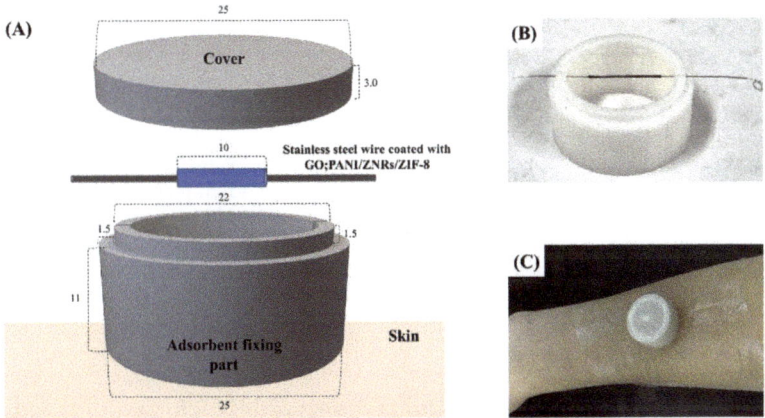

Figure 3. (**A**) A schematic view of adsorption kit for the collection of body odor from human skin (the unit of length: mm). (**B**) Actual image of adsorbent placed on the fixing part made by 3D printer. (**C**) Sampling of body odor at the arm using adsorbent kit covered by a medical tape.

2.6. Optimization Using Response Surface Methodology (RSM)

In the fabrication of adsorbent, the concentrations of GO, aniline, and 2-MI that may affect adsorption amount of the target compounds were optimized using RSM. 3-factor 3-level Box BBD was used with Minitab 19 (Minitab Inc., State College, PA, USA). A total of 17 experiments were conducted randomly, as shown in Table S2. The levels and coded numbers of each factor are as follows: amount of GO (X_1) was 5 mg L^{-1} (−1), 10 mg L^{-1} (0), and 15 mg L^{-1} (1); amount of aniline (X_2) was 9.31 g L^{-1} (−1), 14.0 g L^{-1} (0), and 18.6 g L^{-1} (1); and amount of 2-MI (X_3) was 10.26 g L^{-1} (−1), 102.6 g L^{-1} (0), and 195.0 g L^{-1} (1).

In addition, the 3-factor 3-level BBD was used to optimize the INME analysis condition. As shown in Table S3, total of 17 experiments were conducted randomly. Extraction temperature (X_1), adsorption time (X_2), and desorption time (X_3) were selected as factors. The levels and coded numbers for each factor were as follows: extraction temperature (X_1) was 30 °C (−1), 40 °C (0), and 50 °C (1); the adsorption time (X_2) was 30 min (−1), 45 min (0), and 60 min (1); and the desorption time (X_3) was 1 min (−1), 3 min (0), and 5 min (1).

The data obtained from the experiments for fabrication condition and the INME analysis conditions were analyzed by the response surface regression procedure using second-order polynomial model (Equation (1)).

$$Y = \alpha_0 + \alpha_1 X_1 + \alpha_2 X_2 + \alpha_3 X_3 + \alpha_{11} X_1^2 + \alpha_{22} X_2^2 + \alpha_{33} X_3^2 + \alpha_{12} X_1 X_2 + \alpha_{13} X_1 X_3 + \alpha_{23} X_2 X_3 \quad (1)$$

where Y is the predicted response (amount of adsorption); α_0 is the intercept, α_1, α_2, and α_3 are linear coefficients; α_{11}, α_{22}, and α_{33} are quadratic coefficients; and α_{12}, α_{13}, and α_{23}, are the interactive coefficients.

3. Results and Discussion

3.1. Optimization of Human Body Odor Adsorbent Fabrication Conditions Using RSM

Reactants such as GO (X_1), aniline (X_2), and 2-MI (X_3) added were optimized through RSM. It determines the optimized value of each variable and analyzes the interactions between the variables. Optimized values for each variable were obtained using the regression equation calculated using Minitab (Table 1). A three-dimensional response surface curve is a graphical representation of the results from the regression equation. In each graph, after fixing one of the three independent variables to level 0, the effect of the remaining two independent variables on peak area obtained from the total ion chromatogram (TIC) was investigated (Figure S1) [23]. Confirming the relationship between GO and aniline, the peak area value rapidly increased as aniline increased, while the peak area value increased gently as the amount of GO increased. In the relationship between GO and 2-MI, as the amount of 2-MI increased, the peak area increased rapidly except for *trans*-2-nonenal, and the amount of GO showed a gradual increase. In the relationship between aniline and 2-MI, the peak area increased rapidly as 2-MI increased, except for *trans*-2-nonenal, and the amount of aniline increased gradually. Consequently, the order affecting the peak areas is 2-MI, aniline, and GO. The optimal BBD results for the reactants are determined as 15 mg L^{-1} of GO, 18.3 g L^{-1} of aniline, and 151.6 g L^{-1} of 2-MI. The error of the modeling equation obtained through the BBD method was about 10%. The peak area value for each target compound can be predicted by substituting the concentration of each reactant.

Table 1. Peak area prediction formula of each material obtained through BBD in optimization of the reactant concentration.

Compound	Model Formula
trans-2-Nonenal	1,373,897 + 65,796 X_1 + 71,554 X_2 − 165,521 X_3 + 737,146 X_1^2 − 307,202 X_2^2 + 2692 X_3^2 − 9779 X_1X_2 + 70,330 X_1X_3 − 3443 X_2X_3
Benzothiazole	2,738,952 + 459,936 X_1 + 881,137 X_2 + 255,043 X_3 + 1,078,534 X_1^2 + 520,184 X_2^2 + 591,432 X_3^2 − 221,829 X_1X_2 − 241,702 X_1X_3 + 320,824 X_2X_3
Isopropyl palmitate	55,849 + 10,191 X_1 + 10,199 X_2 + 7497 X_3 − 11,547 X_1^2 − 4503 X_2^2 − 15,959 X_3^2 + 3465 X_1X_2 + 1502 X_1X_3 + 3925 X_2X_3
Hexyl salicylate	667,682 + 81,671 X_1 + 81,140 X_2 + 93,743 X_3 + 47,534 X_1^2 − 18,836 X_2^2 − 27,142 X_3^2 − 888 X_1X_2 + 24,879 X_1X_3 + 34,062 X_2X_3
α-Hexyl cinnamaldehyde	1,127,759 + 173,707 X_1 + 296,075 X_2 + 229,720 X_3 + 53,389 X_1^2 + 84,526 X_2^2 + 26,918 X_3^2 + 150,333 X_1X_2 − 60,256 X_1X_3 + 112,567 X_2X_3

X_1: amount of graphene oxide (mg L^{-1}), X_2: amount of aniline (g L^{-1}), and amount of X_3: 2-methyl imidazole (g L^{-1}).

3.2. Optimization of INME Extraction Conditions Using RSM

HS-INME-GO:PANI/ZNRs/ZIF−8 extraction conditions, including extraction temperature (X_1), adsorption time (X_2), and desorption time (X_3), were optimized using the RSM method. Figure S2 is a response surface graph showing the relationship between the variables and the peak area values. In the relationship of extraction temperature and adsorption time, it was observed that the extraction temperature had a more significant effect on extraction for *trans*-2-nonenal and benzothiazole, while the adsorption time had a greater effect on the remaining compounds. Confirming the relationship of extraction temperature and desorption time, the extraction temperature had a greater effect than the desorption time for all target compounds. The relationship between adsorption time and desorption time was found that desorption time had a greater effect on extraction. The optimal extraction conditions based on the RSM results are as follows. The optimal condition was 40 °C for extraction temperature, 60 min for adsorption time, and 4.3 min for desorption time, respectively. The modeling equation showed an error about 10%. and the value of the peak area of each target compound could be reasonably predicted from the modeling equation (Table 2).

Table 2. Peak area prediction formula of each compound obtained through BBD in optimization of INME-GO:PANI/ZNRs/ZIF−8 analysis conditions.

Compound	Model Formula
trans-2-Nonenal	$6,715,422 + 1,488,188\ X_1 + 990,247\ X_2 - 290,098\ X_3 - 154,172\ X_1^2 - 416,644\ X_2^2 + 858,182\ X_3^2 - 1,371,651\ X_1X_2 + 379,903\ X_1X_3 - 108,883\ X_2X_3$
Benzothiazole	$3,475,053 - 572,871\ X_1 + 578,351\ X_2 + 260,594\ X_3 + 80,741\ X_1^2 + 19,364X_2^2 + 507,335\ X_3^2 - 1,048,215\ X_1X_2 - 390,972\ X_1X_3 + 686,615\ X_2X_3$
Isopropyl palmitate	$70,372 + 50,455\ X_1 + 25,337\ X_2 - 1915\ X_3 + 15,132\ X_1^2 - 7361\ X_2^2 + 20,237\ X_3^2 + 9741\ X_1X_2 - 6146\ X_1X_3 - 18,574\ X_2X_3$
Hexyl salicylate	$1,477,657 + 311,065\ X_1 + 284,017\ X_2 + 104,003\ X_3 - 37,787\ X_1^2 - 246,864\ X_2^2 + 85,773\ X_3^2 + 5355\ X_1X_2 + 18,324\ X_1X_3 + 76,669\ X_2X_3$
α-Hexyl cinnamaldehyde	$2,459,037 + 6,365,250\ X_1 + 452,391\ X_2 - 67,492\ X_3 - 291,829\ X_1^2 - 446,990\ X_2^2 + 118,680\ X_3^2 + 20,695\ X_1X_2 - 27,709\ X_1X_3 + 195,394\ X_2X_3$

X_1: extraction temperature (°C), X_2: adsorption time (min), and X_3: desorption time (min).

3.3. Characterization of the Adsorbent

The products generated at each synthesis process prepared under optimal conditions were characterized through FT-IR, TGA, XRD, BET, and FE-SEM. FT-IR spectra (Figure 4A) show that GO was well-synthesized from graphite by verifying the O-H band at 3389 cm^{-1}, the C=O band at 1721 cm^{-1}, the C=C band in the phenol ring at 1613 cm^{-1}, the C-OH band at 1227 cm^{-1}, and C-O band at 1048 cm^{-1} in the IR spectrum of GO [24]. Through the following peaks in the IR spectrum of GO:PANI, it was confirmed that the GO plate was covered by PANI nanofibers to form GO:PANI identified by peaks at 1561 cm^{-1} from C=C in quinone, 1479 cm^{-1} from benzene ring, 1302 cm^{-1} from C-N in the second amine ring, 1242 cm^{-1} from C-N vibrational band in the protonic acid, 1126 cm^{-1} from C=N vibrational band, and 805 cm^{-1} from the aromatic C-H band [24]. In the IR spectrum of GO:PANI/ZNRs, peaks of amine groups and benzene ring shifted to higher wavenumbers than GO:PANI from the previous synthesis step were observed at 1585 cm^{-1}, 1498 cm^{-1}, 1297 cm^{-1}, 1143 cm^{-1}, and 826 cm^{-1}. This indicates that the amine and imine nitrogen atoms interact with Zn^{2+} via either protonation or complexation [25]. In addition, the O-Zn-O stretching vibration peak was observed at 432 cm^{-1} [26]. Amine groups and a benzene ring were observed at 1596 cm^{-1}, 1500 cm^{-1}, 1308 cm^{-1}, 1148 cm^{-1}, and 827 cm^{-1} in the GO:PANI/ZNRs/ZIF−8 IR spectrum. Additionally, a peak at 748 cm^{-1} from out of plane bending of the 2-MI ring and a 420 cm^{-1} peak from Zn-N stretching confirmed that the nitrogen atom was connected to 2-MI linker to form ZIF−8 [27].

From the XRD analysis, it was confirmed that ZNRs and ZIF−8 were well-formed (Figure 4B). In the GO:PANI/ZNRs, peaks at 2θ = 31.68° (100), 34.35° (002), 36.16° (101), 47.43° (012), and 56.49° (110), which are identical to the peaks from the ZnO standard, were observed. Both ZnO and ZIF−8 peaks in GO:PANI/ZNRs/ZIF−8 were confirmed to compare with peaks from the ZnO and ZIF−8 standards. Peaks observed at 2θ = 7.33° (011), 10.37° (002), 12.71° (112), 14.70° (022), 16.43° (013), 18.52° (222), 22.14° (114), 24.48° (233), and 26.61° (134) were identical to the peaks of the ZIF−8 standard, while peaks at 2θ = 31.71° (100), 34.40° (002), 36.20° (101), 47.48° (012), and 56.53° (110) were observed from the ZnO standard. It was concluded that the adsorbent consists of GO:PANI/ZNRs/ZIF−8 was well-synthesized and expected to adsorb compounds having various polarities due to various functional groups.

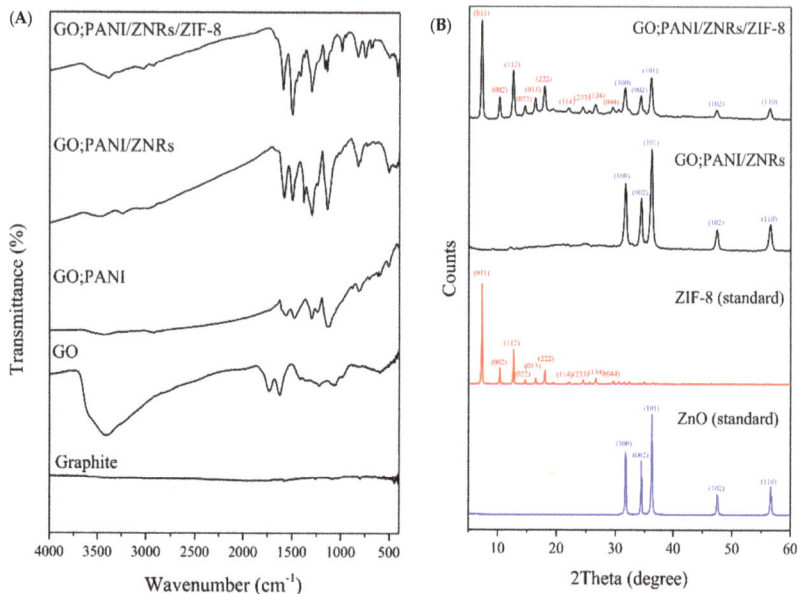

Figure 4. (**A**) FT-IR spectrums of graphite, GO, GO:PANI, GO:PANI/ZNRs, and GO:PANI/ZNRs/ZIF−8 and (**B**) XRD spectrums of the ZnO (standard), ZIF−8 (standard), GO:PANI/ZNRs/ZIF−8, and GO:PANI/ZNRs/ZIF−8.

Thermal stability of the adsorbent synthesized at each step was measured by TGA (Figure S3). Significant weight loss was not observed and confirmed thermal stability up to 273 °C. Mass change at 81 °C is regarded as the dehydration process. The mass change of GO:PANI/ZNRs was not observed until 262 °C. The effect of GO addition on thermal stability was shown clearly to compare the decomposition temperature between PANI/ZNRs/ZIF−8 and GO:PANI/ZNRs/ZIF−8, which was 195 °C and 291 °C, relatively.

The effect of ZIF−8 addition on pore size distribution is shown in Figure 5 as the pore size distribution and N_2 adsorption and desorption isotherm. The BET specific surface area of GO:PANI/ZNRs/ZIF−8 was 4640 m^2 g^{-1}, which was much higher than that of GO:PANI/ZNRs with a specific surface area of 5.35 m^2 g^{-1}. According to the pore size distribution curve, GO:PANI/ZNRs has mesopore (2~50 nm), whereas GO:PANI/ZNRs/ZIF−8 has micropore (<2 nm) and mesopore (2~50 nm) [28]. In addition, N_2 adsorption and desorption isotherm of GO:PANI/ZNRs/ZIF−8 showed type I isotherm indicating that micropore and mesopore coexist [29].

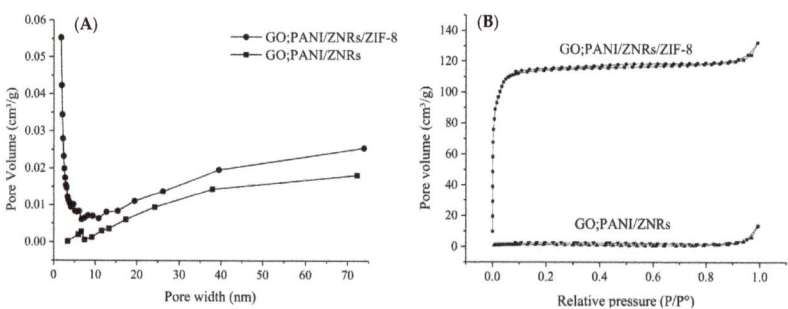

Figure 5. (**A**) Pore width distribution of GO:PANI/ZNRs and GO:PANI/ZNRs/ZIF−8 and (**B**) N_2 adsorption-desorption isotherms of GO:PANI/ZNRs and GO:PANI/ ZNRs/ZIF−8.

The morphology of the adsorbent coated on stainless-steel wire was observed using FE-SEM in Figure 6. The image of GO:PANI showed a rough and porous surface due to the deposition of PANI nanofibers on the GO sheets [24]. In GO:PANI/ZNRs image, flower-shaped ZnO nanostructures were constructed on the porous GO:PANI. Rhombic dodecahedron ZIF−8 was observed in GO:PANI/ZNRs/ZIF−8 image, which indicates that ZIF−8 was well-dispersed on the ZNRs surface of GO:PANI/ZNRs [30].

Figure 6. SEM images of stainless-steel wire coated with (**A**) GO:PANI, (**B**) GO:PANI/ZNRs, and (**C**) GO:PANI/ZNRs/ZIF−8 (1; ×5000, 2; ×20,000).

Comparing the adsorption amount by each synthesized step and PANI/ZNRs/ZIF−8, GO:PANI/ZNRs/ZIF−8 shows the highest adsorption for all target compounds (Figure S4). The final adsorbent has moieties with different polarities from ZNRs and ZIF−8 to enhance the adsorption potentials, and its large specific surface area and micropores might play an important role in increasing the adsorption amounts. Strong π–π stacking interaction and hydrogen bonding between target compounds and adsorbent contribute to the improved extraction of target compounds [22,23]. In comparison between GO:PANI/ZNRs/ZIF−8 and PANI/ZNRs/ZIF−8, GO addition tends to increase the adsorption amount of *trans*-2-nonenal that appears to interact between GO and C=O of *trans*-2-nonenal.

Based on the literature reviews, there are few studies on GO:PANI/ZNRs/ZIF−8 adsorbent for the human body odor compounds. On the other hand, the adsorbents composed of GO, PANI, ZNRs, and ZIF−8 as SPME fibers coupled with chromatography have been studied in water, soil, air, and food samples [21,22,24,27,28]. The GO:PANI was used as an effective adsorbent for Zn ion removal from the water sample [24]. The adsorbent of zinc oxide nanorods directly coated on stainless-steel wire was applied as SPME fiber to extract aldehydes for instant noodle samples [21]. Hexanal, heptanal, octanal, nonanal, and decanal were analyzed with a linear range between 0.08 and 5.0 µg g^{-1} [21]. PANI/ZNRs/ZIF−8 was synthesized to extract 2-methylnaphthalene, 1-methylnaphthalene, pyrene, phenanthrene, cyrysene, and benzo[a]pyrene of polycyclic aromatic hydrocarbons (PAHs) in a sewage water sample [22]. The synthesized ZIFs-ZnO composite as a sorbent for SPME-HPLC-UV was used to remove Sudan dyes in the water sample showing 0.002 ng mL^{-1} LOD and wide linear ranges 0.02–20 ng mL^{-1} [27].

3.4. Comparison of Extraction Efficiency

The Dynamic INME and static INME method were compared by enrichment factor (EF). EF was calculated as follows:

$$EF = A_1/A_0 \qquad (2)$$

A_1 is the peak area from dynamic INME method that sucks the target compounds inside the needle using a reciprocating pump, and A_0 is the peak are from the static INME method that was performed without using a pump. The EF value of *trans*-2-nonenal was 1062.29 ± 33.72, benzothiazole was 390.12 ± 73.14, hexyl salicylate was 29.71 ± 5.41, α-hexyl cinnamaldehyde was 36.79 ± 1.30, and isopropyl palmitate was 1.54 ± 0.59, respectively. It was concluded that all target compounds were more efficiently extracted than the static INME- method.

3.5. Feasibility Test Using an Adsorption Kit

The body odor adsorption kit developed in this study was applied to in vivo sampling. A feasibility test was performed on petri dish spiking 30 µg target compounds to place an adsorption kit fixed with the medical tape.

After 2 h exposure to compounds, the wire coated with GO:PANI/ZNRs/ZIF−8 adsorbent inserted inside the adsorption kit was removed, put into the Hamilton needle, and connected to the plunger to desorb the compounds at GC/MS injector. The TIC obtained from this process is shown in Figure 7. From TIC, all five target compounds were well-separated and measured. *trans*-2-Nonenal, benzothiazole, hexyl salicylate, α-hexyl cinnamaldehyde, and isopropyl palmitate were successfully separated at the retention time of 12.45 min, 14.42 min, 24.15 min, 25.40 min, and 29.92 min, respectively. In conclusion, the body odor adsorption kit developed in this study could be used as an in vivo sampling method for human body odor analysis.

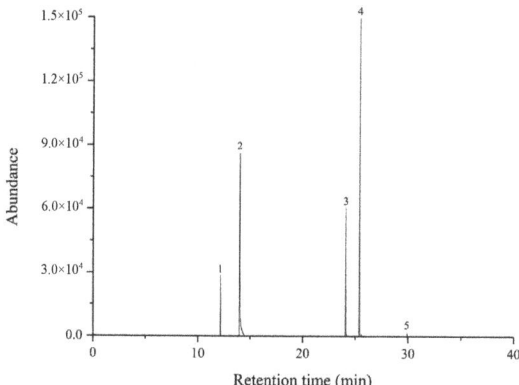

Figure 7. Total ion chromatogram of aging-related body odor using adsorption kit ((1) *trans*-2-noenal, (2) benzothiazole, (3) hexyl salicylate, (4) a-hexyl cinnamaldehyde, and (5) isopropyl palmitate).

3.6. Method Validation

The INME method for in vitro body odor extraction was validated by analyzing five target compounds repeatedly. For all target compounds, the linearity of the seven-points calibration curves had r^2 values greater than 0.99 in the dynamic range. Limit of detection (LOD) and limit of quantitation (LOQ) were calculated according to ISO definition [31]. The calculated LOD and LOQ were 4.89 ng~128 ng and 14.8 ng~38.8 ng, respectively. The dynamic range was 4.89 ng~15,000 ng.

$$\text{Recovery} = (A_C + A_I) - A_0/A_S \times 100 \tag{3}$$

In addition, a recovery test was performed by analyzing a 100% cotton T-shirt with an analyte spiked by the proposed method (Figure 8). The recovery rate was calculated using Equation (3), that A_C is the amount of target compounds adsorbed on the 100% cotton T-shirt, A_I is the amount of target compounds adsorbed on GO:PANI/ZNRs/ZIF−8 adsorbent using HS-INME method, A_0 is the amount of target compounds in the unspiked sample, and A_S is the amount of target compounds spiked on the 100% cotton T-shirt [19]. The recovery rate shows acceptable values, ranging from 91.27% to 103.47%.

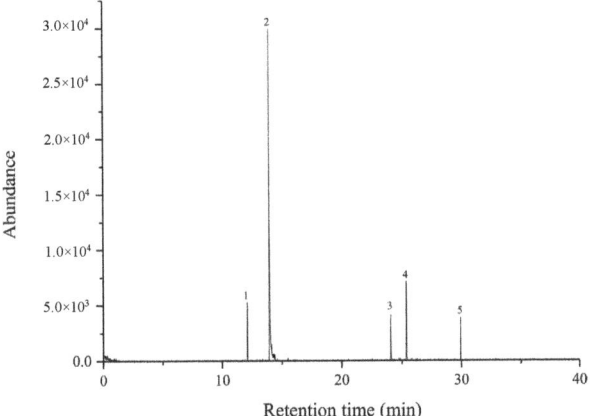

Figure 8. Total ion chromatogram of aging-related body odor spiked on 100% cotton T-shirt using INME method ((1) *trans*-2-nonenal, (2) benzothiazole, (3) hexyl salicylate, (4) a-hexyl cinnamaldehyde, and (5) isopropyl palmitate).

The reproducibility of the proposed method was confirmed by obtaining the relative standard deviation of intra assay (run to run, $n = 5$) and inter assay (needle to needle, $n = 5$). As a result, in the intra and inter assays, all target compounds showed about 10% and 15% precision, respectively.

4. Conclusions

In this study, in vitro and in vivo methods using GO:PANI/ZNRs/ZIF−8 adsorbent that were suitable for body odor adsorption was developed. The amounts of GO, aniline, and 2-MI added during the adsorbent fabrication process were simply and accurately optimized through the RSM method determined as 15 mg L^{-1} of GO, 18.3 g L^{-1} of aniline, and 151.6 g L^{-1} of 2-MI, respectively. The physicochemical characterization of the synthesized adsorbent was confirmed by FT-IR, XRD, TGA, BET, and SEM. The large surface area and micropore of the adsorbent increased the adsorption amount of the target compounds, and the thermal stability was increased due to the addition of GO. In in vivo sampling, a wire coated with adsorbent could be placed directly to the human skin using an adsorbent kit developed in this study. In in vitro method, a wire coated with GO:PANI/ZNRs/ZIF−8 to be used as dynamic HS-INME method to adsorb human body odor VOCs saturated in the headspace of the vial. As a result of the validation of the in vitro method, analysis of aging-related human body odor compounds was successfully performed with good precision and recovery. GO:PANI/ZNRs/ZIF−8 adsorbent could be used as INME method that could adsorb five aging-related human body odor compounds successfully and demonstrate over 100 applications after the conditioning process.

Supplementary Materials: The following supporting information can be downloaded at https://www.mdpi.com/article/10.3390/molecules27154795/s1: Figure S1: Response surface diagrams of target compounds by factors.; Figure S2: Response surface diagrams of target compounds by factors. Figure S3: TGA curves of GO:PANI, GO:PANI/ZNRs, GO:PANI/ZNRs/ZIF−8, and PANI/ZNRs/ZIF−8. Figure S4: Comparison of the adsorption amount of five target compounds according to the adsorbent synthesis steps, and without GO. Table S1: Physicochemical properties of target compounds. Table S2: Experimental design matrix for optimization of the reactant concentrations in fabrication of adsorbent. Table S3: Experimental design matrix for optimization of HS-INME conditions.

Author Contributions: Conceptualization, S.B.; methodology, S.K.; validation, S.K. and S.B.; investigation, S.K.; writing—original draft preparation, S.K.; writing—review and editing, S.B.; and supervision, S.B. All authors have read and agreed to the published version of the manuscript.

Funding: This research was funded by the Korean National Research Foundation, grant number (2020R1A2C1014112), and Seoul Women's University (2022-0081).

Institutional Review Board Statement: Not applicable.

Informed Consent Statement: Informed consent was obtained from all subjects involved in the study.

Data Availability Statement: Not applicable.

Conflicts of Interest: The authors declare no conflict of interest.

Sample Availability: Samples of the compounds are not available from the authors.

References

1. Haze, S.; Gozu, Y.; Nakamura, S.; Kohno, Y.; Sawano, K.; Ohta, H.; Yamazaki, K. 2-NonenalNewly Found in Human Body Odor Tends to Increase with Aging. *J. Investig. Dermatol.* **2001**, *116*, 520–524. [CrossRef] [PubMed]
2. Yamazaki, S.; Hoshino, K.; Kusuhara, M. Odor Associated with Aging. *J. Anti-Aging Med.* **2010**, *7*, 60–65. [CrossRef]
3. Ackerl, K.; Atzmueller, M.; Grammer, K. The scent of fear. *Neuroendocr. Lett.* **2002**, *23*, 79–84.
4. Singh, D.; Bronstad, P.M. Female body odour is a potential cue to ovulation. *Proc. R. Soc. B* **2001**, *268*, 797–801. [CrossRef]
5. Abaffy, T.; Möller, M.; Riemer, D.D.; Milikowski, C.; Defazio, R.A. A Case Report-olatile Metabolomic Signature of Malignant Melanoma using Matching Skin as a Control. *J. Cancer Sci. Ther.* **2011**, *3*, 140–144. [CrossRef]
6. Jha, S.K. Characterization of human body odor and identification of aldehydes using chemical sensor. *Rev. Anal. Chem.* **2017**, *36*, 20160028. [CrossRef]
7. Gallagher, M.; Wysocki, C.J.; Leyden, J.J.; Spielman, A.I.; Sun, X.; Preti, G. Analyses of volatile organic compounds from human skin. *Br. J. Derm.* **2008**, *159*, 780–791. [CrossRef]
8. Kimura, K.; Sekine, Y.; Furukawa, S.; Takahashi, M.; Oikawa, D. Measurement of 2-nonenal and diacetyl emanating from human skin surface employing passive flux sampler—GCMS system. *J. Chromatogr. B Anal. Technol. Biomed. Life Sci.* **2016**, *1028*, 181–185. [CrossRef]
9. Doo, H.; Won, B.; Han, J.P.; Hwang, S.J.; Kim, J.; Park, J.; Park, S.N. Antioxidative Effect and Tyrosinase Inhibitory Activity of Lindera obtusiloba Blume Extracts. *J. Soc. Cosmet. Sci. Korea* **2012**, *38*, 297.
10. Curran, A.; Rabin, S.; Prada, P.; Furton, K. Comparison of the Volatile Organic Compounds Present in Human Odor Using Spme-GC/MS. *J. Chem. Ecol.* **2005**, *31*, 1607–1619. [CrossRef]
11. Curran, A.M.; Ramirez, C.F.; Schoon, A.A.; Furton, K.G. The frequency of occurrence and discriminatory power of compounds found in human scent across a population determined by SPME-GC/MS. *J. Chromatogr. B Analt. Technol. Biomed. Life Sci.* **2007**, *846*, 86–97. [CrossRef]
12. Penn, D.J.; Oberzaucher, E.; Grammer, K.; Fischer, G.; Soini, H.A.; Wiesler, D.; Novotny, M.V.; Dixon, S.J.; Xu, Y.; Brereton, R.G. Individual and gender fingerprints in human body odour. *J. R. Soc. Interface* **2007**, *4*, 331–340. [CrossRef]
13. Jiang, R.; Cudjoe, E.; Bojko, B.; Abaffy, T.; Pawliszyn, J. A non-invasive method for in vivo skin volatile compounds sampling. *Anal. Chim. Acta* **2013**, *804*, 111–119. [CrossRef]
14. Lee, S.Y.; Yoon, J.H.; Bae, S.; Lee, D.S. In-needle microextraction coupled with gas chromatography/mass spectrometry for the analysis of phthalates generating from food containers. *Food Anal. Methods* **2018**, *11*, 2767–2777.
15. Jeon, H.L.; Son, H.H.; Bae, S.; Lee, D.S. Use of polyacrylic acid and polydimethylsiloxane mixture for in-needle microextraction of volatile aroma compounds in essential oils. *Bull. Korean Chem. Soc.* **2015**, *36*, 2730–2739. [CrossRef]
16. Hwang, Y.R.; Lee, Y.L.; Ahn, S.Y.; Bae, S. Electrochemically polyaniline-coated microextraction needle for phthalates in water. *Anal. Sci. Technol.* **2020**, *33*, 76–85.
17. Bang, Y.J.; Hwang, Y.R.; Lee, S.Y.; Park, S.M.; Bae, S. Sol–gel-adsorbent-coated extraction needles to detect volatile compounds in spoiled fish. *J. Sep. Sci.* **2017**, *40*, 3839–3847. [CrossRef]
18. Kim, S.; Bae, S.; Lee, D.-S. Characterization of scents from Juniperus chinensis by headspace in-needle microextraction using graphene oxide-polyaniline nanocomposite coated wire followed by gas chromatography-mass spectrometry. *Talanta* **2022**, *245*, 123463. [CrossRef]
19. Jeong, H.Y. Analysis of Age-related Volatile Organic Compounds Using INME Coated with PDMS/MWCNT followed by GC/MS. Master's Thesis, Seoul Women's University, Seoul, Korea, 2021.
20. Son, H.; Bae, S.; Lee, D. New needle packed with polydimethylsiloxane having a micro-bore tunnel for headspace in-needle microextraction of aroma components of citrus oils. *Anal. Chim. Acta* **2012**, *751*, 86–93. [CrossRef]
21. Ji, J.; Liu, H.; Chen, J.; Zeng, J.; Huang, J.; Gao, L.; Wang, Y.; Chen, X. ZnO nanorod coating for solid phase microextraction and its applications for the analysis of aldehydes in instant noodle samples. *J. Chromatogr. A* **2012**, *1246*, 22–27. [CrossRef]

22. Zeng, J.; Li, Y.; Zheng, X.; Li, Z.; Zeng, T.; Duan, W.; Li, Q.; Shang, X.; Dong, B. Controllable Transformation of Aligned ZnO Nanorods to ZIF−8 as Solid-Phase Microextraction Coatings with Tunable Porosity, Polarity, and Conductivity. *Anal. Chem.* **2019**, *91*, 5091–5097. [CrossRef]
23. Park, H.E.; Ho, K. Optimization of Synthesis Condition of Monolithic Sorbent Using Response Surface Methodology. *Appl. Chem. Eng.* **2013**, *24*, 299–304.
24. Ramezanzadeh, M.; Asghari, M.; Ramezanzadeh, B.; Bahlakeh, G. Fabrication of an efficient system for Zn ions removal from industrial wastewater based on graphene oxide nanosheets decorated with highly crystalline polyaniline nanofibers (GO-PANI): Experimental and ab initio quantum mechanics approaches. *Chem. Eng. J.* **2018**, *337*, 385–397. [CrossRef]
25. Brožová, L.; Holler, P.; Kovářová, J.; Stejskal, J.; Trchová, M. The stability of polyaniline in strongly alkaline or acidic aqueous media. *Polym. Degrad. Stab.* **2008**, *93*, 592–600. [CrossRef]
26. Andres Verges, M.; Serna, C.J. Morphological characterization of ZnO powders by X-ray and IR spectroscopy. *J. Mater. Sci. Lett.* **1988**, *7*, 970–972. [CrossRef]
27. Ling, X.; Chen, Z. Immobilization of zeolitic imidazolate frameworks with assist of electrodeposited zinc oxide layer and application in online solid-phase microextraction of Sudan dyes. *Talanta* **2019**, *192*, 142–146. [CrossRef]
28. Storck, S.; Bretinger, H.; Maier, W.F. Characterization of micro- and mesoporous solids by physisorption methods and pore-size analysis. *Appl. Catal. A* **1998**, *174*, 137–146. [CrossRef]
29. Xing, T.; Lou, Y.; Bao, Q.; Chen, J. Surfactant-assisted synthesis of ZIF−8 nanocrystals in aqueous solution via microwave irradiation. *CrystEngComm* **2014**, *16*, 8994–9000. [CrossRef]
30. He, L.; Li, L.; Zhang, L.; Xing, S.; Wang, T.; Li, G.; Wu, X.; Su, Z.; Wang, C. ZIF−8 templated fabrication of rhombic dodecahedron-shaped ZnO@SiO2, ZIF−8@SiO2 yolk–shell and SiO2 hollow nanoparticles. *CrystEngComm* **2014**, *16*, 6534. [CrossRef]
31. Wenzl, T.; Haedrich, J.; Schaechtele, A.; Robouch, P.; Stroka, J. *Guidance Document on the Estimation of LOD and LOQ for Measurements in the Field of Contaminants in Feed and Food*; Publications office of the European Union: Geel, Belgium, 2016; Volume 28099. [CrossRef]

Article

Metabolic Evaluation of Urine from Patients Diagnosed with High Grade (HG) Bladder Cancer by SPME-LC-MS Method

Kamil Łuczykowski [1,†], Natalia Warmuzińska [1,†], Sylwia Operacz [1], Iga Stryjak [1], Joanna Bogusiewicz [1], Julia Jacyna [2], Renata Wawrzyniak [2], Wiktoria Struck-Lewicka [2], Michał J. Markuszewski [2] and Barbara Bojko [1,*]

[1] Department of Pharmacodynamics and Molecular Pharmacology, Faculty of Pharmacy, Collegium Medicum in Bydgoszcz, Nicolaus Copernicus University in Toruń, 85-089 Bydgoszcz, Poland; k.luczykowski@cm.umk.pl (K.Ł.); n.warmuzinska@cm.umk.pl (N.W.); Sylwia_1292@op.pl (S.O.); i.stryjak@cm.umk.pl (I.S.); j.bogusiewicz@cm.umk.pl (J.B.)

[2] Department of Biopharmacy and Pharmacodynamics, Faculty of Pharmacy, Medical University of Gdańsk, 80-416 Gdańsk, Poland; julia.jacyna@gumed.edu.pl (J.J.); renata.wawrzyniak@gumed.edu.pl (R.W.); wiktoria.struck-lewicka@gumed.edu.pl (W.S.-L.); michal.markuszewski@gumed.edu.pl (M.J.M.)

* Correspondence: bbojko@cm.umk.pl
† These authors contributed equally to this work.

Abstract: Bladder cancer (BC) is a common malignancy of the urinary system and a leading cause of death worldwide. In this work, untargeted metabolomic profiling of biological fluids is presented as a non-invasive tool for bladder cancer biomarker discovery as a first step towards developing superior methods for detection, treatment, and prevention well as to further our current understanding of this disease. In this study, urine samples from 24 healthy volunteers and 24 BC patients were subjected to metabolomic profiling using high throughput solid-phase microextraction (SPME) in thin-film format and reversed-phase high-performance liquid chromatography coupled with a Q Exactive Focus Orbitrap mass spectrometer. The chemometric analysis enabled the selection of metabolites contributing to the observed separation of BC patients from the control group. Relevant differences were demonstrated for phenylalanine metabolism compounds, i.e., benzoic acid, hippuric acid, and 4-hydroxycinnamic acid. Furthermore, compounds involved in the metabolism of histidine, beta-alanine, and glycerophospholipids were also identified. Thin-film SPME can be efficiently used as an alternative approach to other traditional urine sample preparation methods, demonstrating the SPME technique as a simple and efficient tool for urinary metabolomics research. Moreover, this study's results may support a better understanding of bladder cancer development and progression mechanisms.

Keywords: bladder cancer (BC); metabolomics; solid phase microextraction (SPME); liquid chromatography; mass spectrometry; urine

1. Introduction

Bladder cancer (BC) is mainly diagnosed among people over 50 years of age. It is diagnosed three times more often in men and represents the fourth most common cancer in this group. Approximately 90% of bladder cancers are transitional cell carcinoma, while the other 10% are squamous cell carcinoma and adenocarcinoma [1,2]. Among the recognized risk factors—contributing to the progress of neoplastic changes in the bladder—smoking is the most significant [3]. Active smokers have a four times higher risk of developing the disease compared to persons who have never smoked. Other causes include genetic disorders, exposure to specific chemical compounds, or chronic bladder irritation [2]. Hematuria is the most common symptom of bladder cancer. Other possible symptoms include weight loss and pain in the abdomen and in the kidneys. However, these symptoms are not specific to bladder cancer, making diagnosis challenging since it requires exclusion of other diseases [2]. Commonly performed tests for cases where there is a suspicion of bladder cancer include cystoscopy (the "gold standard"), cytology of urinary sediment, and imaging diagnostics. However, these diagnostic methods have significant drawbacks: cystoscopy is invasive and uncomfortable for the patient, whereas laboratory tests have low sensitivity and specificity for early stages of bladder cancer [1]. These limitations have fostered research on alternative, more effective, minimally invasive approaches for bladder cancer diagnosis. As a first step in the creation of new diagnostic tools, untargeted metabolomics/metabolomic profiling of biological fluids can be used to identify biomarkers of bladder cancer [4], opening new paths towards the creation of novel biomarker-based diagnostic tools.

Human urine is a biological matrix that is easy to obtain in large volumes. In addition to its easy availability, urine is an important biological matrix due to the type of information it may hold in relation to the various metabolic processes occurring throughout the human body. Urine is produced by the kidneys, which filter unwanted substances from the body; as such, urine contains many unnecessary and harmful by-products of metabolism. The main components of urine are water, urea, creatinine, ammonia, inorganic salts, and other water-soluble substances [5]. Given that urine is largely free from interfering lipids or proteins, sample preparation for this matrix is significantly less time-consuming in comparison to other biological matrices. Hence, metabolomic studies using urine as a biological matrix are promising [5]. Metabolites present in the urine, which are the final products of processes occurring in healthy and cancerous cells, can be of particular importance in the diagnosis of bladder cancer. Differences observed between metabolomic profiles of BC patients compared to healthy persons may allow us to find potential BC indicators [6].

Even though urine is a relatively simple biological matrix, it still requires careful sample preparation considerations with respect to issues such as matrix effects and sample-to-sample variations. The ideal sample preparation method should be simple (with the smallest number of steps) and reproducible. Besides, it must ensure sample purification, recovery of a wide range of analytes, as well as inhibition of metabolism, which protects compounds from degradation [7]. The simplest and most common preparation method of urine samples prior to MS analysis is dilution ("dilute and shoot"). However, this simplistic strategy compromises sensitivity and does not ensure metabolism quenching, thus spurring the development of alternative approaches to urine sample preparation to address these shortcomings. Of the variety of methods developed for urine analysis, solid-phase microextraction has been demonstrated to largely fulfill the abovementioned conditions, given that in addition to integrating various steps of the sample preparation and extraction process, it additionally provides efficient sample clean-up, making it compatible with liquid chromatography and mass spectrometry [7,8]. In addition, the technology offers flexibility in terms of device geometry, chemistry of the extraction phase, and high-throughput capabilities. With regards to the latter, the most common format of the device i.e., the microfiber, is widely used for low-throughput analysis of tissues or in animal studies [9,10], as well as for high throughput analyses where sample volumes are limited [11]. When sample volume is of no concern, the thin film format of SPME (TFME) is preferable, as it increases recovery of analytes and is compatible with commercial automated or semi-automated high throughput robotic systems [10].

In the current study, thin film SPME in high throughput semi-automated mode was used for metabolomics screening of urine of patients with bladder cancer and healthy controls. Following experimental data acquisition, statistical analysis, and a biological pathway analysis were carried out to identify compounds that might be important in the identification of BC as well as in the identification of metabolic pathways involved in the pathogenesis and progression of the disease.

2. Results

2.1. Subject's Characteristics

Urine samples used for the study were obtained from advanced-stage bladder cancer patients and from healthy volunteers. Both groups consisted of 24 patients at a similar age, with similar BMI values, and with a similar number of smokers. The average age of patients in both groups fell within the age bracket known for carrying the highest risk of developing bladder cancer. A summary of patient demographics for the two groups is presented in Table 1. The group of healthy volunteers did not undergo any treatment during the sample collection period, and they were declared to be in good health condition, which was also confirmed with laboratory results. The enrollment of BC patients depended on confirmation of high grade, muscle-invasive BC during the histopathological evaluation of biopsies collected during the diagnostic procedure.

Table 1. Demographic characteristics of the studied groups.

Studied Group	Group Size		Age [years]	BMI [kg/m^2]	Smokers [%]
	Men	Women			
BC patients	18	6	65 ($\pm$12.0)	26.03 ($\pm$4.1)	67
Healthy Volunteers	18	6	64 ($\pm$10.4)	25.87 ($\pm$2.2)	75

2.2. Untargeted Metabolomics Analysis

The attained data was subjected to principal component analysis (PCA) in order to assess data quality as well as determine differences between the metabolic profiles of control and BC patients. As shown in Figure 1A,B, QC samples formed a tight cluster, confirming the quality of the obtained results. Additionally, the two studied groups achieved relatively good separation in both positive and negative ionization modes. Based on the PCA score plots and a 95% confidence ellipse using Hotelling T-squared, five outliers were removed: three from the group of BC patients and two from the healthy group for positive ionization mode, and two from the group of BC patients and three from the healthy group for negative ionization mode. Thereafter, supervised multivariate analysis: orthogonal projections to latent structures discriminant analysis (OPLS-DA) was performed to achieve maximum separation among the groups (Figure 1C,D). The Q2 and R2 values for the model were 46% and 85.2% for positive ionization mode and 43.3% and 88.2% for negative ionization mode, respectively.

Based on the obtained model, metabolites found to contribute the most in differentiating the two groups according to their VIP scores were selected for further analysis. A VIP score value >1 was selected as cut-off value. Table 2 shows the metabolites meeting the abovementioned criteria. Additional information regarding the annotation of these compounds is presented in Supplementary Information Table S1.

Figure 2 presents Box Whisker plots for selected annotated metabolites exhibiting significant differences in urinary levels between BC patients and healthy controls. Chromatograms for each of these compounds are shown in Supplementary Information Figure S2. Box Whisker plots represent peak areas for a selected component on the y axis as a rectangle against equally spaced sample groups on the x axis. The height of the rectangle represents the peak areas in the interquartile range. The following equations were used to calculate the upper and lower whiskers: Interquartile range (IQR) = Quartile 3 (Q3) − Quartile 1 (Q1), Upper whisker = Q3 + IQR × 1.5, Lower whisker = Q1 − IQR × 1.5.

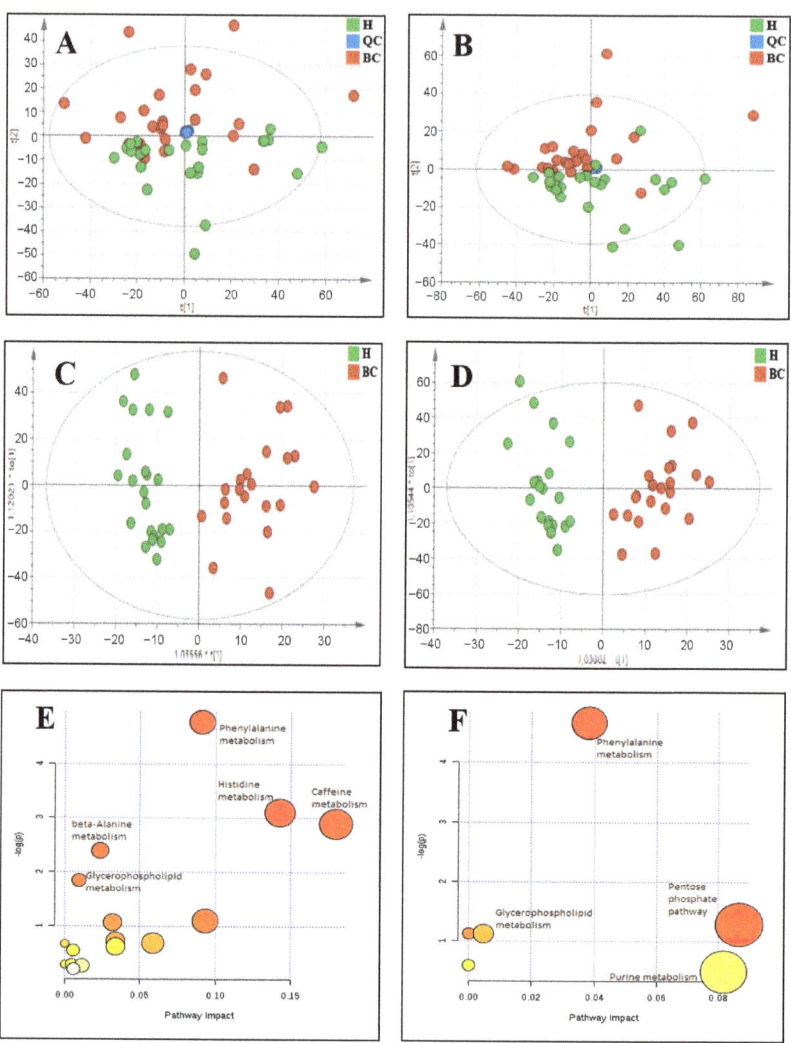

Figure 1. Observed differences in metabolic profiles of control and BC patients: (**A**) PCA score plot for positive ionization mode, with PC1 describing 12.8% of the variation and PC2 describing 5.5%, (**B**) PCA plot for negative ionization mode, with PC1 describing 13.6% of the variation and PC2 describing 5.5%, (**C**) Score plot of OPLS-DA model for positive ionization mode, with R2 = 85.2% and Q2 = 46%, (**D**) Score plot of OPLS-DA model for negative ionization mode, with R2 = 88.2% and Q2 = 43.3%, (**E**) Pathway analysis of differential metabolites in positive ionization mode, (**F**) Pathway analysis of differential metabolites in negative ionization mode.

Table 2. Differential metabolites in positive and negative ionization mode. References indicate BC's previous clinical studies in which the metabolite was selected as discriminant and/or dysregulated (MW—molecular weight; ↓ indicates down-regulation and ↑ indicates up-regulation in BC).

Metabolites	MW	RT	VIP Score	Trend
Positive ionization mode				
2-Acetyl-1-alkyl-sn-glycero-3-phosphocholine [12]	523.3638	17.98	1.66212	↑
3-Dehydroxycarnitine	145.1103	3.51	1.13186	↓
3-Methylxanthine [13]	166.0491	1.38	1.80968	↓
4-Hydroxycinnamic acid	164.0475	4.17	1.80279	↓
5-Hydroxyindoleacetic acid [13]	191.0582	8.42	1.17858	↓
Adenine	135.0545	3.08	2.03098	↑
Benzoic acid [14]	122.0370	4.17	1.38913	↓
Carnosine	226.1064	3.95	1.06014	↓
Epinephrine	183.0896	8.35	1.52508	↓
Hippuric acid [13–15]	179.0582	7.83	2.37849	↓
Histidine	155.0695	2.14	1.30363	↓
Isoniazid	137.0589	7.97	1.43638	↓
LysoPE(18:1)	479.3014	17.55	1.66856	↑
N-Acetyl-phenylalanine	207.0896	9.17	1.73877	↓
p-Aminobenzoic acid [13]	137.0477	1.44	1.45154	↓
Retinol	286.2295	12.93	1.07027	↓
Theophylline	180.0648	6.76	2.21588	↓
Negative ionization mode				
3-(3-sulfooxyphenyl)propanoic acid	246.0195	7.18	1.72228	↓
Adenosine monophosphate *	347.0631	1.18	1.60799	↑
Gluconic acid [16,17]	196.0587	1.39	1.74004	↓
Hippuric acid [13–15]	179.0583	7.81	2.21443	↓
Indolelactic acid [13]	205.0739	11.00	1.31510	↓

* Fragmentation spectrum not confirmed with online databases.

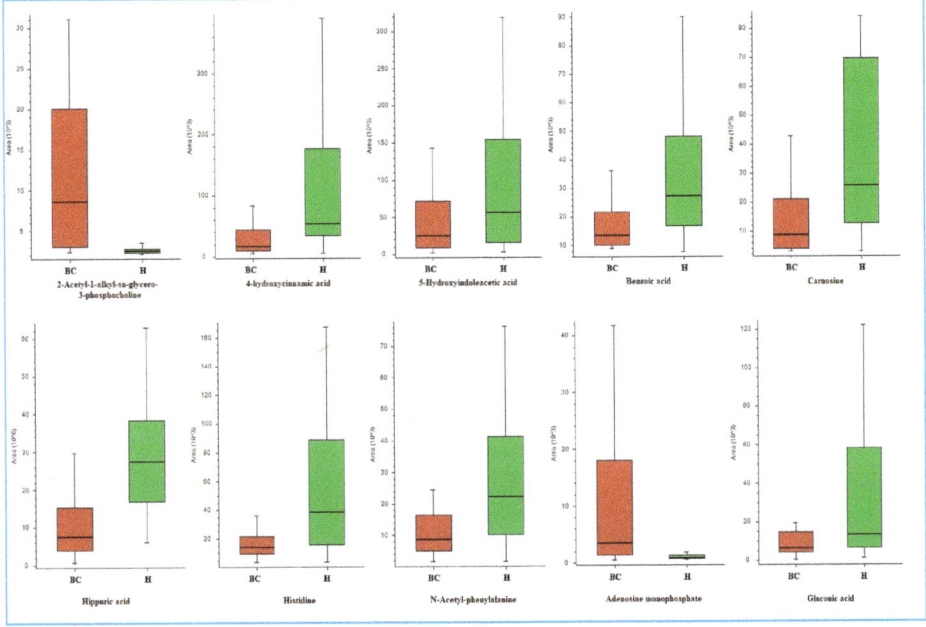

Figure 2. Box Whisker charts for selected compounds differentiating the studied groups (red (BC)—bladder cancer patients, green (H)—healthy volunteers).

2.3. Pathway Analysis

A pathway analysis was performed to obtain biological information about relevant networks of metabolic pathways that undergo changes in patients with bladder cancer in comparison to the healthy population. Our pathway analysis was carried out using the list of significantly differential metabolic features previously determined as described above (VIP score > 1). Results obtained from both ionization modes are presented in Figure 1E,F. The attained results indicate that the identified compounds may be mainly associated with phenylalanine, histidine, and caffeine metabolism. Table 3 shows the full list of identified metabolites and associated pathways in both positive and negative ionization modes.

Table 3. Metabolic pathways and associated metabolites in positive and negative ionization mode.

Pathway Name	Metabolites
Positive Ionization Mode	
Phenylalanine metabolism	Benzoic acid; Hippuric acid; 4-Hydroxycinnamic acid; N-Acetyl-phenylalanine
Histidine metabolism	Histidine; Carnosine;
Caffeine metabolism	Theophylline; 3-Methylxanthine;
beta-Alanine metabolism	Carnosine; Histidine
Glycerophospholipid metabolism	LysoPE(18:1); Phosphatidyl-N-dimethylethanolamine;
Retinol metabolism	Retinol;
Ether lipid metabolism	2-Acetyl-1-alkyl-sn-glycero-3-phosphocholine;
Ubiquinone and other terpenoid-quinone biosynthesis	4-Hydroxycinnamic acid;
Drug metabolism—other enzymes	Isoniazid;
Nitrogen metabolism	Histidine;
Folate biosynthesis	p-Aminobenzoic acid
Aminoacyl-tRNA biosynthesis	Histidine
Tyrosine metabolism	Epinephrine;
Tryptophan metabolism	5-Hydroxyindoleacetic acid
Purine metabolism	Adenine;
Negative Ionization Mode	
Phenylalanine metabolism	Hippuric acid;
Pentose phosphate pathway	Gluconic acid;
Nitrogen metabolism	Adenosine monophosphate;
Glycerophospholipid metabolism	Phosphatidyl-N-dimethylethanolamine;
Tryptophan metabolism	Indolelactic acid;
Purine metabolism	Adenosine monophosphate;

3. Discussion

In recent years, the urology field has seen rapid development in terms of urine-based metabolomic analysis. Given that urine comes in direct contact with bladder epithelial cells and tumor tissue, it may therefore contain metabolites shed from cancerous cells in the bladder [6]. In metabolomics, sample preparation has appreciable influence on the quality of the obtained results, particularly for biofluid samples such as urine [18]. In the current study, a technology integrating sample preparation and extraction of metabolites, namely solid phase microextraction, was presented for metabolomics analysis of urine samples of bladder cancer patients. It has been already demonstrated that SPME is an effective tool for metabolomics studies in plasma and tissue [8,10], but no direct immersion urinary metabolomics analyses have been reported to date to the best of our knowledge. However, extensive anti-doping multi-residue analyses in urine have been successfully carried out via the thin film SPME automated high throughput system [19], showing the potential of the technology for simultaneous determination of over 100 compounds with no matrix effect and good sensitivity in urine, all while affording a simple analytical protocol. Herein, thin film SPME followed by LC-HRMS analysis was used for the first time for urinary untargeted metabolomics in cancer patients.

The metabolites selected as discriminant compounds and, subsequently, the results of the pathways analysis, are consistent with previous reports [13]. Compounds found present at decreased levels in BC patients in comparison to healthy controls included benzoic acid (BA) and hippuric acid (HA), which are known to be involved in the phenylalanine metabolism pathway. Hippuric acid is a normal component of urine as a metabolite of aromatic compounds from food [20]. While the serum concentration of HA is increased in uremic patients as a result of reduced renal clearance [21], lower levels of hippuric acid in urine samples from BC patients are in agreement with previous studies, making it characteristic for this group of patients [14,15]. Benzoic acid, a compound commonly used as a food preservative, conjugates to glycine in the liver and is excreted as hippuric acid [20]. Benzoic acid and hippuric acid are also believed to play essential roles in gut microbial pathways [14,22]. The important role of gut microbiota in health has been gradually gaining recognition in science; although studies regarding the physiological functions of BA have mainly involved pigs as a research model for humans [22], benzoic acid administration has been found to improve gut functions including digestion and absorption, and also could improve gut barrier function, including nonspecific barrier mechanisms and specific immunological responses [22]. The lower concentrations of hippuric acid and benzoic acid found in the bladder cancer group might be associated with gut microbial pathways, decreased activity of the hippurate hydrolase, or increased activity of the glycine N-benzoyltransferase [14].

Another down-regulated metabolite was N-Acetyl-phenylalanine, which also participates in the phenylalanine metabolism pathway. Acetylphenylalanine is a product of enzyme phenylalanine N-acetyltransferase [20,23], and lower concentrations of this compound in saliva have been previously reported in oral squamous cell carcinoma patients [23]. Histidine and carnosine also exhibited decreased levels in BC patients. The latter plays a role in histidine, beta-alanine, and nitrogen metabolism as well as in the aminoacyl-tRNA biosynthesis pathway. Histidine is an α-amino acid that is used in the biosynthesis of proteins. One of the proteins significantly related to tumor progression is fragile histidine triad protein (FHIT), a protein that has been proven to have suppressor properties. The lack of FHIT protein or its reduced level is observed in many types of cancer as well as in various cancer cell lines [24–26]. Reduced concentration of histidine in bladder cancer patients may correlate with a lower activity of FHIT protein [24]. FHIT protein deficiency has been also reported in bladder cancer, mainly in high-grade tumors. This phenomenon correlates with poorer prognosis in BC patients [26–28]. Histidine is also a precursor for carnosine biosynthesis. This dipeptide links histidine and the beta-alanine metabolism pathway. While carnosine has not been reported in the literature as a metabolite of significant importance in bladder cancer, numerous studies have shown that carnosine exhibits some antioxidant effects [29–31]. Carnosine's ability to scavenge reactive oxygen species (ROS) may be important in the prevention of cancers. In this study, carnosine levels were found to be higher in urine of the control group, which would be in line with these assumptions.

The 2-Acetyl-1-alkyl-sn-glycero-3-phosphocholine levels in urine exhibited relevant differences in comparisons of metabolomic profiles of bladder cancer patients and healthy volunteers. This compound is also known as a platelet-activating factor (PAF), playing a role in platelet aggregation, inflammation, and allergic reactions. PAF and its receptor have been implicated in malignant processes such as tumor development, growth, and metastatic angiogenesis [32]. Kispert et al. observed significant increases in PAF accumulation in bladder cancer cells following cigarette smoke extract exposure [12]. The increased levels of PAF in BC patients are in accordance with the current study; however, the number of smokers in both studied groups was similar, which might suggest that smoking is not the only factor influencing PAF accumulation in the bladder.

A higher level of 4-hydroxycinnamic acid was found in the urine of healthy volunteers compared to patients with bladder cancer. This organic compound, also known as p-Coumaric acid, is a naturally occurring phenolic acid present in most plants, including commonly consumed vegetables and fruits [33]. There are numerous reports describing its antioxidant and anticancer properties [34–36]. Kong et al. have shown that *p*-coumaric acid inhibits tumor growth through reduction of angiogenesis within the tumor [36]. While our current findings cannot directly establish a relationship between 4-hydroxycinnamic acid and bladder cancer, the higher levels of this compound in healthy volunteers may be evidence of its role in cancer prevention.

We have also identified changes in the pentose phosphate pathway demonstrated by the observed decrease in gluconic acid levels. Gluconic acid, a naturally occurring carboxylic acid, as well as its derivatives are used in food, pharmaceuticals, and cosmetics as additives or buffer salts; however, the mechanisms of their biological activity are still not completely understood [20,37]. Recent research in this area has indicated that gluconic acid and its derivatives may contain antioxidative properties [37]. Lower concentrations of this compound in the BC group as compared to the healthy control might be related to oxidative stress, which has been implicated in the pathogenesis of numerous diseases [38]. Decreased concentration of gluconic acid in the urine of the bladder cancer patients has been reported previously [16,17]. Interestingly, gluconic acid such as carnosine and *p*-coumaric acid has antioxidative properties and was also found to be present in lower quantities in the urine of the BC group.

We acknowledge that there are limitations to our study. The analysis was performed with a relatively small number of patients without differentiation of cancer type. However, as emphasized throughout the manuscript, the results do not provide a well-validated panel of biomarkers mainly due to the small number of participants and larger cohort study to confirm or deny this work's results. On the other hand, based on this study, we were able to identify some interesting metabolites not described previously in the view of BC that might contribute to the pathogenesis of bladder cancer or serve as protective compounds in cancer development, and this information can be a good starting point for targeted analysis of these compounds and the pathway they are involved in. Moreover, our work has demonstrated that TFME hyphenated to LC-HRMS can be an alternative platform for untargeted urinary analysis toward biomarker discovery.

4. Materials and Methods

Analytical grade sodium chloride, potassium chloride, potassium phosphate monobasic, sodium phosphate dibasic, hydrochloric acid, and sodium hydroxide, used for the phosphate-buffered saline solution (PBS pH 7.4) preparation, as well as the LC-MS grade chromatographic solvents water, methanol, acetonitrile, and formic acid, were purchased form Sigma-Aldrich (Poznań, Poland).

Urine was obtained from advanced-stage bladder cancer patients and from healthy volunteers (control group). The urine samples were collected from the first urination in the morning. Samples were provided by the Department of Biopharmaceutics and Pharmacodynamics, Medical University of Gdańsk.

Solid phase microextraction (SPME) in its thin film format was used for preparation of samples, and each step of the process was carried out on a high throughput 96-semi-automated SPME system (Professional Analytical System (PAS) Technology, Magdala, Germany), which allowed for simultaneous analysis of all samples (Figure 3). Extraction was performed by using steel blades coated with a polystyrene divinylbenzene (PS-DVB) sorbent (Alchem, Toruń, Poland). Coating preparation procedures were based on the spraying method described by Mirnaghi et al. [39]. Steel blades were purchased form Professional Analytical System (PAS) Technology (Magdala, Germany) and polypropylene Nunc 96 DeepWell plates were purchased from Sigma-Aldrich (Poznań, Poland).

Prior to the analysis, urine was diluted in PBS (1:1; v/v) so as to establish a matrix pH of 7.4. Before extraction, coatings were conditioned for 30 min with 1.0 mL of a methanol: water (1:1; v/v) solution in 96-well-plates with agitation set at 750 rpm to improve sorbent surface activation and prepare the sorbent to retain the analytes, then subsequently submitted to a 10 s wash step to remove residual methanol. The extraction process was executed from 0.5 mL of urine samples for 1 h at 37 °C. After extraction, the blades were placed in 1 mL of nanopure water for 10 s to remove particulates, salts, and other contaminants loosely attached on the coating surface by non-specific interactions, which could potentially cause matrix effects and instrument contamination. Following this wash step, desorption was conducted in 1 mL of acetonitrile: water (1:1; v/v) solution with agitation (750 rpm) for 90 min.

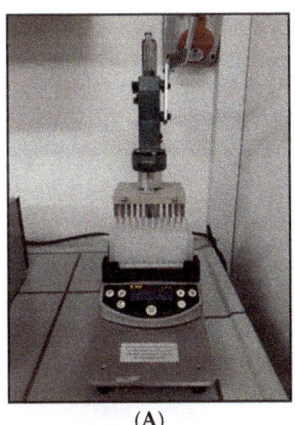 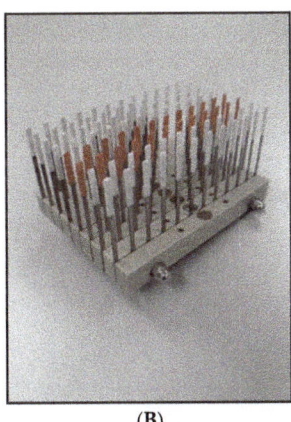

(A) (B)

Figure 3. The high throughput 96-semi-automated SPME system used for analysis of urine samples (A) and a set of 96 SPME devices coated with different types of extraction phases (the PS-DVB coating selected for the study is represented in rows 3 and 4 from the right) (B).

4.1. LC-MS Conditions

Chromatographic separation was performed on the Dionex UHPLC system. Urine extracts, obtained as described above, were injected at a volume of 10 µL on a reversed phase pentafluorophenyl (PFP) column (Discovery HS F5 100 mm × 2.1 mm, 3 µm (Sigma-Aldrich, Poznań, Poland)). Autosampler and column temperatures were set to 4 °C and 25 °C, respectively. The flow rate was 0.3 mL/min. The mobile phase A was water with formic acid (99.9:0.1; v/v) and mobile phase B was acetonitrile with formic acid (99.9:0.1; v/v). The total time of analysis per sample was 40 min. The starting mobile phase conditions were 0% B from 0 to 3.0 min, followed by a linear gradient to 90% B from 3.0 to 25.0 min, an isocratic hold at 90% B until 34.0 min, and a 6 min column re-equilibration time [40]. Total ion chromatograms of QC samples in both ionization modes are shown in Supplementary Information Figure S1.

The analyses were performed in both positive and negative electrospray ionization modes in separate runs on a Q Exactive Focus Orbitrap mass spectrometer (Thermo Fisher Scientific). In positive ionization mode, the following HESI ion source parameters were used: spray voltage 1500 V, capillary temperature 300 °C, sheath gas 40 a.u., aux gas flow rate 15 a.u., probe heater temperature 300 °C, and S-Lens RF level 55%. For negative ionization mode, HESI ion source parameters were as follows: spray voltage 2500 V, capillary temperature 256 °C, sheath gas 48 a.u., aux gas flow rate 11 a.u., probe heater temperature 413 °C, and S-Lens RF level 55%. Scan range was set on m/z 80–1000 with resolution 70,000. The instrument was calibrated using external calibration immediately before the analysis and every 48 h, resulting in mass accuracy <2 ppm. Data acquisition was performed with Xcalibur software v. 4.0.

All samples were analyzed in one randomized sequence and QC samples were run periodically (8–10 injections) to verify instrument performance. QC samples were prepared by mixing 20 µL of each of the 48 urine extracts.

The putative identification of compounds was confirmed based on Full MS/dd-MS2 mode. Fragmentation parameters were as follows. Mass resolution: 35,000 full width at half maximum (FWHM), AGC target: 2E4, minimum AGC: 8E3, intensity threshold: auto, maximum IT: auto, isolation window: 3.0 m/z, stepped collision energy: 10 V, 20 V, 40 V, loop count: 2, dynamic exclusion: auto. Fragmentation spectra were confirmed with online databases such as LIPID MAPS, HMDB, METLIN, and mzCloud.

4.2. Data Processing and Statistical Analysis

Raw data was processed by the Compound Discoverer 3.0 (Thermo Fisher Scientific) software with aims to identify metabolites present in the samples. Detected metabolites with signal-to-noise >3 and peak intensity >1,000,000 were subjected to analysis. Intensity tolerance was set as 30%, and RT tolerance as 0.2 min. QC-based area was used for correction (min 50% coverage, max 30% RSD in QC, normalization by constant mean). A data filtration step removed 49% of the 7907 features and 29% of the 6114 features in positive and negative ionization modes, respectively. After peak alignment, gap filling was applied to fill missing values by a very small peak at the level of spectrum noise for the compound. The obtained table with accurate masses of annotated compounds was inserted into SIMCA 15 (Umetrics), where principal component analysis (PCA) and orthogonal projections to latent structures discriminant analysis (OPLS-DA) were performed using unit variance (UV) scaling. Statistically significant compounds in the projection used in OPLS-DA models were selected using Variable Importance in Projection (VIP) scores >1 and then filtered, taking into account the biological relevance and consistency of the fragmentation spectrum with databases. Pathway analysis was carried out with MetaboAnalyst 4.0, using Homo sapiens Kyoto Encyclopedia of Genes and Genomes (KEGG) metabolic pathway database. The analysis used the hypergeometric test as an enrichment method and relative-betweenness centrality for topology analysis, and the results were presented in the form of scatter plots.

5. Conclusions

The main goals of this study were to test applicability of TFME for untargeted screening of urinary metabolites and the identification of compounds involved in tumor development and progression in bladder cancer patients. Multivariate OPLS-DA analysis was performed to differentiate BC patients from healthy individuals. This profiling has not only identified a group of metabolites that may contribute to bladder cancer development, but also compounds that show potential in the prevention of cancer. Pathway analysis allowed for the integration of metabolomics data and biological information to enhance knowledge about the biological links between selected metabolites and BC pathogenesis. Future quantitative targeted studies on a larger number of patients are required to validate current findings and to evaluate the predictive value of the selected metabolites. Moreover, the results of the present study demonstrated that thin film SPME can be efficiently used as an alternative approach to other traditional urine sample preparation methods, demonstrating the SPME technique as a simple and efficient tool for urinary metabolomics research.

Supplementary Materials: The following are available online. Figure S1: Total ion chromatogram of QC sample in positive (A) and negative (B) ionization mode. Figure S2: Chromatograms for selected compounds differentiating the studied groups: 2-Acetyl-1-alkyl-sn-glycero-3-phosphocholine (A), 4-Hydroxycinnamic acid (B), 5-Hydroxyindoleacetic acid (C), Adenosine monophosphate (D), Benzoic acid (E), Hippuric acid (F), Gluconic acid (G), Carnosine (H), Histidine (I), N-Acetyl-phenylalanine (J), Table S1: Differential metabolites in positive and negative ionization mode with information on the annotation in the ChemSpider database.

Author Contributions: Conceptualization, M.J.M. and B.B.; formal analysis, data curation, K.Ł., N.W. and J.B.; investigation, methodology, S.O. and I.S.; writing—original draft preparation, K.Ł. and N.W.; writing—review and editing, J.J., R.W., W.S.-L., M.J.M. and B.B.; visualization, K.Ł. and N.W.; supervision, B.B.; project administration, B.B. All authors have read and agreed to the published version of the manuscript.

Funding: This research received no external funding.

Institutional Review Board Statement: The study was conducted according to the guidelines of the Declaration of Helsinki, and approved by Independent Bioethics Commission for Research of Medical University of Gdańsk (NKBBN/49/2013) and the Bioethics Committee of Collegium Medicum in Bydgoszcz at Nicolaus Copernicus University in Toruń (KB 628/2015).

Informed Consent Statement: Informed consent was obtained from all subjects involved in the study.

Data Availability Statement: Not applicable.

Acknowledgments: The authors would like to thank Thermo Fisher Scientific for access to the Q-Exactive Focus Orbitrap mass spectrometer. We thank the Department of Biopharmaceutics and Pharmacodynamics, Medical University of Gdańsk for urine samples.

Conflicts of Interest: The authors declare no conflict of interest.

Sample Availability: Extracts of the urine samples are available from the authors.

References

1. Turo, R.; Cross, W.; Whelan, P. Bladder cancer. *Medicine (Baltimore)* **2012**, *40*, 14–19. [CrossRef]
2. Kaufman, D.S.; Shipley, W.U.; Feldman, A.S. Bladder cancer. *Lancet* **2009**, *374*, 239–249. [CrossRef]
3. Chan, E.C.Y.; Pasikanti, K.K.; Hong, Y.; Ho, P.C.; Mahendran, R.; Mani, L.R.N.; Chiong, E.; Esuvaranathan, K. Metabonomic profiling of bladder cancer. *J. Proteome Res.* **2015**, *14*, 587–602. [CrossRef] [PubMed]
4. Cheng, X.; Liu, X.; Liu, X.; Guo, Z.; Sun, H.; Zhang, M.; Ji, Z.; Sun, W. Metabolomics of Non-muscle Invasive Bladder Cancer: Biomarkers for Early Detection of Bladder Cancer. *Front. Oncol.* **2018**, *8*, 1–11. [CrossRef] [PubMed]
5. Bouatra, S.; Aziat, F.; Mandal, R.; Guo, A.C.; Wilson, M.R.; Knox, C.; Bjorndahl, T.C.; Krishnamurthy, R.; Saleem, F.; Liu, P.; et al. The Human Urine Metabolome. *PLoS ONE* **2013**, *8*, e73076. [CrossRef] [PubMed]
6. Rodrigues, D.; Jerónimo, C.; Henrique, R.; Belo, L.; De Lourdes Bastos, M.; De Pinho, P.G.; Carvalho, M. Biomarkers in bladder cancer: A metabolomic approach using in vitro and ex vivo model systems. *Int. J. Cancer* **2016**, *139*, 256–268. [CrossRef]
7. Fernández-Peralbo, M.A.; Luque de Castro, M.D. Preparation of urine samples prior to targeted or untargeted metabolomics mass-spectrometry analysis. *TrAC—Trends Anal. Chem.* **2012**, *41*, 75–85. [CrossRef]
8. Bojko, B.; Reyes-Garce's, N.; Bessonneau, V.; Goryński, K.; Mousavi, F.; Souza Silva, E.A.; Pawliszyn, J. Solid-phase microextraction in metabolomics. *TrAC—Trends Anal. Chem.* **2014**, *61*, 168–180. [CrossRef]
9. Cudjoe, E.; Bojko, B.; Togunde, P.; Pawliszyn, J. In vivo solid-phase microextraction for tissue bioanalysis. *Bioanalysis* **2012**, *4*, 2605–2619. [CrossRef]
10. Reyes-Garcés, N.; Gionfriddo, E.; Gomez-Rios, G.A.; Alam, M.N.; Boyaci, E.; Bojko, B.; Singh, V.; Grandy, J.; Pawliszyn, J. Advances in Solid Phase Microextraction and Perspective on Future Directions. *Anal. Chem.* **2018**, *90*, 302–360. [CrossRef]
11. Jaroch, K.; Boyaci, E.; Pawliszyn, J.; Bojko, B. The use of solid phase microextraction for metabolomic analysis of non-small cell lung carcinoma cell line (A549) after administration of combretastatin. *Sci. Rep.* **2019**, *9*, 1–9. [CrossRef] [PubMed]
12. Kispert, S.; Marentette, J.; McHowat, J. Cigarette smoking promotes bladder cancer via increased platelet-activating factor. *Physiol. Rep.* **2019**, *7*, 1–10. [CrossRef] [PubMed]
13. Loras, A.; Trassierra, M.; Sanjuan-Herráez, D.; Martínez-Bisbal, M.C.; Castell, J.V.; Quintás, G.; Ruiz-Cerdá, J.L. Bladder cancer recurrence surveillance by urine metabolomics analysis. *Sci. Rep.* **2018**, *8*, 1–10. [CrossRef]
14. Mpanga, A.Y.; Siluk, D.; Jacyna, J.; Szerkus, O.; Wawrzyniak, R.; Markuszewski, M.; Matuszewski, M.; Kaliszan, R.; Markuszewski, M.J. Targeted metabolomics in bladder cancer: From analytical methods development and validation towards application to clinical samples. *Anal. Chim. Acta* **2018**, *1037*, 188–199. [CrossRef]
15. Huang, Z.; Chen, Y.; Hang, W.; Gao, Y.; Lin, L.; Li, D.Y.; Xing, J.; Yan, X. Holistic metabonomic profiling of urine affords potential early diagnosis for bladder and kidney cancers. *Metabolomics* **2013**, *9*, 119–129. [CrossRef]
16. Pasikanti, K.K.; Esuvaranathan, K.; Ho, P.C.; Mahendran, R.; Kamaraj, R.; Wu, Q.H.; Chiong, E.; Chun, E.; Chan, E.C.Y. Noninvasive Urinary Metabonomic Diagnosis of Human Bladder Cancer. *J. Proteome Res.* **2010**, *9*, 2988–2995. [CrossRef]
17. Pasikanti, K.K.; Esuvaranathan, K.; Hong, Y.; Ho, P.C.; Mahendran, R.; Mani, L.R.N.; Chiong, E.; Chan, E.C.Y. Urinary Metabotyping of Bladder Cancer Using Two-Dimensional Gas Chromatography Time-of-Flight Mass Spectrometry. *J. Proteome Res.* **2013**, *12*, 3865–3873. [CrossRef]
18. Mousavi, F.; Bojko, B.; Pawliszyn, J. High-Throughput Solid-Phase Microextraction–Liquid Chromatography–Mass Spectrometry for Microbial Untargeted Metabolomics. *Microb. Metabolomics Methods Mol. Biol.* **2019**, *1859*, 133–152.
19. Boyacı, E.; Goryński, K.; Rodriguez-Lafuente, A.; Bojko, B.; Pawliszyn, J. Introduction of solid-phase microextraction as a high-throughput sample preparation tool in laboratory analysis of prohibited substances. *Anal. Chim. Acta* **2014**, *809*, 69–81. [CrossRef]
20. Pubchem. U.S. National Library of Medicine National Center for Biotechnology Information. Available online: https://pubchem.ncbi.nlm.nih.gov/ (accessed on 15 October 2019).

21. Yu, T.H.; Tang, W.H.; Lu, Y.C.; Wang, C.P.; Hung, W.C.; Wu, C.C.; Tsai, I.T.; Chung, F.M.; Houng, J.Y.; Lan, W.C.; et al. Association between hippuric acid and left ventricular hypertrophy in maintenance hemodialysis patients. *Clin. Chim. Acta* **2018**, *484*, 47–51. [CrossRef]
22. Mao, X.; Yang, Q.; Chen, D.; Yu, B.; He, J. Benzoic Acid Used as Food and Feed Additives Can Regulate Gut Functions. *BioMed Res. Int.* **2019**, *2019*, 1–6. [CrossRef] [PubMed]
23. Wang, Q.; Gao, P.; Wang, X.; Duan, Y. The early diagnosis and monitoring of metabolomics. *Sci. Rep.* **2014**, *4*, 1–9.
24. Hassan, M.I.; Naiyer, A.; Ahmad, F. Fragile histidine triad protein: Structure, function, and its association with tumorogenesis. *J. Cancer Res. Clin. Oncol.* **2010**, *136*, 333–350. [CrossRef] [PubMed]
25. D'Arca, D.; Lenoir, J.; Wildemore, B.; Gottardo, F.; Bragantini, E.; Shupp-Byrne, D.; Zanesi, N.; Fassan, M.; Croce, C.M.; Gomella, L.G.; et al. Prevention of urinary bladder cancer in the FHIT knock-out mouse. *Urol. Oncol.* **2010**, *28*, 189–194. [CrossRef]
26. Liu, X.P.; Yin, X.H.; Yan, X.H.; Zeng, X.T.; Wang, X.H. The Clinical Relevance of Fragile Histidine Triad Protein (FHIT) in Patients with Bladder Cancer. *Med. Sci. Monit.* **2018**, *24*, 3113–3118. [CrossRef]
27. Zhang, C.T.; Lu, R.; Lin, Y.L.; Liu, R.L.; Zhang, Z.H.; Yang, K.; Dang, R.F.; Zhang, H.T.; Shen, Y.G.; Kong, P.Z.; et al. The Significance of Fragile Histidine Triad Protein as a Molecular Prognostic Marker of Bladder Urothelial Carcinoma. *J. Int. Med. Res.* **2012**, *40*, 507–516. [CrossRef]
28. Ye, F.; Wang, L.; Castillo-Martin, M.; McBride, R.; Galsky, M.D.; Zhu, J.; Boffetta, P.; Zhang, D.Y.; Cordon-Cardo, C. Biomarkers for bladder cancer management: Present and future. *Am. J. Clin. Exp. Urol.* **2014**, *2*, 1–14.
29. Noori, S.; Mahboob, T. Antioxidant effect of carnosine pretreatment on cisplatin-induced renal oxidative stress in rats. *Indian J. Clin. Biochem.* **2010**, *25*, 86–91. [CrossRef]
30. Gaunitz, F.; Hipkiss, A.R. Carnosine and cancer: A perspective. *Amino Acids* **2012**, *43*, 135–142. [CrossRef]
31. Metwally, N.S.; Ali, S.A.; Mohamed, A.M.; Khaled, H.M.; Ahmed, S.A. Levels of certain tumor markers as differential factors between bilharzial and non-biharzial bladder cancer among Egyptian patients. *Cancer Cell Int.* **2011**, *11*, 1–11. [CrossRef] [PubMed]
32. Tsoupras, A.B.; Iatrou, C.; Frangia, C.; Demopoulos, C.A. The Implication of Platelet Activating Factor in Cancer Growth and Metastasis: Potent Beneficial Role of PAF-Inhibitors and Antioxidants. *Infect. Disord.—Drug Targets* **2009**, *9*, 390–399. [CrossRef] [PubMed]
33. Sharma, S.H.; Rajamanickam, V.; Nagarajan, S. Antiproliferative effect of p-Coumaric acid targets UPR activation by downregulating Grp78 in colon cancer. *Chem. Biol. Interact.* **2018**, *291*, 16–28. [CrossRef] [PubMed]
34. Bouzaiene, N.N.; Jaziri, S.K.; Kovacic, H.; Chekir-Ghedira, L.; Ghedira, K.; Luis, J. The effects of caffeic, coumaric and ferulic acids on proliferation, superoxide production, adhesion and migration of human tumor cells in vitro. *Eur. J. Pharmacol.* **2015**, *766*, 99–105. [CrossRef]
35. Rosa, L.S.; Silva, N.J.A.; Soares, N.C.P.; Monteiro, M.C.; Teodoro, A.J. Anticancer Properties of Phenolic Acids in Colon Cancer—A Review. *J. Nutr. Food Sci.* **2016**, *6*, 1–7.
36. Kong, C.S.; Jeong, C.H.; Choi, J.S.; Kim, K.J.; Jeong, J.W. Antiangiogenic Effects of P-Coumaric Acid in Human Endothelial Cells. *Phytother. Res.* **2013**, *27*, 317–323. [CrossRef] [PubMed]
37. Kolodziejczyk, J.; Saluk-Juszczak, J.; Wachowicz, B. In vitro study of the antioxidative properties of the glucose derivatives against oxidation of plasma components. *J. Physiol. Biochem.* **2011**, *67*, 175–183. [CrossRef] [PubMed]
38. Reuter, S.; Gupta, S.C.; Chaturvedi, M.M.; Aggarwal, B.B. Oxidative stress, inflammation, and cancer: How are they linked? *Free Radic. Biol. Med.* **2010**, *49*, 1603–1616. [CrossRef]
39. Mirnaghi, F.S.; Chen, Y.; Sidisky, L.M.; Pawliszyn, J. Optimization of the Coating Procedure for a High-Throughput 96-Blade Solid Phase Microextraction System Coupled with LC À MS/MS for Analysis of Complex Samples. *Anal. Chem.* **2011**, *83*, 6018–6025. [CrossRef] [PubMed]
40. Vuckovic, D.; Pawliszyn, J. Systematic Evaluation of Solid-Phase Microextraction Coatings for Untargeted Metabolomic Profiling of Biological Fluids by Liquid Chromatography—Mass Spectrometry. *Anal. Chem.* **2011**, *83*, 1944–1954. [CrossRef]

Article

Magnetic Micro-Solid-Phase Extraction Using a Novel Carbon-Based Composite Coupled with HPLC–MS/MS for Steroid Multiclass Determination in Human Plasma

Andrea Speltini [1,*], Francesca Merlo [2], Federica Maraschi [2], Giorgio Marrubini [1], Anna Faravelli [2] and Antonella Profumo [2,*]

1. Department of Drug Sciences, University of Pavia, 27100 Pavia, Italy; giorgio.marrubini@unipv.it
2. Department of Chemistry, University of Pavia, 27100 Pavia, Italy; francesca.merlo02@universitadipavia.it (F.M.); federica.maraschi@unipv.it (F.M.); anna.faravelli01@universitadipavia.it (A.F.)
* Correspondence: andrea.speltini@unipv.it (A.S.); antonella.profumo@unipv.it (A.P.); Tel.: +39-0382-987349 (A.S.)

Abstract: A micron-sized sorbent, Magn-Humic, has been prepared by humic acids pyrolysis onto silica-coated magnetite. The material was characterized by scanning electron microscopy (SEM), transmission electron microscopy (TEM), energy dispersive spectroscopy (EDS), thermogravimetric analysis (TGA), and Brunauer, Emmett, and Teller (BET) surface area measurements and applied for simultaneous magnetic solid-phase extraction (MSPE) of glucocorticoids, estrogens, progestogens, and androgens at ng mL^{-1} levels from human plasma followed by high-performance liquid chromatography coupled with mass spectrometry (HPLC–MS/MS). Due to the low affinity for proteins, steroids extraction was done with no need for proteins precipitation/centrifugation. As highlighted by a design of experiments, MSPE was performed on 250 µL plasma (after 1:4 dilution) by 50 mg Magn-Humic (reusable for eight extractions) achieving quantitative recovery and satisfying clean-up. This was improved by washing (2 mL 2% v/v formic acid) prior to analytes elution by 0.5 mL 1:1 v/v methanol-acetonitrile followed by 0.5 mL methanol; eluate reduction to 0.25 mL compensated the initial sample dilution. The accuracy was assessed in certified blank fetal bovine serum and in human plasma, gaining satisfactory recovery in the range 65–122%, detection limits in the range 0.02–0.3 ng mL^{-1} (0.8 ng mL^{-1} for 17-β-estradiol) and suitable inter-day precision (relative standard deviation (RSD) <14%, n = 3). The method was evaluated in terms of selectivity, sensitivity, matrix-effect, instrumental carry-over, and it was applied to human plasma samples.

Keywords: bioanalysis; carbon materials; HPLC–MS; biological matrices; solid-phase extraction; sample preparation

1. Introduction

The search for new sample preparation procedures with improved throughput, simple workflow, and reduced use of organic solvents is nowadays one of the most desired goals in analytical sample treatment. This is especially important in the case of complex matrices such as environmental, food, and biological samples, wherein in most cases the target analytes are present at very low concentrations together with huge amounts of other matrix constituents [1–4]. These are major interfering species in various steps of the analytical protocol, from analyte isolation to final instrumental quantitation, thus working procedures for extraction, clean-up and, possibly, pre-concentration, are increasingly required.

With regard to sample treatment for biological matrices, proteins (up to 80 g L^{-1}) are the major interferents in plasmatic steroids determination, calling for pretreatments such as sample dilution and proteins precipitation [5–13] before extraction that is done, for instance, by solid-phase extraction (SPE) [5,7–9,12,13]. Quantitation is today mainly performed by high-performance liquid chromatography coupled with mass spectrometry detection (HPLC–MS), which ensures selective and sensitive determination [7–9,12,14–16]; some methods also involving gas chromatography coupled with MS detection have been proposed, necessarily requiring analytes derivatization before analysis [13,17].

In this context, our previous works showed the versatility of the mixed-mode HA-C@silica sorbent, employed in conventional SPE cartridges for enrichment/clean-up of various compounds in environmental and biological matrices prior to HPLC–MS [18–21]. With regard to the biomatrix, HA-C@silica proved to be advantageous due to its protein exclusion [21].

Besides conventional column SPE, magnetic solid-phase extraction (MSPE) has emerged in the last decades as a promising and a straightforward sample preparation technique due to simple and quick extraction [1,2]. In dispersive MSPE, the magnetic sorbent is dispersed under agitation in the sample solution, providing high surface contact and full interaction between the analytes and the sorbent particles, and then it is easily isolated from the solution using a small magnet [2].

Based on these advantages, MSPE has been adopted in the last years as sample treatment for determination of various pharmaceuticals in biological matrices, mostly antibiotics, antidepressants, narcotic analgesics, benzodiazepines, and anti-inflammatory drugs [2]. However, it should be noted that only very few MSPE-based methods, entailing use of gold-modified nanoparticles, have been reported to extract some selected steroids from human plasma [6] and urine [6,22]. Based on the above discussion and in light of our earlier research [21], in this study a novel magnetic sorbent (Magn-Humic) has been prepared by pyrolysis of humic acids (HAs) onto silica-coated magnetite. The siliceous shell, grown up on magnetite (Fe_3O_4) by a sol-gel procedure, was conveniently exploited to support HAs before pyrolysis, yielding a micron-sized magnetic sorbent relying on the sorption properties of the HA-derived carbon phase, to be easily used in human plasma for batch-extraction of steroids, namely prednisolone (PREDLO), prednisone (PRED), hydrocortisone (H-CORT), cortisone (CORT), betamethasone (BETA), dexamethasone (DEXA), triamcinolone (TRIAM), 17-β-estradiol (E2), testosterone (TST), epitestosterone (EPI), 17-α-ethynylestradiol (EE2), estrone (E1), hydroxyprogesterone (H-PROG), fluocinolone (FLUO), progesterone (PROG), and medroxyprogesterone acetate (M-PROG).

The material was characterized by various techniques, namely thermogravimetric analysis (TGA), scanning electron microscopy (SEM), transmission electron microscopy (TEM), energy dispersive spectroscopy (EDS), and surface area measurements by Brunauer, Emmett, and Teller (BET) method. Preliminary protein exclusion tests and extractions were done in bovine serum albumin (BSA) solution, and then the MSPE was moved in a real biological sample, i.e., certified hormone-free fetal bovine serum (FBS). A simple MSPE was developed and optimized by a design of experiments (DoE) to extract steroids while achieving sample clean-up and preconcentration prior to HPLC–MS/MS. The proposed method, assessed by the main figures of merit and compared with the currently available procedures based on SPE before instrumental analysis, was tested in human plasma and applied to multiclass steroids determination in blind plasma samples.

2. Results and Discussion

2.1. Magn-Humic Characterizations

The morphology of the prepared materials was appraised by SEM. As shown in Figure 1, the Fe_3O_4 nanoparticles characterized by definite edges (Figure 1a) have been coated after sol-gel by sphere-like silica particles, with a better homogeneity in shape and dimension observed on $SiO_2@Fe_3O_4$ (Figure 1(b1,b2)) compared to that air-calcined after sol-gel, c-$SiO_2@Fe_3O_4$ (Figure 1(c1,c2)).

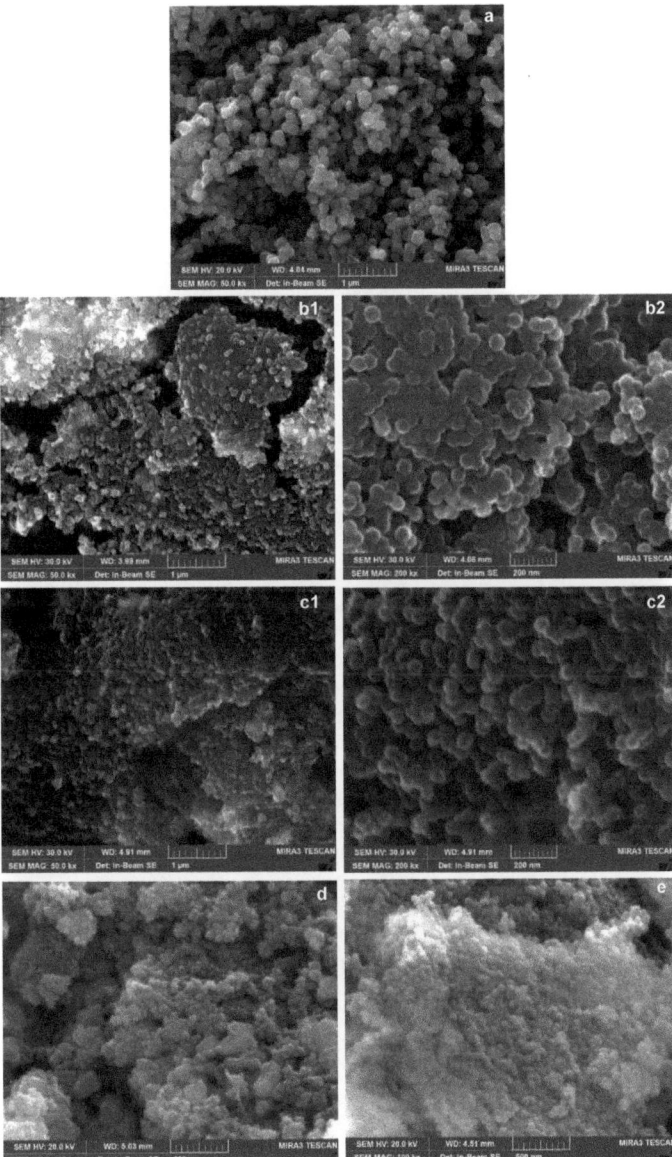

Figure 1. Representative scanning electron microscopy (SEM) images acquired on (**a**) pristine Fe_3O_4, (**b1,b2**) $SiO_2@Fe_3O_4$, (**c1,c2**) c-$SiO_2@Fe_3O_4$, (**d**) Magn-Humic, and (**e**) c-Magn-Humic.

The siliceous coating on the magnetic core was better evidenced by TEM (average thickness 10–20 nm) and confirmed by the Si/O and Si/Fe ratios from compositional EDS analysis, which showed a homogeneous distribution of the elements (see Supplementary Information). The SEM images of Magn-Humic (Figure 1d) and c-Magn-Humic (Figure 1e) evidenced a more compact structure of the latter—in agreement with the lower surface area, showed hereafter—and some carbon structures more visible in Magn-Humic. The overall procedure yielded micrometric materials, with particles ranging from few to some tens of microns, as shown by the additional SEM images in Supplementary Material.

The amount of carbon phase in the composites, determined by TGA, resulted to be 2.0 and 3.3 wt% for c-Magn-Humic and Magn-Humic, respectively. To achieve accurate quantitation of the carbonaceous fraction, the weight losses of c-SiO$_2$@Fe$_3$O$_4$ and SiO$_2$@Fe$_3$O$_4$—used as "blank" samples—were subtracted to those of the respective sorbents. In the case of SiO$_2$@Fe$_3$O$_4$, prepared with no calcination, an isothermal pretreatment of the sample (320 °C, 12 h) was necessary to remove cetyltrimethylammonium bromide (CTAB) entrapped in the silica shell. This was necessary because the great weight loss due to the surfactant release during the sample heating overlapped the weight loss between 320 °C and 600 °C of Magn-Humic related to the pyrolyzed HAs, making the calculation of the actual carbon phase wt% impracticable. TGA profiles are shown in Supplementary Information.

Surface area data, mean values from three measurements on each sample, are shown in Table 1.

Table 1. Surface area values determined by Brunauer, Emmett, and Teller (BET) method (relative standard deviation (RSD) < 5%, n = 3).

Material	Surface Area (m^2 g^{-1})
c-SiO$_2$@Fe$_3$O$_4$	305
c-Magn-Humic	169
SiO$_2$@Fe$_3$O$_4$	81
Magn-Humic	183

As apparent, surface area was enlarged compared to pristine magnetite (20–50 m^2 g^{-1}) due to the formation of the silica shell by sol-gel, and this increase is more evident performing calcination (see c-SiO$_2$@Fe$_3$O$_4$), which removes CTAB [23,24]. Instead, the deposition of pyrolyzed HAs on c-SiO$_2$@Fe$_3$O$_4$ induces a decrease of surface area because of carbon structures growing in the silica pores. This turns into agreement with the preparation of HA-C@silica [18,19]. For SiO$_2$@Fe$_3$O$_4$, after pyrolysis surface area showed a remarkable increase justifiable considering that CTAB is anyhow released during the pyrolytic treatment (600 °C). These findings fit with the TGA results above discussed and underline that the preparation of Magn-Humic is doubly advantageous as calcination after sol-gel can be avoided obtaining, in any case, a carbon-based magnetic material with higher surface area.

2.2. Protein Exclusion and Explorative Extraction Tests

In the first part of this study, the prepared materials were investigated for their affinity toward proteins, according to the studies of restricted access carbon nanotubes (RACNTs) for clean-up of biological matrices [25–27]. Protein exclusion tests were here performed in batch (rotating plate, 170 rpm, 3 min) by contacting 50 mg sorbent with 1 mL PBS containing 7 mg BSA [21], a quantity below saturation [26,27] and however lower compared to those of biological samples (see Section 2.4). The excluded protein—not retained on the solid phase—was quantified by UV-Vis spectrophotometry (spectra acquisition 200–800 nm, quantification at λ_{max} 280 nm) [21,26], and results are shown in Table 2.

Table 2. Protein exclusion (%) observed in bovine serum albumin (BSA) solution (RSD < 6 %, n = 3).

Sorbent	% BSA Exclusion [1]	Ref.
Magn-Humic	90(5)	This work
c-Magn-Humic	95(3)	This work
RACNTs	90(3)	[21]
HA-C@silica	86(2)	[21]

[1] in parentheses the standard deviation.

As apparent, up to 90–95% of the sample BSA is excluded by the two magnetic materials, in good agreement with the behavior of HA-C@silica and with a performance similar to that experimentally observed on RACNTs [21]. As discussed more in-depth in previous work [21], also in line with Mullet and Pawlyszin [28], the low affinity of the sorbents for proteins is essentially due to the small amount of carbon phase (2–3%, by TGA) joined to the low hydrophobicity imparted by oxygenated groups embedded in the carbon phase [18,19] that hamper protein retention. Predictably, protein exclusion was almost quantitative (97%) on Fe_3O_4 as control sample.

Both magnetic composites were tested for explorative MSPE experiments (in duplicate) by contacting 50 mg of sorbent with 2 mL BSA solution (10 g L^{-1}, 0.01 M phosphate buffer solution, PBS, pH 7.2 [21]), spiked with 2 mg L^{-1} of CORT, E2, TST, and PROG as probes. After extraction (3 min vortex, 1400 rpm), the sorbent was washed with 2 mL 2% formic acid (FA) followed by 2 mL 30% methanol (MeOH) aqueous solutions, and analytes were eluted in vortex by 2 × 1 mL MeOH-acetonitrile (ACN) (1:1) [21] and quantified by HPLC–UV (see Appendix A). Higher recoveries (in the range 40–76%) were observed for Magn-Humic compared to c-Magn-Humic (between 18 and 52%). At the same time, control tests on the intermediate materials (recovery 16–56% and 1–13% for $SiO_2@Fe_3O_4$ and c-$SiO_2@Fe_3O_4$, respectively) proved the major role of the carbonaceous phase deriving from HAs pyrolysis, able to retain steroids by a mixed-mode mechanism relying on π stacking and polar–apolar balanced interactions [18,20]. As expected, pristine magnetite did not show retention capability for the steroids, which were not quantifiable in the MSPE eluate (<0.2 mg L^{-1}).

In light of these explorative recovery tests, protein exclusion data and results from physical-chemical characterization, Magn-Humic prepared with no calcination after sol-gel was selected for in-depth investigation.

2.3. Development of the MSPE Procedure in BSA Solution Using Magn-Humic

For the MSPE development, experiments were undertaken in solution of BSA (10 g L^{-1}, 0.01 M PBS pH 7.2) as model protein focusing on extraction, clean-up, and elution. All experiments were run using 50 mg Magn-Humic and 2 mL samples spiked with 2 mg L^{-1} of each compound (CORT, E2, TST, and PROG) and, besides recovery, residual protein in the washing and in the eluate was monitored by the conventional Bradford assay (see Supplementary Material).

Concerning analytes adsorption on Magn-Humic, rotating plate shaker proved to favor extraction more consistently than vortex (data not shown), and not significantly different results were observed in going from 3 to 30 min contact; thus, 3 min was selected as the extraction time. In the washing step, 30% MeOH [21] caused a significant release of sorbed analytes, especially CORT, the steroid with the lowest partition coefficient (Log$P$) among the four probes (see Supplementary Material). Analytes recovery increased by reducing both the volume of the washing solution (from 2 to 1 mL) and the % of MeOH (from 30 to 5%, v/v). Considering that the residual protein in the eluate did not vary significantly, just the acidic washing (2 mL 2% v/v FA) was performed, affording removal of about 800 µg (40%) of adsorbed protein, with no loss of analytes.

Elution by 1 mL MeOH-ACN (1:1) allowed for the collection of a consistent fraction of steroids (65–80%) and a second elution, performed using the same eluent or 1 mL of MeOH, evidenced that the latter provided good elution and lower release of protein from the sorbent compared to the binary mixture. In the final eluate, obtained combining the two fractions, the residual protein was around 65 µg (against ca. 105 µg of the double elution with the mixture), corresponding to 0.3% of the BSA in the sample submitted to MSPE, as a result of the high protein exclusion (ca. 90%) joined to the acidic washing. At the same time, under these conditions, recovery was in the range 85–101% for all compounds. These findings account for a sorption process wherein the interaction with the sorbent displaces steroid–protein association [28], and elution by organic solvents induces the release of the potential fraction of protein-associated analytes [28].

To assess batch-to-batch reproducibility, additional recovery tests were done over non-consecutive days employing three independently synthesized batches of Magn-Humic. The observed RSD < 12% for the analytes recoveries is proof of reproducibility.

2.4. Optimization and Evaluation of MSPE in Biological Matrices

With the aim of maximizing recovery and method sensitivity in a real biological matrix, a DoE was planned to specifically focus on the performance of Magn-Humic in relation to the sample amount. Two factors were accordingly studied, namely sample volume (x_1) and sorbent amount (x_2) working on 1:4 diluted FBS samples (10 g L^{-1} proteins), spiked with 200 ng mL^{-1} of each analyte, in line with the experimental domain included in Supplementary Material. The mean multiclass recoveries observed under the different conditions are presented in Table 3, together with the residual protein in the MSPE eluate.

Table 3. Mean multiclass recovery ($n = 3$), average of all analytes recoveries of each experiment, and residual protein in the magnetic solid-phase extraction (MSPE) eluate obtained in the conditions of the experimental plan.

Exp.	FBS Volume (µL), x_1	Magn-Humic Amount (mg), x_2	Recovery (%)	Residual Proteins (µg)
1	250	10	55	57
2	1250	10	28	123
3	250	50	81	133
4	1250	50	65	237

Recovery values were used as the experimental response (y) relative to each variable (x_i), and they were modeled by the CAT software (Chemometric Agile Tool, available freely on the site of the Italian Group of Chemometrics) [29]) according to the following equation:

$$y = b_0 + b_1 x_1 + b_2 x_2 + b_{12} x_1 x_2 \tag{1}$$

The plot of the coefficients (b_i) of the model and the response surface are gathered in Figure 2 (part a, and part b, respectively).

As shown in Figure 2a, both x_1 and x_2 proved to significantly affect recovery (*** $p < 0.001$); in particular, recovery was favored working with the lowest volume of biological matrix (x_1) and the highest amount of magnetic material (x_2). As well, interaction between the two factors ($x_1 x_2$) resulted statistically relevant (*** $p < 0.001$), and, in line with the surface response graph (Figure 2b), recovery from about 80% upwards can be reached keeping x_1 and x_2 at the lowest and the highest level, respectively.

The model elaborated on the results from the MSPE tests, which yielded $y = 57 - 11x_1 + 16x_2 + 3x_1 x_2$, was validated by the experiment at the test point $x_1 = 0$ and $x_2 = 0$, which means working with 750 µL FBS and 30 mg sorbent; indeed, the experimental recovery (65%) well matched the theoretical one predicted by the model (relative error 12%).

Under optimal conditions (50 mg Magn-Humic and 1 mL sample containing 250 µL FBS), which were also convenient in terms of clean-up (Table 3), recovery at 100 ng mL^{-1} was quantitative for all analytes, as shown in Table 4. To improve method sensitivity, additional recovery tests were done using a smaller volume of eluent, i.e., 2 × 0.5 mL, observing unchanged recovery.

To better cover the steroids concentration range typical of human plasma, further MSPE trials were undertaken in FBS to verify accuracy also at lower concentrations, ranging from 1 to 25 ng mL^{-1}, and representative chromatograms are shown in Supplementary Material. As shown in Table 4, all compounds were quantified also at the lowest spike (1 ng mL^{-1}, except E2, the steroid with the lowest instrumental sensitivity), in this case by evaporating to dryness the eluate and reconstituting the residue in 0.25 mL MeOH.

In this way, enrichment factor (EF) 4 was achieved with a substantial sensitivity gain compared with our earlier report [21] and a within-laboratory inter-day precision RSD < 14% ($n = 3$). As can be seen, quantitative recovery was gained for all compounds and only slightly lower (65%) for PREDLO just at the lowest spiking level.

The MSPE was then moved on 1:4 diluted plasma (~20 g L^{-1} proteins) spiked with 25 ng mL^{-1} with unchanged recovery, highlighting that the procedure works well also in biological fluids with a protein content higher than that of FBS, thus representing a simplified and effective alternative to the intensive sample treatment workflow required in bioanalysis [4].

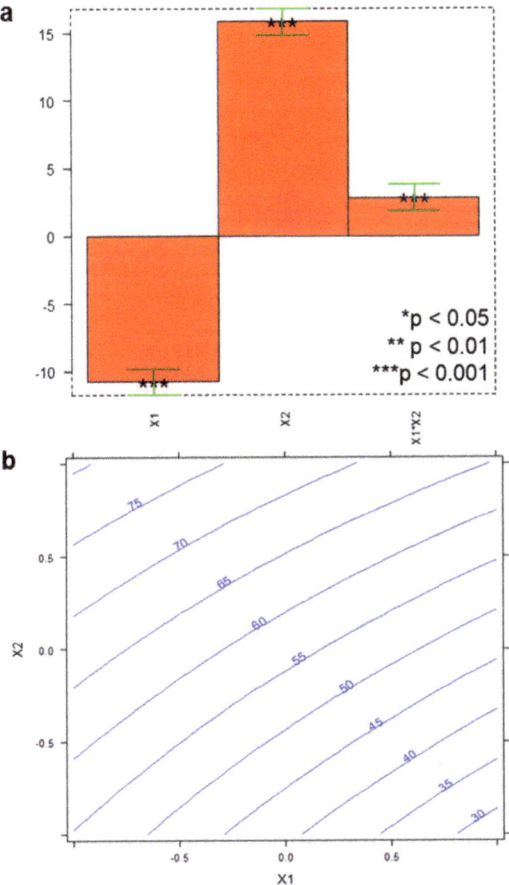

Figure 2. Design of experiments (DoE) results: (**a**) plot of the coefficients of the model (stars indicate the significance level of the coefficients, error bars indicate the confidence interval of the mean values for the coefficients, computed for $\alpha = 0.05$); (**b**) response surfaces of predicted recovery as function of x_1 (FBS volume) and x_2 (sorbent amount).

Table 4. Recoveries obtained in FBS by the optimized MSPE procedure followed by high-performance liquid chromatography coupled with mass spectrometry (HPLC–MS/MS) (n = 3).

		Mean Recovery (%) [1]		
Spike (ng mL^{-1})	100	25	5	1 [2]
PREDLO	95	84	87	65
PRED	109	98	97	107
H-CORT	80	80	71	70
CORT	91	87	97	70
BETA	97	94	104	122
DEXA	97	96	80	75
TRIAM	100	95	110	104
E2	106	115	109 [2]	n.q. [3]
TST	84	81	95	92
EPI	84	82	94	94
EE2	89	82	88	97
E1	86	89	105	90
H-PROG	105	90	84	96
FLUO	75	91	97	106
PROG	88	89	80	84
M-PROG	98	101	99	82

[1] RSDs < 14%, n = 3. [2] evaporation of the MSPE eluate and reconstitution in 0.25 mL MeOH (EF 4) before analysis. [3] n.q., not quantifiable at this concentration level.

2.5. Analytical Evaluation of the Method and Application to Bioanalysis

Selectivity is guaranteed by LC–MS with multiple-reaction monitoring (MRM) detection, which allows identification/quantification of the target compounds by using the two most intense transitions of each compound. Actually, no peaks of isobaric compounds or interfering species potentially co-extracted by Magn-Humic were evidenced at the steroids retention times in the chromatograms of FBS (blank matrix) MSPE eluate (Figure S5b).

The matrix-matched calibration, for quantitation of the concentrations expected after MSPE, was performed in the range 1–100 ng mL^{-1} (5–100 ng mL^{-1} for E2) by three independent calibration curves in the MSPE eluate from blank FBS and provided good linearity (r^2 0.9938–0.9999). Matrix effect (ME) resulted in an average signal suppression between 27 and 58% compared to the responses observed in pure solvent (MeOH), and it was quite well compensated by standard additions to the MSPE eluate. With regard to sensitivity, method detection and quantification limits (MDLs and MQLs) were in the range 0.02–0.3 ng mL^{-1} (0.8 ng mL^{-1} for E2) and 0.07–1 ng mL^{-1} (2.5 ng mL^{-1} for E2), respectively.

Trueness was assessed both in FBS and in human plasma at the ng mL^{-1} levels, obtaining satisfactory recoveries and good within-laboratory inter-day precision (Table 4) in agreement with criteria for analytical methods development at ng mL^{-1} levels [30,31].

No instrumental carry over was observed in the chromatograms of pure MeOH injected after each MSPE eluate, hence excluding cross-contamination. Carry over did not occur also in the MSPE step when recycling the sorbent, and reusability tests proved that Magn-Humic preserves its extraction performance for eight consecutive MSPE (recovery 65–107% at the eighth extraction).

The method was applied to the analysis of three clinical human blind samples, and representative HPLC–MS/MS chromatographic profiles are reported in Figure 3.

Five steroids were quantified at concentrations from few to tens nanograms per milliliter, i.e., CORT (7–8 ng mL^{-1}), H-CORT (34–71 ng mL^{-1}), BETA (2 ng mL^{-1}), EPI, and TST (1–2 ng mL^{-1}), with RSD < 10% (n = 3). These plasmatic levels fall within the typical intervals reported in human plasma for such compounds [32–34]. The synthetic glucocorticoid BETA was found just in one of the three samples, and its presence is usually correlated to recent drug intake [32], while CORT and H-CORT were determined in all samples. The TST/EPI plasmatic concentrations ratio, strictly related to the urinary concentrations monitored in antidoping controls, resulted around 1 in the samples analyzed here.

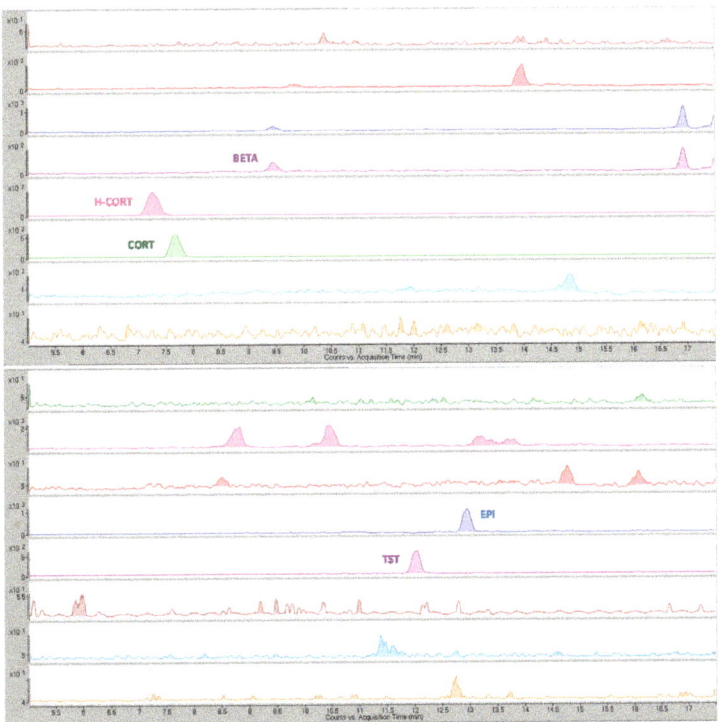

Figure 3. Representative multiple-reaction monitoring (MRM) chromatograms of the MSPE eluate from human plasma sample.

2.6. Comparison with Literature and Critical Discussion

The sorbent here proposed, Magn-Humic, is attractive compared to the new materials recently used for (M)SPE ([6,11,33], Table 5), and the final extraction procedure, coupled to chromatographic separation, is a valid tool for multiclass steroids determination in human plasma.

Magn-Humic MSPE is easily done by a common-laboratory equipment and provides simultaneous extraction, clean-up, and pre-concentration in complex biological matrices avoiding protein precipitation [5,6,13], which is a cause of analyte loss [35] or large sample dilution [5,6]. Although the explorative 3D-printed sorbents newly proposed [11,33] can offer enhanced sample throughput in the 96-well plate format [11], the extraction here obtained is quantitative and definitely quicker. Compared to our earlier report on HA-C@silica [21], quantitation is here gained in plasma after 1:4 dilution (instead of 1:8); moreover, dilution is fully compensated by the EF. For its sensitivity, the method is suitable for therapeutic drug monitoring and pharmacokinetic studies.

The sorbent is a micron-sized composite suitable for MSPE thus avoiding use of packed cartridges, often affected by bed blockage or reduced flow rate [36], and vacuum systems/peristaltic pumps necessary for traditional column SPE. Meanwhile, the phase separation with an external magnetic field is rapid compared to centrifugation and filtration required for dispersive SPE using non-magnetic sorbents [36]. The sample preparation is carried out with just 50 mg sorbent and thus can be defined as micro-MSPE [37,38], and, at the same time, it requires smaller amounts of plasma (250 µL) than those (1–4 mL) generally used for MSPE of drugs in biological matrices [2]. An additional advantage is the reusability of the sorbent material; thus, 50 mg can be conveniently used for extraction of eight plasma samples.

Table 5. Comparison with current analytical methods involving (M)SPE followed by chromatographic separation for steroids multiclass determination in human plasma.

Steroids, Analysed Number and Classes	Plasma Volume (μL)	Protein Precipitation	Centrifugation	Dilution	Extraction Technique	Sorbent (amount, mg)	Elution	Derivatization	Recovery (%)	RSD (%)	MQLs (ng mL^{-1})	Analysis	Ref.
10: 4 estrogens, 1 androgen, 3 progestagens, 2 glucocorticoids	2000	MeOH	n.a.[1]	H$_2$O (+36 mL)	SPE	C18 (500 mg)	2 mL MeOH	-	85.3–99.9	0.2–8.3	4–157 (MDLs)	HPLC-UV	[5]
19: 3 estrogens, 6 androgens, 4 progestagens, 6 glucocorticoids	400	-	-	H$_2$O (+4 mL)	SPE	C18 (500 mg)	5 mL MeOH-H$_2$O (80:20)	-	93.9–137.3	1.5–15.6	0.055–0.530	HPLC-MS	[8]
7: 2 estrogens, 4 androgens, 1 progestagen	495	0.1 % FA	20,220× g, 10 min, 4 °C	H$_2$O (to 2 mL)	SPE	C18 (500 mg)	3 mL ethylacetate	Step 1. 30 min, 30 °C Step 2. 30 min, 40 °C	69.2–100	1.6–35.5	0.01–5	GC-MS	[13]
16: 3 estrogens, 2 androgens, 3 progestagens, 8 glucocorticoids	250	-	-	PBS (to 2 mL)	SPE	HA-C@silica[2] (100 mg)	1 mL MeOH-ACN (1:1)	-	64–118	<15	2–10 (15 for E2)	HPLC-MS	[21]
7: 1 estrogen, 3 androgens, 2 glucocorticoids, 1 mineralcorticoid	n.a.	-	-	n.a.	dispersive SPE	3D-printed LayFOMM 60®	ACN-H$_2$O (80:20), 75 min, 750 rpm	-	19.3–84.9	1.44–9.46	3–10	HPLC-MS	[33]
5: 1 estrogen, 2 androgens, 1 glucocorticoid, 1 mineralcorticoid	300	-	-	PBS (to 1.5 mL)	96-well plate SPE	3D-printed LayFOMM 60®	ACN-H$_2$O (80:20), 75 min	-	2.05–38.07	3.02–18.14	n.a.	HPLC-MS	[11]
2: 1 androgen, 1 progestagen	n.a.	ACN	3000 rpm, 30 min	H$_2$O (to 50 mL)	MSPE	TMSPT-MNP@Au[3] (50 mg)	1 mL MeOH, 3 min	-	94.5–99.1	3.49–4.19	0.05–0.07 (MDLs)	HPLC-UV	[6]
16: 3 estrogens, 2 androgens, 3 progestagens, 8 glucocorticoids	250	-	-	PBS (to 1 mL)	MSPE	Magn-Humic (50 mg)	0.5 mL MeOH-ACN (1:1) + 0.5 mL MeOH vortex, 1400 rpm, 3 min)	-	65–122	5–14	0.07–1 (2.5 for E2)	HPLC-MS	This work

[1] n.a., not available. [2] HA-C@silica, silica-supported carbon from humic acids pyrolysis. [3] TMSPT-MNP@Au, Au nanoparticles grafted on 3-(trimethoxysilyl)-1-propanethiol modified Fe$_3$O$_4$ magnetic nanoparticles.

3. Materials and Methods

3.1. Chemicals and Materials

Fe_3O_4 (50–100 nm, 20–50 m^2 g^{-1}), triethanolamine (TEOA, >99%), CTAB ($\geq$99%), ethanol (96% v/v), HAs sodium salt (technical grade), BSA (> 98%), Bradford reagent (for micro and standard assays, 1–10 mg L^{-1} and 50–1400 mg L^{-1} proteins, respectively), nylon filters (0.2 µm), charcoal-stripped FBS, and high purity steroids standards were purchased from Sigma-Aldrich (Milan, Italy). Analytical grade H-PROG was supplied by Steroids (Cologno Monzese, Italy), and TRIAM and FLUO by Farmabios (Gropello Cairoli, Italy). Molecular structures and Log$P$ values are shown in Supplementary Material. Technical grade acetone, HPLC gradient grade MeOH, ACN, and ultrapure water were provided by VWR (Milan, Italy). Tetraethyl orthosilicate (TEOS, 98%), FA (99%), NH_4F ($\geq$ 98%), Na_2HPO_4 (99%), and $NaH_2PO_4 \cdot H_2O$ (99%) were acquired from Carlo Erba Reagents (Milan, Italy).

MeOH steroids stock solutions (1000 µg mL^{-1}) were stored in the dark (4 °C). Working solutions $\leq$ 1 µg mL^{-1} were prepared weekly in MeOH by dilution from a 10 µg mL^{-1} solution.

3.2. Preparation and Characterization of HA-C@SiO$_2$@Fe$_3$O$_4$ (Magn-Humic)

Based on the results of our previous works [18–21] and the great advantages offered by MSPE [1,2], the idea of this study was to prepare a magnetic sorbent to be employed in a simplified extraction procedure in batch, i.e., dispersive MSPE. Considering two recent papers on magnetic porous silica prepared via sol-gel [23,24], this route was chosen to obtain an intermediate material (SiO_2@Fe_3O_4) as the support for HAs (Appendix A) to prepare the magnetic sorbent HA-C@SiO_2@Fe_3O_4, named Magn-Humic in the paper. In detail, 200 mg HAs were dissolved in 100 mL distilled water in a round-bottom flask, and 2 g SiO_2@Fe_3O_4 or c-SiO_2@Fe_3O_4 (air-calcined after sol-gel) were added, and the suspension was stirred for 2 min. Water was removed by rotary evaporator, and the obtained solid was pyrolyzed in an alumina combustion boat inside a quartz tube (600 °C, 1 h, N_2 flow, heating 10 °C min^{-1}, cooling 10 °C min^{-1}) to convert HAs into a hydrophilic–lipophilic balanced carbonaceous phase [18,19]. Before use, Magn-Humic and c-Magn-Humic were washed in a filtering flask with plenty of distilled water until neutrality of the eluate. The batch-to-batch reproducibility was checked by recovery tests on three independent Magn-Humic preparations.

Microstructural characterizations were performed by a high-resolution scanning electron microscope (TESCAN Mira 3, Brno, Czech Republic), operating at 20.0 kV. Images were acquired on the powders after carbon (for SiO_2@Fe_3O_4 and c-SiO_2@Fe_3O_4) or platinum (for Magn-Humic and c-Magn-Humic) coatings, which were performed by either a Cressington 208C carbon coater or a Cressington HR 208, respectively (Watford, England, UK). The same instrumentation was used for compositional EDS analysis. TEM images were acquired by a JEOL JEM-1200EXIII instrument provided with a Mega View III CCD camera. Few milligrams samples were dispersed by sonication in about 3 mL water (Fe_3O_4), MeOH (SiO_2@Fe_3O_4), or acetone (Magn-Humic), and then 10 µL of each suspension were deposited on grids and left to dry (room temperature).

TGA was performed using a Q5000 Instrument (TA Instruments Inc., New Castle, DE, USA). Each sample (10 mg) was heated (20 °C min^{-1}) into a Pt pan from 25 up to 900 °C, using 100 mL min^{-1} air flow.

Surface areas were measured by the BET single point method using a Flowsorb II 2300 (Micromeritics, Norcross, GA, USA) apparatus. The sample was weighed and degassed at 80 °C (1.5 h) under a continuous stream of a N_2-He (30:70) mixture, and then it was put in liquid N_2 for gas adsorption.

3.3. Biological Samples

Being certified hormone-free FBS the recommended surrogate matrix [8,15,21], it was used as the blank for recovery tests at concentrations in the range 1–100 ng mL^{-1}. Human plasma blind samples were provided by IRCCS Policlinico San Matteo (Pavia, Italy).

Aliquots of the samples were frozen and stored at −20 °C. Before extraction, sub-samples were left to thaw at room temperature and vortexed for 20 s at 1400 rpm. In the case of recovery tests, samples were spiked, and after 30 min equilibration at room temperature, re-vortexed before MSPE.

3.4. MSPE Procedure for Simultaneous Extraction, Clean-Up, and Pre-concentration of Multiclass Steroids in Human Plasma

The MSPE procedure was carried out using 50 mg Magn-Humic in a self-standing 2 mL screw-cap glass vial. The material was conditioned using 2 mL phosphate buffer solution (PBS, 0.01 M, pH 7.2) by vortex (1400 rpm, 3 min), and then fast sedimentation of the solid was achieved by a neodymium magnet (Ø 10 mm, h 4 mm) placed under the vial. The liquid was withdrawn by a pipette, and the sample (1 mL from 1:4 plasma dilution, in PBS [11,21,25]) was added in the vial. Extraction was done on a rotating plate shaker (170 rpm, 3 min), the liquid was removed as above described, and washing was performed with 2 mL 2% (v/v) FA aqueous solution (vortex, 1400 rpm, 3 min). Analytes were eluted by 0.5 mL MeOH-ACN (1:1, v/v) and 0.5 mL MeOH, sequentially (vortex, 1400 rpm, 3 min). The two eluates were merged, filtered (0.2 μm), and evaporated to dryness under gentle N_2 flow. The residue was re-dissolved in 0.25 mL MeOH for the HPLC–MS/MS analysis (see Appendix A). The overall time required for the extraction procedure is approximately 15 min. After use, the sorbent was contacted with 2 mL eluting solutions to avoid potential carryover, washed with 2 mL 2% (v/v) FA aqueous solution, and conditioned with 2 mL PBS for reusability tests.

3.5. MSPE Followed by HPLC–ESI-MS/MS: Analytical Parameters

Method selectivity was checked by analysis of blank samples (FBS) processed by all the steps of the analytical procedure described in the section above.

Linearity was assessed by ordinary linear least squares regression (OLLSR) on five-point calibration curves (1–100 ng mL^{-1}) generated in both neat solvent, i.e., MeOH, and in FBS MSPE eluate after evaporation to dryness and reconstitution in 0.25 mL MeOH.

Matrix-matched calibration in the MSPE eluate was selected for quantification [18,20,21], at the same time compensating ME. This was calculated as:

$$ME\ (\%) = \frac{b_m}{b_s} \times 100 \qquad (2)$$

where b_m and b_s are the slopes of the matrix-matched calibration curve and the calibration line obtained in pure solvent, respectively [18,20,21]. MDLs and MQLs were calculated from the matrix-matched calibration curves, obtained in FBS MSPE extracts after N_2 evaporation, as 3 and 10 times, respectively, the ratio between the baseline noise away from the peak tail and the regression line slope [18,20,21], considering that the pre-concentration (EF 4) compensated the initial sample dilution (1:4).

Accuracy was assessed in terms of trueness and precision. Due to the unavailability of certified reference materials (CRMs), trueness was verified by recovery tests (1–100 ng mL^{-1}) in spiked FBS and human plasma samples by independent MSPE trials ($n = 3$), and the within-laboratory inter-day precision was evaluated based on RSD%. Instrumental carry-over was monitored by injections of MeOH, as control blank, after each chromatographic run.

4. Conclusions

The novel carbon-based magnetic material Magn-Humic has been prepared, characterized by various techniques, and successfully applied as sorbent for micro-MSPE of steroids in serum/plasma samples. Coupling the high sample protein exclusion and quantitative extraction afforded by using Magn-Humic to LC-MS^2 analysis, satisfactory clean-up and multianalyte determination were possible with high selectivity. The sample treatment procedure, optimized by DoE, allows one to avoid large sample dilution and protein precipitation, requires small amount of sample, is simple, quick (around 15 min) and effective for multiclass determination of steroid hormones. The sorbent is reusable for repeated extractions, and it could be extended to environmental and food matrices.

Supplementary Materials: The following are available online, Figure S1: Representative TEM images acquired on (a) pristine Fe_3O_4 (50 kx) and (b1, b2) $SiO_2@Fe_3O_4$ (150 kx); Figure S2: Representative SEM images acquired on (a) Magn-Humic and (b) c-Magn-Humic; Figure S3: TGA profiles recorded on (a) c-$SiO_2@Fe_3O_4$, (b) c-Magn-Humic, (c) $SiO_2@Fe_3O_4$, and (d) $SiO_2@Fe_3O_4$ after isothermal pretreatment for 12 h at 320 °C, (e) Magn-Humic; Figure S4: Mean calibration curve ($n = 6$) for the Bradford assay; Figure S5: MRM chromatograms of MSPE eluates from (a) FBS spiked with 5 ng mL^{-1} of each compound before extraction and (b) unspiked FBS; Table S1: Molecular structures and Log$P$ values of the studied steroids (data from https://hmdb.ca/, accessed on 26 February 2021); Table S2: Compositional results collected on the materials obtained after sol-gel compared to pristine magnetite; Table S3: MRM conditions for HPLC–ESI-MS/MS analysis of the steroids; Table S4: Experimental domain of the variables selected for the 2^2 factorial design.

Author Contributions: Conceptualization, A.S. and A.P.; software, G.M.; validation, A.S., F.M. (Francesca Merlo), F.M. (Federica Maraschi), and A.P.; formal analysis, G.M. and A.F.; investigation, A.F., F.M. (Francesca Merlo) and A.S.; resources, F.M. (Francesca Merlo) and F.M. (Federica Maraschi); data curation, A.S., F.M. (Francesca Merlo), and F.M. (Federica Maraschi); writing—original draft preparation, A.S.; writing—review and editing, A.S., F.M. (Francesca Merlo), F.M. (Federica Maraschi), and A.P.; visualization, A.S. and F.M. (Francesca Merlo); supervision, A.S. and A.P.; funding acquisition, A.S. and A.P. All authors have read and agreed to the published version of the manuscript.

Funding: This research was funded by "Fondo Ricerca Giovani" (FR&G) by University of Pavia.

Institutional Review Board Statement: The study was conducted according to the guidelines of the Declaration of Helsinki. The plasma samples are obtained from blood donations obtained and validated according to the provisions of the current national law on the subject: "Disposizioni relative ai requisiti di qualità e sicurezza del sangue e degli emocomponenti" (Supplemento ordinario n. 69 to GAZZETTA UFFICIALE Serie generale - n. 300). For this type of supply without sensitive data and randomized between products not intended for clinical use, there is no intervention by the Ethics Committee. Approval from the Ethics Committee would be required in case donation or additional blood sample are specifically obtained for a defined research.

Informed Consent Statement: Informed consent is defined and collected for each donation according to the current national law: "Disposizioni relative ai requisiti di qualità e sicurezza del sangue e degli emocomponenti" (Article 8 and Annex II), and it is allowed to be used for research purposes in the medical and biomedical fields. For this type of supply without sensitive data and randomized between products not intended for clinical use, there is no dedicated informed consent: the blood donor signs a form borrowed from the text of the law as each donation is intended to be used for research purposes. Dedicated consent would be required in case donation or additional blood sample are specifically obtained for a defined research.

Data Availability Statement: The data presented in this study are available on request from the corresponding author.

Acknowledgments: Ilenia Tredici is acknowledged for the SEM/EDS analyses performed at the CISRiC (Centro Interdipartimentale di Studi e Ricerche per la Conservazione del Patrimonio Culturale) of Pavia. We are grateful to Chiara Milanese and to Lidia Romani (Department of Chemistry, University of Pavia) for TGA and TEM acquisitions, respectively. We acknowledge Giuseppina Sandri (Department of Drug Sciences, University of Pavia) for surface area measurements. We thank Paola Isernia (IRCCS Policlinico San Matteo of Pavia) for providing human plasma blind samples.

Conflicts of Interest: The authors declare no conflict of interest.

Sample Availability: Samples of the compounds are not available from the authors.

Appendix A

A.1. Preparation of $SiO_2@Fe_3O_4$

1 g Fe_3O_4 was suspended in 90 mL water and sonicated for 30 min. Subsequently, 1.7 g of CTAB and 1 g of TEOA were added, and the mixture was continuously mixed

by an overhead mechanical stirrer for 1 h at 80 °C using a thermostatically controlled bath (in a fumehood). Then, 14 mL TEOS were rapidly added, and the reaction mixture was maintained under stirring at 80 °C for 2 h. The obtained $SiO_2@Fe_3O_4$ was recovered by filtration, washed with 50 mL ethanol, dried in oven (60 °C, 24 h), and used in the subsequent step, eventually after calcination (c-$SiO_2@Fe_3O_4$) at 540 °C, 7 h [23].

A.2 HPLC–UV

The chromatographic apparatus consisted of a Shimadzu (Milan, Italy) LC-20AT solvent delivery module equipped with a DGU-20A3 degasser and interfaced with an SPD-20A detector. A Scharlab Kroma Phase 100 C18 (250 × 4.6 mm, 5 μm) column coupled with a Supelco Supelguard Ascentis C18 (20 × 2.1 mm, 5 μm) guard-column was used. After an equilibration period of 3 min, 20 μL of each sample was manually injected in the system. The mobile phase was (A) water and (B) ACN, flow rate of 1 mL min^{-1}. Elution program: linear gradient from 30 to 90% B until 12 min, then to 95% B until 15 min, finally to 100% B until 18 min (kept for 5 min). The detection wavelengths were 225 nm for estrogens and 242 nm for the other compounds. Calibration standard solutions (1–9 mg L^{-1}) were prepared in MeOH-ACN (1:1, v/v) obtaining good linearity ($r^2 > 0.9923$).

A.3 HPLC–ESI-MS/MS

The target substances were analyzed with a HPLC apparatus Agilent 1260 Infinity coupled with an Agilent 6460C MS spectrometer ESI-MS/MS system (Cernusco sul Naviglio, Italy). The MS operating parameters, optimized by Agilent Mass Hunter Source Optimizer Software (Agilent, USA), were the following: drying gas (N_2) temperature 350 °C; drying gas flow 12 L min^{-1}; nebulizer 50 psi; sheath gas temperature 400 °C; sheath gas flow 12 L min^{-1}; capillary voltage 4000 V positive, 3000 V negative; nozzle voltage 0 V positive, 1500 V negative; electron multiplier voltage (EMV) 200 V positive, 0 V negative; and cell accelerated voltage 1 and 4 V for negative and positive mode, respectively. Quantitative analysis was performed in MRM mode, using the most intense transitions from precursor ion to product ions for each analyte (see Table S3).

References

1. Speltini, A.; Sturini, M.; Maraschi, F.; Profumo, A. Recent trends in the application of the newest carbonaceous materials for magnetic solid-phase extraction of environmental pollutants. *Trends Environ. Anal. Chem.* **2016**, *10*, 11–23. [CrossRef]
2. Vasconcelos, I.; Fernandes, C. Magnetic solid phase extraction for determination of drugs in biological matrices. *Trends Anal. Chem.* **2017**, *89*, 41–52. [CrossRef]
3. Jiang, H.L.; Li, N.; Cui, L.; Wang, X.; Zhao, R.S. Recent application of magnetic solid phase extraction for food safety analysis. *Trends Anal. Chem.* **2019**, *120*, 115632. [CrossRef]
4. Abdel-Rehim, M.; Pedersen-Bjergaard, S.; Abdel-Rehim, A.; Lucena, R.; Moein, M.M.; Cárdenas, S.; Miró, M. Microextraction approaches for bioanalytical applications: An overview. *J. Chromatogr. A* **2020**, *1616*, 460790. [CrossRef] [PubMed]
5. Zarzycki, P.K.; Kulhanek, K.M.; Smith, R.; Clifton, V.L. Determination of steroids in human plasma using temperature-dependent inclusion chromatography for metabolomics investigations. *J. Chromatogr. A* **2006**, *1104*, 203–208. [CrossRef]
6. Beiraghi, A.; Pourghazi, K.; Amoli-Diva, M. Au nanoparticle grafted thiol modified magnetic nanoparticle solid phase extraction coupled with high performance liquid chromatography for determination of steroid hormones in human plasma and urine. *Anal. Methods* **2014**, *6*, 1418–1426. [CrossRef]
7. Gaudl, A.; Kratzsch, J.; Bae, Y.J.; Kiess, W.; Thiery, J.; Ceglarek, U. Liquid chromatography quadrupole linear ion trap mass spectrometry for quantitative steroid hormone analysis in plasma, urine, saliva and hair. *J. Chromatogr. A* **2016**, *1464*, 64–71. [CrossRef]
8. Weisser, J.J.; Hansen, C.H.; Poulsen, R.; Weber Larsen, L.; Cornett, C.; Styrishave, B. Two simple clean-up methods combined with LC-MS/MS for quantification of steroid hormones in in vivo and in vitro assays. *Anal. Bioanal. Chem.* **2016**, *408*, 4883–4895. [CrossRef]
9. Nair, S.G.; Patel, D.P.; Sanyal, M.; Singhald, P.; Shrivastav, P.S. Simultaneous analysis of glucocorticosteroid fluticasone propionate and its metabolite fluticasone propionate 17β-carboxylic acid in human plasma by UPLC–MS/MS at sub pg/mL level. *J. Pharm. Biomed. Anal.* **2017**, *135*, 1–7. [CrossRef]
10. Khataei, M.M.; Yamini, Y.; Nazaripour, A.; Karimi, M. Novel generation of deep eutectic solvent as an acceptor phase in three phase hollow fiber liquid phase microextraction for extraction and preconcentration of steroidal hormones from biological fluids. *Talanta* **2018**, *178*, 473–480. [CrossRef]

11. Belka, M.; Konieczna, L.; Okonska, M.; Pyszka, M.; Ulenberg, A.; Bączek, T. Application of 3D-printed scabbard-like sorbent for sample preparation in bioanalysis expanded to 96-wellplate high-throughput format. *Anal. Chim. Acta* **2019**, *1081*, 1–5. [CrossRef] [PubMed]
12. Denver, N.; Khan, S.; Stasinopoulos, I.; Church, C.; Homer, N.Z.M.; MacLean, M.R.; Andrew, R. Derivatization enhances analysis of estrogens and their bioactive metabolites in human plasma by liquid chromatography tandem mass spectrometry. *Anal. Chim. Acta* **2019**, *1054*, 84–94. [CrossRef] [PubMed]
13. Luque-Córdoba, D.; López-Bascón, M.A.; Priego-Capote, F. Development of a quantitative method for determination of steroids in human plasma by gas chromatography-negative chemical ionization-tandem mass spectrometry. *Talanta* **2020**, *220*, 121415. [CrossRef]
14. van der Veen, A.; van Faassen, M.; de Jong, W.H.A.; van Beek, A.P.; Dijck-Brouwer, D.A.J.; Kema, I.P. Development and validation of a LC-MS/MS method for the establishment of reference intervals and biological variation for five plasma steroid hormones. *Clin. Biochem.* **2019**, *68*, 15–23. [CrossRef] [PubMed]
15. van Nuland, M.; Venekamp, N.; Wouters, W.M.E.; van Rossum, H.H.; Rosing, H.; Beijnen, J.H. LC–MS/MS assay for the quantification of testosterone, dihydrotestosterone, androstenedione, cortisol and prednisone in plasma from castrated prostate cancer patients treated with abiraterone acetate or enzalutamide. *J. Pharm. Biomed. Anal.* **2019**, *170*, 161–168. [CrossRef]
16. Tircova, B.; Bosakova, Z.; Kozlik, P. Development of an ultra-high performance liquid chromatography–tandem mass spectrometry method for the determination of anabolic steroids currently available on the black market in the Czech Republic and Slovakia. *Drug Test. Anal.* **2019**, *11*, 355–360. [CrossRef]
17. Van Renterghem, P.; Viaene, W.; Van Gansbeke, W.; Barrabin, J.; Iannone, M.; Polet, M.; T'Sjoen, G.; Deventer, K.; Van Eenoo, P. Validation of an ultra-sensitive detection method for steroid esters in plasma for doping analysis using positive chemical ionization GC-MS/MS. *J. Chromatogr. B* **2020**, *1141*, 122026. [CrossRef]
18. Speltini, A.; Merlo, F.; Maraschi, F.; Sturini, M.; Contini, M.; Calisi, N.; Profumo, A. Thermally condensed humic acids onto silica as SPE for effective enrichment of glucocorticoids from environmental waters followed by HPLC-HESI-MS/MS. *J. Chromatogr. A* **2018**, *1540*, 38–46. [CrossRef]
19. Speltini, A.; Pastore, M.; Merlo, F.; Maraschi, F.; Sturini, M.; Dondi, D.; Profumo, A. Humic acids pyrolyzed onto silica microparticles for solid-phase extraction of benzotriazoles and benzothiazoles from environmental waters. *Chromatographia* **2019**, *82*, 1275–1283. [CrossRef]
20. Merlo, F.; Speltini, A.; Maraschi, F.; Sturini, M.; Profumo, A. HPLC-MS/MS multiclass determination of steroid hormones in environmental waters after preconcentration on the carbonaceous sorbent HA-C@silica. *Arab. J. Chem.* **2020**, *13*, 4673–4680. [CrossRef]
21. Speltini, A.; Merlo, F.; Maraschi, F.; Villani, L.; Profumo, A. HA-C@silica sorbent for simultaneous extraction and clean-up of steroids in human plasma followed by HPLC-MS/MS multiclass determination. *Talanta* **2021**, *221*, 121496. [CrossRef] [PubMed]
22. An, X.; Chai, W.; Deng, X.; Chen, H.; Ding, G. A bioinspired polydopamine approach toward the preparation of gold-modified magnetic nanoparticles for the magnetic solid-phase extraction of steroids in multiple samples. *J. Sep. Sci.* **2018**, *41*, 2774–2782. [CrossRef] [PubMed]
23. Tian, Z.; Yu, X.; Ruan, Z.; Zhu, M.; Zhu, Y.; Hanagata, N. Magnetic mesoporous silica nanoparticles coated with thermo-responsive copolymer for potential chemo—and magnetic hyperthermia therapy. *Microporous Mesoporous Mater.* **2018**, *256*, 1–9. [CrossRef]
24. Fang, X.; Yao, J.; Hu, X.; Li, Y.; Yan, G.; Wu, H.; Deng, C. Magnetic mesoporous silica of loading copper metal ions for enrichment and LC-MS/MS analysis of salivary endogenous peptides. *Talanta* **2020**, *207*, 120313. [CrossRef] [PubMed]
25. Barbosa, A.F.; Barbosa, V.M.P.; Bettini, J.; Luccas, P.O.; Figueiredo, E.C. Restricted access carbon nanotubes for direct extraction of cadmium from human serum samples followed by atomic absorption spectrometry analysis. *Talanta* **2015**, *131*, 213–220. [CrossRef]
26. Dipe de Faria, H.; Tosin Bueno, C.; Krieger, J.E.; Moacyr Krieger, E.; Costa Pereira, A.; Lima Santos, P.C.J.; Figueiredo, E.C. Online extraction of antihypertensive drugs and their metabolites from untreated human serum samples using restricted access carbon nanotubes in a column switching liquid chromatography system. *J. Chromatogr. A* **2017**, *1528*, 41–52. [CrossRef]
27. Dipe de Faria, H.; Azevedo Rosa, M.; Thalison Silveira, A.; Figueiredo, E.C. Direct extraction of tetracyclines from bovine milk using restricted access carbon nanotubes in a column switching liquid chromatography system. *Food Chem.* **2017**, *225*, 98–106. [CrossRef]
28. Mullett, W.M.; Pawliszyn, J. Direct LC analysis of five benzodiazepines in human urine and plasma using an ADS restricted access extraction column. *J. Pharmaceut. Biomed.* **2001**, *26*, 899–908. [CrossRef]
29. Leardi, R.; Melzi, C.; Polotti, G. CAT, Cchemometric Agile Tool. Freely. Available online: http://gruppochemiometria.it/index.php/software (accessed on 18 February 2021).
30. González, A.G.; Herrador, M.A. A practical guide to analytical method validation, including measurement uncertainty and accuracy profiles. *Trend Anal. Chem.* **2007**, *26*, 227–238. [CrossRef]
31. SANCO/12571/2013, Guidance Document on Analytical Quality Control and Validation Procedures for Pesticide Residues Analysis in Food and Feed. Available online: https://www.eurl-pesticides.eu/library/docs/allcrl/AqcGuidance_Sanco_2013_12571.pdf (accessed on 26 March 2021).
32. Rochester 2021 Interpretive Handbook; Mayo clinic laboratories: Rochester, USA. Available online: https://www.mayocliniclabs.com/test-catalog/pod/MayoTestCatalog-Rochester--SortedByTestName-duplex-interpretive.pdf (accessed on 18 February 2021).

33. Koniecza, L.; Belka, M.; Okonska, M.; Pyszka, M.; Bączec, T. New 3D-printed sorbent for extraction of steroids from human plasma preceding LC-MS analysis. *J. Chromatogr. A.* **2018**, *1545*, 1–11. [CrossRef]
34. Márta, Z.; Bobály, B.; Fekete, J.; Magda, B.; Imre, T.; Mészáros, K.V.; Bálint, M.; Szabó, P.T. Simultaneous determination of thirteen different steroid hormones using micro UHPLC-MS/MS with on-line SPE system. *J. Pharmaceut. Biomed. Anal.* **2018**, *150*, 258–267. [CrossRef] [PubMed]
35. Li, F.; Lu, L.; Gao, D.; Wang, M.; Wang, D.; Xia, Z. Rapid synthesis of three-dimensional sulfur-doped porous graphene via solid-state microwave irradiation for protein removal in plasma sample pretreatment. *Talanta* **2018**, *185*, 528–536. [CrossRef] [PubMed]
36. Manousi, N.; Rosenberg, E.; Deliyanni, E.; Zachariadis, G.A.; Samanidou, V. Magnetic solid-phase extraction of organic compounds based on graphene oxide nanocomposites. *Molecules* **2020**, *25*, 1148. [CrossRef] [PubMed]
37. Ghorbani, M.; Aghamohammadhassan, M.; Chamsaz, M.; Akhlaghi, H.; Pedramrad, T. Dispersive solid phase microextraction. *Trends Anal. Chem.* **2019**, *118*, 793–809. [CrossRef]
38. Moreda-Piñeiro, J.; Moreda-Piñeiro, A. Combined assisted extraction techniques as green sample pre-treatments in food analysis. *Trends Anal. Chem.* **2019**, *118*, 1–18. [CrossRef]

Article

Online In-Tube Solid-Phase Microextraction Coupled with Liquid Chromatography–Tandem Mass Spectrometry for Automated Analysis of Four Sulfated Steroid Metabolites in Saliva Samples

Hiroyuki Kataoka * and Daiki Nakayama

School of Pharmacy, Shujitsu University, Nishigawara, Okayama 703-8516, Japan; s1916073se@shujitsu.jp
* Correspondence: hkataoka@shujitsu.ac.jp

Abstract: Accurate measurement of sulfated steroid metabolite concentrations can not only enable the elucidation of the mechanisms regulating steroid metabolism, but also lead to the diagnosis of various related diseases. The present study describes a simple and sensitive method for the simultaneous determination of four sulfated steroid metabolites in saliva, pregnenolone sulfate (PREGS), dehydroepiandrosterone sulfate (DHEAS), cortisol sulfate (CRTS), and 17β-estradiol-3-sulfate (E2S), by online coupling of in-tube solid-phase microextraction (IT-SPME) and stable isotope dilution liquid chromatography–tandem mass spectrometry (LC–MS/MS). These compounds were extracted and concentrated on Supel-Q PLOT capillary tubes by IT-SPME and separated and detected within 6 min by LC–MS/MS using an InertSustain swift C18 column and negative ion mode multiple reaction monitoring systems. These operations were fully automated by an online program. Calibration curves using their stable isotope-labeled internal standards showed good linearity in the range of 0.01–2 ng mL^{-1} for PREGS, DHEAS, and CRTS and of 0.05–10 ng mL^{-1} for E2S. The limits of detection (S/N = 3) of PREGS, DHEAS, CRTS, and E2S were 0.59, 0.30, 0.80, and 3.20 pg mL^{-1}, respectively. Moreover, intraday and interday variations were lower than 11.1% ($n = 5$). The recoveries of these compounds from saliva samples were in the range of 86.6–112.9%. The developed method is highly sensitive and specific and can easily measure sulfated steroid metabolite concentrations in 50 µL saliva samples.

Keywords: sulfated steroid metabolites; saliva; online automated analysis; in-tube solid-phase microextraction (IT-SPME); liquid chromatography–tandem mass spectrometry (LC–MS/MS); stable isotope dilution

1. Introduction

Steroid hormones include glucocorticoids, which regulate carbohydrate and lipid metabolism and have inflammatory and immunosuppressive effects; mineralocorticoids, which regulate blood pressure through ionic equilibrium in body fluids; and sex hormones, which control reproductive functions and secondary sexual characteristics [1–3]. Starting from cholesterol, all of these compounds are biosynthesized in the adrenal cortex, gonads, and brain by various enzymes, but their biosynthetic and metabolic pathways are complex, and the molecular roles of these compounds have not been fully determined [1,2]. After their biosynthesis, these steroid hormones are metabolized by Phase I reactions involving oxidation and reduction and Phase II reactions involving conjugation [1,3].

Most steroids are sulfated by sulfotransferases (SULTs), which increase their solubility in aqueous solution and their excretion into urine. In addition, some of these sulfated steroids are desulfated by steroid sulfatases (STSs), with these steroids returning to their free forms [1,3]. This cycle plays an important role in the regulation of total steroid content and bioactivity in vivo [4,5], and the lack of STSs is a major factor in the pathogenesis of

STS deficiency, recessive X-linked ichthyosis (RXLI), and the metabolic syndrome [6,7]. The ability to accurately measure the concentrations of sulfated steroid metabolites in vivo can not only provide insight into the regulation of steroid metabolism by the balance between SULTs and STSs, but also lead to the diagnosis of various diseases.

The sulfated steroid metabolite dehydroepiandrosterone-3-sulfate (DHEAS) [8,9] has been reported to be involved in responses to stress [10], reproductive function [5], the onset of age-related diseases, and human longevity [4,11]. During acute periods of stress, salivary DHEAS concentrations increase, but their levels decrease in long-term situations [10,12]. In addition, low DHEAS concentrations have been observed in subjects with aging-related diseases, such as sarcopenia, Alzheimer's disease, depression, cardiovascular disease, and low libido [4,11].

Sulfated steroid metabolites can be analyzed by immunoassays or by mass spectrometry coupled to chromatography. Immunoassays, however, are not suitable for simultaneous analysis of a series of steroids due to their cross-reactivity, difficulties distinguishing among steroids with similar structures, and the need to generate specific antibodies against each compound [2,3,10,13]. Gas chromatography–mass spectrometry (GC–MS) cannot directly detect conjugated metabolites and requires chemical or enzymatic cleavage of sulfate groups and volatile derivatization steps [1,2,6,13–15]. Chemical cleavage of sulfated steroid metabolites also results in the hydrolysis of other conjugates, and commercially available sulfatase enzymes, which enzymatically cleave these sulfated compounds, also possess glucuronidase activity, making it difficult to distinguish between steroid glucuronides and sulfated steroids [6,14]. In contrast, liquid chromatography–mass spectrometry (LC–MS) and LC–tandem mass spectrometry (MS/MS) exhibit excellent ionization properties via electrospray ionization and do not require derivatization of compounds, and therefore the specific fragmentation patterns of these compounds allow selective and sensitive analysis by MS/MS [1–4,6,7,10,13–25].

Sulfated steroids have been measured in plasma, serum, and urine samples. However, blood collection is invasive and may itself induce stress, whereas urine collection is simple, but urinary concentrations of compounds are affected by the volume of urine excreted. In contrast, saliva can be easily collected non-invasively from subjects of all ages, from children to the elderly, and collection devices are relatively inexpensive. Furthermore, the concentrations of sulfated steroids in saliva are highly correlated with their concentrations in plasma or serum [10,23]. Because sulfated steroid content is lower in saliva than in serum, tedious and laborious pretreatment operations such as organic solvent extraction [23] and solid-phase extraction [10,20] are essential to separate and extract target analytes from the samples.

In-tube solid-phase microextraction (IT-SPME) is a method by which samples can be easily extracted and concentrated using open-tube fused silica capillaries with coated inner surfaces as extraction devices, followed by online coupling to LC and LC–MS online using column switching technique (Figure S1). The entire process, from sample extraction/concentration to separation, detection, and data analysis, can be fully automated [26–28]. This method has been used to develop online analytical systems for a variety of compounds [28]. In addition, highly sensitive analytical methods were developed to determine the concentrations of non-sulfated steroid hormones in urine and saliva samples [29–32].

The aim of this study was to establish a fully automated online simultaneous analysis system, consisting of IT-SPME coupled with stable isotope dilution LC–MS/MS, for four sulfated steroid metabolites (Figure S2), pregnenolone sulfate (PREGS), DHEAS, 17β-estradiol 3-sulfate (E2S), and cortisol 21-sulfate (CRTS), which act as neuroactive steroids [4,8,11,33], and apply this system to the non-invasive analysis of these compounds in saliva samples.

2. Results and Discussion

2.1. Optimization of IT-SPME and Desorption of Sulfated Steroid Metabolites

An IT-SPME system that uses a capillary column as an extraction device involves the online extraction of the compounds of interest on the capillary column and the online desorption of these compounds by switching the draw/ejection flow of the sample solution and mobile phase flow (Figure S1). Extraction efficiency is mainly affected by the type of capillary coating, the number and flow rate of draw/eject cycles, and sample pH. Based on previous findings [31,32], these IT-SPME conditions were optimized for 1 ng mL^{-1} each of PREGS, DHEAS, and CRTS and 5 ng mL^{-1} of E2S. As PLOT columns including Supel-Q and Carboxen 1006 have a larger adsorption surface area and thicker film layer, the amounts extracted were greater than those for other liquid-phase columns (Figure 1A). Supel-Q has a higher affinity for sulfated steroids with cyclic skeletons than Carboxen 1006, which is a carbon molecular sieve, due to its divinylbenzene structure. All four sulfated steroid metabolites were efficiently extracted into a Supel-Q PLOT capillary by more than 25 repeated draw/eject cycles of 40 µL sample at a flow rate of 0.2 mL min^{-1} (Figure 1B). The length of the capillary is dependent on the draw/eject volume of the sample and is an important factor affecting extraction efficiency and time. However, capillaries that are too long and sample volumes that are too large will increase band width and require more time. Comparisons showed that, for a draw/eject volume of 40 µL of sample, a 60 cm long capillary with an inner diameter of 0.32 mm was optimal. In contrast, adjustment of sample pH can improve the distribution coefficient of compounds by suppressing their ionization. The optimal pH for non-sulfated steroid hormones in the previous reports [29,31,32] was 4, but PREGS, DHEAS, CRTS, and E2S contain sulfate groups, and therefore all four should be extracted into the capillary stationary phase at a more acidic pH. A pH that is too low, however, may cause damage to the extraction coating, affecting its service life and enrichment effect. Among the pH 2–9 buffers tested, potassium hydrogen phthalate–HCl buffer (pH 3) was found to be the most effective (Figure 2).

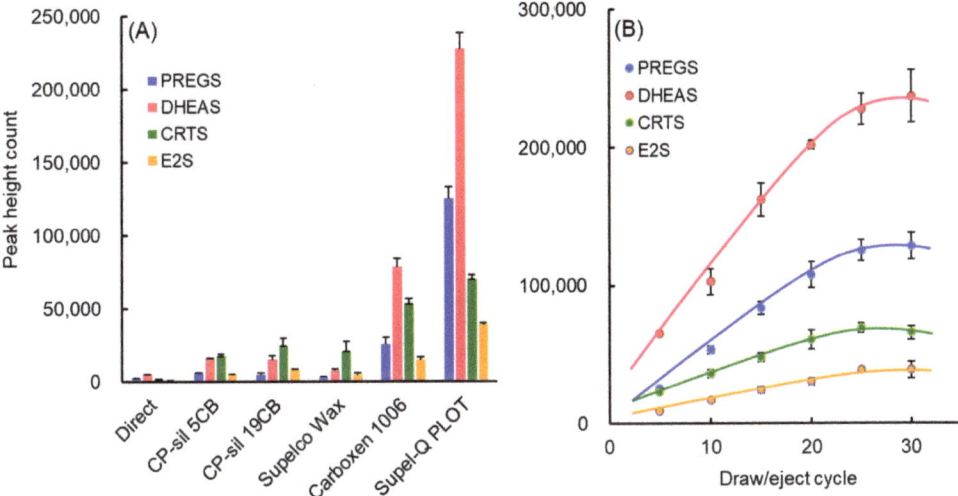

Figure 1. Effects of (**A**) capillary coatings and (**B**) number of draw/eject cycles on the IT-SPME of sulfated steroid metabolites. Standard solution containing 1 ng mL^{-1} each of PREGS, DHEAS, and CRTS and 5 ng mL^{-1} of E2S were extracted by (**A**) 25 draw/eject cycles of 40 µL of standard solution at a flow rate of 200 µL min^{-1}, and (**B**) the indicated number of draw/eject cycles of 40 µL of standard solution on a Supel-Q PLOT capillary at a flow rate of 200 µL min^{-1}.

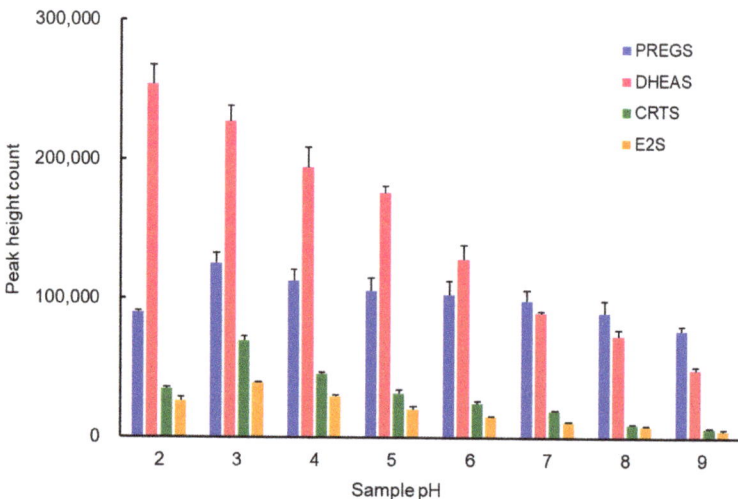

Figure 2. Effects of sample pH on the IT-SPME of sulfated steroid metabolites. Standard solution (40 µL) containing 1 ng mL^{-1} each of PREGS, DHEAS, and CRTS and 5 ng mL^{-1} of E2S were extracted by 25 draw/eject cycles on a Supel-Q PLOT capillary at a flow rate of 200 µL min^{-1}.

The absolute extractable amounts of sulfated steroid metabolites onto the capillary column were calculated by comparing peak area counts with the corresponding amounts in standard solution directly injected onto the LC columns. Using a 1 mL standard solution containing 1.0 ng mL^{-1} of each compound, the average extraction yields (n = 3) of PREGS, DHEAS, CRTS, and E2S onto the Supel-Q PLOT capillary column were found to be 56.6 ± 3.5%, 45.0 ± 2.2%, 47.3 ± 2.4%, and 39.7 ± 0.8%, respectively. The sulfated steroid metabolites extracted into the capillary column were dynamically desorbed and introduced directly into the LC column by online mobile phase flow using a column switching system. After analysis, the capillary column was cleaned and conditioned with methanol and mobile phase flow prior to the next analysis, thereby allowing its repeated use without carryover. All of these operations were programmed and automated (Table S1).

2.2. LC–MS/MS Analysis of Sulfated Steroid Metabolites and Their Stable Isotope-Labeled Compounds

Four sulfated steroid metabolites, along with their stable isotope-labeled compounds as internal standards (ISs), exhibited abundant deprotonated ions [M−H]$^-$ (Q1 mass) in the ESI-negative ionization mode. The [HSO$_4$]$^-$ ion (m/z 97) and [SO$_3$]$^-$ ion (m/z 80) formed by cleavage of the [M−H]$^-$ of each compound were detected as fragment ions (Q3 mass). For each precursor ion [M−H]$^-$, the m/z 97 for PREGS, DHEAS, CRTS, and their stable isotope-labeled compounds and the m/z 80 for E2S and E2S-d$_4$ were selected as product ions, and the MS/MS operating parameters were optimized. The MRM transitions and MS/MS parameters set for each compound are shown in Table S2. These findings were in good agreement with previously reported results [13,19,23].

The four sulfated steroid metabolites and their IS compounds were separated by LC on an InertSustain Swift C18 column. Chromatographic conditions were optimized by focusing on short retention times, paying special attention to matrix effects as well as peak shapes. Optimal separation was achieved using water/acetonitrile (55/45, v/v), with a flow rate of 0.2 mL min^{-1} resulting in good peak shapes and selective detection in MRM mode with a runtime of 6 min (Figure 3). The CV% of the retention time for each compound was within 5%. The analysis time per sample was about 23 min, allowing automated analysis of about 60 samples per day by operating overnight.

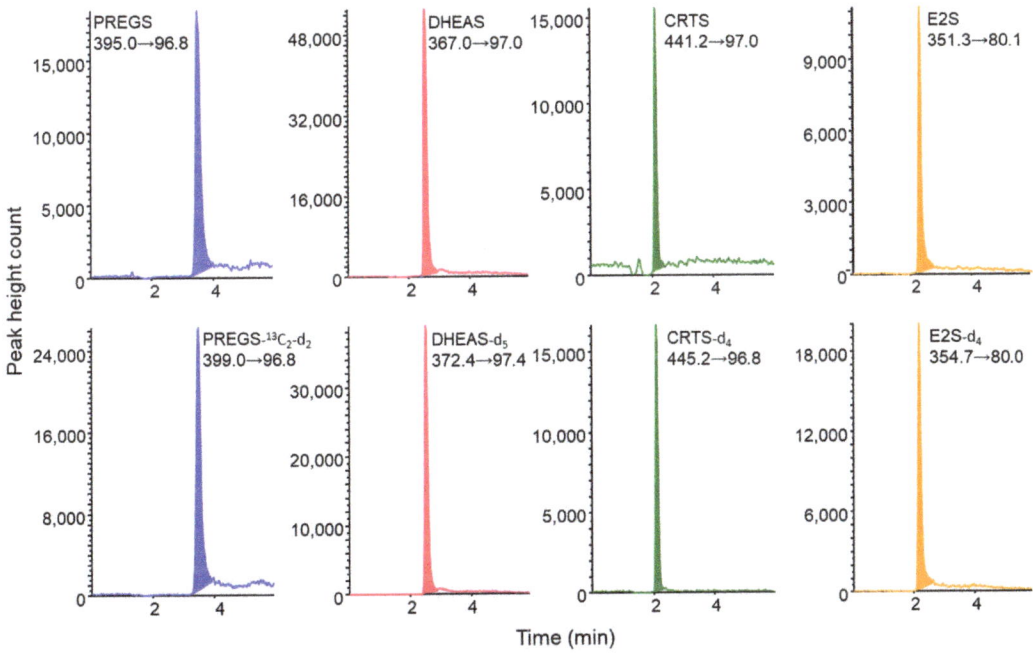

Figure 3. Multiple reaction monitoring (MRM) chromatograms obtained from standard solution containing 0.2 ng mL^{-1} each of PREGS, DHEAS, and CRTS and 1 ng mL^{-1} of E2S, and their stable isotope-labeled compounds. IT-SPME LC–MS/MS conditions are described in Section 3.

2.3. Analytical Method Validation and Advantages of IT-SPME LC–MS/MS Method

The analytical method was validated based on generally accepted validation criteria recommended in the ICH guidelines [34]. Linearity for PREGS, DHEAS and CRTS was validated by triplicate analyses of each compound at eight concentrations (0.01, 0.02, 0.05, 0.1, 0.2, 0.5, 1.0, and 2.0 ng mL^{-1}), in the presence of 0.2 ng mL^{-1} each of PREGS-^{13}C$_2$-d$_2$, DHEAS-d$_5$, and CRTS-d$_4$, respectively, whereas linearity for E2S was validated at concentrations of 0.05, 0.1, 0.2, 0.5, 1.0, 2.0, 5.0, and 10.0 ng mL^{-1} in the presence of 1.0 ng mL^{-1} E2S-d$_4$. Calibration curves for each compound were linear, with correlation coefficients above 0.9998 (Table 1). The CVs of the peak height ratios at each concentration ranged from 0.7% to 12.6% (n = 3).

Table 1. Linearity and sensitivity of the IT-SPME LC–MS/MS method for sulfated steroid metabolites.

Compound	Linearity		LOD [2] (pg mL^{-1})		LOQ [3] (pg mL^{-1})
	Range (ng mL^{-1})	CC [1]	Direct Injection	IT-SPME	IT-SPME
PREGS	0.01–2	0.99991	50.6	0.59	28
DHEAS	0.01–2	0.99994	23.9	0.30	16
CRTS	0.01–2	0.99987	68.4	0.80	47
E2S	0.05–10	0.99995	245.4	3.20	172

[1] Correlation coefficient (n = 24); [2] limits of detection: pg mL^{-1} sample solution (signal-to-noise ratio of 3); [3] limits of quantification: pg mL^{-1} saliva sample (signal-to-noise ratio of 10).

PREGS, DHEAS, CRTS, and E2S gave superior responses in MRM mode detection, with the LODs ($S/N = 3$) of each sulfated steroid metabolite in standard solutions ranging from 0.30 to 3.20 pg mL^{-1} (Table 1). The IT-SPME method was about 76-fold more sensitive than the direct injection method (10 µL injections).

Precision was assessed at concentrations of 0.05, 0.2, and 1.0 ng mL^{-1} for PREGS, DHEAS, and CRTS and of 0.25, 1.0, and 5.0 ng mL^{-1} for E2S. The precision, expressed as CV (%), was validated by performing five independent analyses on the same day and on five different days. The intraday and interday variations of these analyses were found to be 2.1–7.7% and 4.0–11.1%, respectively (Table 2).

Table 2. Precision of the IT-SPME LC–MS/MS method for sulfated steroid metabolites.

Compound	Concentration (ng mL^{-1})	Precision (CV [1] %), (n = 5)	
		Intra-Day	Inter-Day
PREGS	0.05	2.3	9.0
	0.2	2.1	4.0
	1	3.0	6.2
DHEAS	0.05	6.9	10.7
	0.2	3.1	7.8
	1	3.1	6.1
CRTS	0.05	7.7	11.1
	0.2	2.4	4.2
	1	2.6	6.8
E2S	0.25	5.6	7.7
	1	3.6	7.9
	5	2.7	5.2

[1] CV, coefficient variation.

These results obtained based on the ICH guidelines showed that the IT-SPME LC–MS/MS method has good linearity and precision. The method is simpler, more sensitive, and more specific than previously reported methods [6,7,16–21,25] and can be fully automated from extraction and concentration of the four sulfated steroid metabolites to their separation and analysis.

2.4. Application to the Analysis of Saliva Samples

Because the concentrations of steroid hormones and their metabolites in saliva have been reported to reflect their free concentrations in plasma or serum [10,23], saliva is regarded as an excellent physiological medium for non-invasive sampling [10,35,36]. Therefore, saliva analysis can allow, in particular cases (i.e., patients with difficult blood collection), the replacement of blood analysis [36]. In general, however, procedures for the collection, handling, and storage of salivary samples should be considered, because the levels of biomarkers in saliva are affected by factors such as sex, age, smoking, diet, circadian rhythm, and more [10,35,36]. The most commonly used saliva collection procedures are passive drooling, the use of cotton swabs, aspiration through soft devices positioned under the tongue, and chewing of paraffin gum [35,36], but the choice of sampling tube type (glass or polystyrene) has no effect on salivary steroid concentration [10]. Since some biomarkers in saliva are unstable compounds, they need to be cooled after collection and frozen if not analyzed immediately [10,35,36]. In this study, saliva samples were collected into Salisoft tubes containing polypropylene–polyethylene swab, followed by ultrafiltration with Amicon Ultra to remove high-molecular-weight components in saliva, such as mucins and other coexisting proteins. The polymeric saliva collectors, such as Salisoft, have a better recovery rate of steroid hormones compared to cotton saliva collectors [10,32,37]. Furthermore, the 30K (molecular weight 30,000 cutoff) centrifugal filter used for ultrafiltration of saliva samples in the previous report [32] was replaced with a 3K (molecular weight 3000 cutoff) filter to reduce matrix effects.

Stable isotope-labeled compounds as ISs were added to saliva samples prior to extraction to correct the influence of matrix effects on the analysis of sulfated steroid metabolites in the samples. As shown in Figure 4, the saliva samples were successfully analyzed without interference peaks by the IT-SPME LC–MS/MS method with MRM mode detection. The LOQs (S/N = 10) of these sulfated steroid metabolites ranged from 16 to 172 pg mL^{-1} saliva (Table 1). In comparison, the LOQs for DHEAS by previously reported LC–MS/MS methods ranged from 0.06 to 1.14 ng mL^{-1} saliva [20,23], indicating that the IT-SPME LC–MS/MS method was 3.7 times more sensitive than these methods.

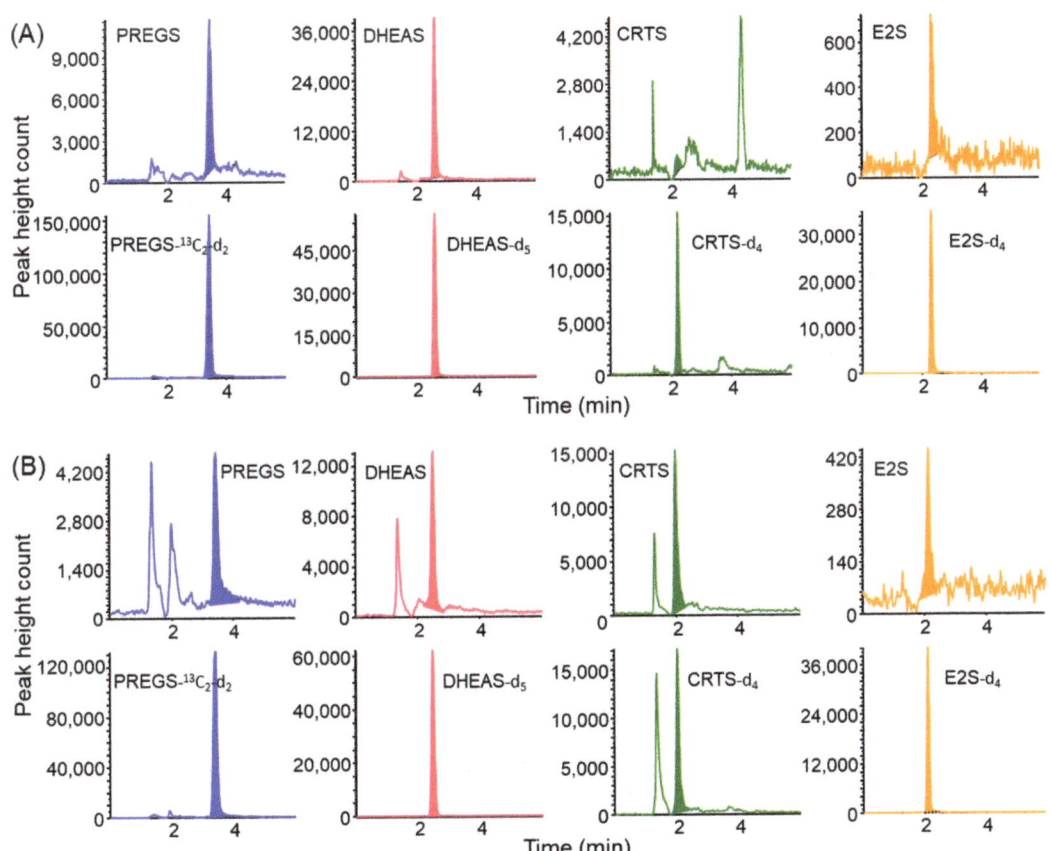

Figure 4. MRM chromatograms obtained from saliva samples of (**A**) a non-smoker and (**B**) a smoker by IT-SPME LC–MS/MS. Analytical conditions are described in Section 3.

To confirm the validity and accuracy of this method, known amounts of PREGS, DHEAS, CRTS, and E2S were spiked into saliva samples, and their recoveries were calculated. The overall recoveries of PREGS, DHEAS, CRTS, and E2S were above 86%, with relative standard deviations of 0.8–9.7% (Table 3). These results show that the IT-SPME LC–MS/MS method has good accuracy and precision and is fully applicable to saliva samples.

Table 3. Recoveries of sulfated steroid metabolites spiked into saliva samples.

Compound	Spiked (ng mL^{-1} Saliva)	Recovery ± SD (%), (n = 3)
PREGS	1.0	87.6 ± 5.3
	4.0	90.1 ± 4.6
	20	112.9 ± 5.8
DHEAS	1.0	86.3 ± 2.4
	4.0	91.4 ± 5.2
	20	93.0 ± 0.7
CRTS	1.0	98.7 ± 9.5
	4.0	96.5 ± 7.0
	20	98.9 ± 3.4
E2S	5.0	86.6 ± 3.8
	20	105.5 ± 2.7
	100	106.5 ± 5.3

The developed method was used to analyze the concentrations of the four sulfated steroid metabolites in saliva samples from 10 male and 10 female subjects. PREGS and DHEAS were detected in all saliva samples, with DHEAS being present at relatively high concentrations (Table 4). In contrast, CRTS and E2S were often below the LOQ because of interference by coexisting peaks, even if peaks were detected. Salivary DHEAS concentrations ranged from 0.36 to 11.9 ng mL^{-1} in males and from 0.05 to 4.8 ng mL^{-1} in females, with lower concentrations in children than in adults. These results are similar to findings showing that DHEAS concentrations peak at around ages 20 to 30 years in both men and women and decrease subsequently with age [4,6].

Table 4. Contents of sulfated steroid metabolites in saliva samples.

Subject			Content (pg mL^{-1} Saliva), (n = 3)			
No.	Sex [1]	Age	PREGS	DHEAS	CRTS	E2S
1	M	6	45 ± 6	1068 ± 79	<LOQ	<LOQ
2	M	7	33 ± 1	869 ± 9	<LOQ	<LOQ
3	M	23	170 ± 14	1914 ± 146	187 ± 15	509 ± 1
4	M	24	128 ± 2	3894 ± 229	114 ± 5	306 ± 7
5	M	25	149 ± 15	5139 ± 71	314 ± 10	466 ± 12
6	M	35	48 ± 6	8272 ± 334	<LOQ	<LOQ
7	M	38	52 ± 2	6022 ± 25	<LOQ	<LOQ
8	M	40	72 ± 1	11,908 ± 730	873 ± 37	<LOQ
9	M	57	86 ± 1	570 ± 35	3215 ± 306	276 ± 21
10	M	67	32 ± 3	365 ± 38	<LOQ	174 ± 12
11	F	4	40 ± 3	47 ± 2	295 ± 18	184 ± 32
12	F	6	44 ± 3	69 ± 5	369 ± 14	179 ± 30
13	F	27	64 ± 0	1729 ± 85	<LOQ	<LOQ
14	F	29	109 ± 19	1244 ± 75	<LOQ	174 ± 13
15	F	30	46 ± 3	4415 ± 8	<LOQ	<LOQ
16	F	33	44 ± 2	4783 ± 120	200 ± 12	<LOQ
17	F	34	41 ± 4	129 ± 8	<LOQ	175 ± 9
18	F	36	90 ± 4	4607 ± 73	<LOQ	<LOQ
19	F	62	46 ± 1	1418 ± 26	<LOQ	<LOQ
20	F	64	28 ± 2	1217 ± 102	<LOQ	<LOQ

[1] M, male; F, female.

3. Materials and Methods

3.1. Reagents and Standard Solutions

PREGS sodium salt and E2S sodium salt were purchased from Sigma-Aldrich Japan (Tokyo, Japan), DHEAS hydrate was from Tokyo Kasei Kogyo (Tokyo, Japan), and CRTS potassium salt was from Toronto Research Chemicals Inc. (TRC, North York, ON, Canada).

Their stable isotope-labeled compounds, PREGS-$^{13}C_2$-d_2 (isotopic purity >98%, Sigma-Aldrich, St. Louis, MA, USA), DHEAS-d_5 (isotopic purity >90%, Sigma-Aldrich), CRTS-d_4 (isotopic purity >95%, TRC), and E2S-d_4 (isotopic purity 95%, TRC), were used as internal standards (IS). These standard and IS compounds (Figure S2) were dissolved in LC–MS-grade methanol to a concentration of 0.1 mg mL^{-1} and diluted with LC–MS-grade distilled water to the required concentration prior to use. The mixed standard solution consisted of 20 ng mL^{-1} each of PREGS, DHEAS, and CRTS and 100 ng mL^{-1} E2S, whereas the mixed IS solution consisted of 2 ng mL^{-1} each of PREGS-$^{13}C_2$-d_2, DHEAS-d_5, and CRTS-d_4 and 10 ng mL^{-1} of E2S-d_4. All of these solutions were stored at 4 °C. LC–MS-grade acetonitrile and distilled water used as mobile phases were purchased from Kanto Chemical (Tokyo, Japan); all other chemicals were of analytical reagent grade.

3.2. Preparation of Saliva Samples

Saliva samples were obtained from 20 healthy volunteers (10 men and 10 women). The experimental protocol was approved by the ethics committee of Shujitsu University (approval code 207; 14 October 2020), and all volunteers provided written informed consent. Saliva samples were collected in Salisoft tubes (Assist, Tokyo, Japan). The tubes were centrifuged at 2500× g for 1 min to elute the saliva solution. If not immediately used for analysis, the samples were stored frozen at −20 °C and thawed spontaneously just before analysis. A 0.1 mL aliquot of mixed IS solution was added to 0.05 mL of each saliva sample, followed by ultrafiltration using an Amicon Ultra 0.5 mL 3K (Millipore, Tullagreen, Ireland) regenerated cellulose 3000 molecular weight cutoff centrifugal filter device at 15,000 rpm for 20 min. Each filtrate was pipetted into a 2.0 mL autosampler vial with septa, to which was added 0.05 mL of 0.2 M potassium hydrogen phthalate–HCl buffer (pH 3). The total volume was made up to 1.0 mL with distilled water, and the vials were set into the autosampler for IT-SPME LC–MS/MS analysis. The concentrations of the sulfated steroid metabolites in saliva were calculated using calibration curves constructed from the ratios of the peak heights of each sulfated steroid metabolite to the peak height of their IS compounds.

3.3. LC–MS/MS Analysis

LC–MS/MS analysis was performed using an Agilent Technologies (Boeblingen, Germany) Model 1100 series LC system and an Applied Biosystems (Foster City, CA, USA) API 4000 triple quadrupole mass spectrometer, with LC separation on an InertSustain swift C18 column (100 mm × 2.1 mm, particle size 5 µm; GL Sciences, Tokyo, Japan). The LC conditions included a column temperature of 30 °C, a mobile phase consisting of distilled water/acetonitrile (55/45, v/v), and a flow rate of 0.2 mL min^{-1}. Electrospray ionization (ESI)–MS/MS conditions included: a turbo ion spray voltage of −4500 V; a turbo ion spray temperature of 450 °C; ion source gas GS1 and GS2 flows of 20 and 11 L min^{-1}, respectively; a curtain gas (CUR) flow of 10 L mL^{-1}; and a collision gas (CAD) flow of 4.0 L min^{-1}. Multiple reaction monitoring (MRM) transitions in negative ion mode and other parameters, including dwell time, declustering potential (DP), entrance potential (EP), collision energy (CE), and collision cell exit potential (CXP), are shown in Supplementary Table S2. Quantification was performed by MRM of the deprotonated precursor molecular ions $[M-H]^-$ and the related product ions for each compound. Quadrupoles Q1 and Q3 were set at unit resolution (Table S2). Analyst Software 1.6.2 (Applied Biosystems) was used for LC–MS/MS data analysis.

3.4. In-Tube SPME

IT-SPME was essentially performed as described in our previous works [31,32]. A GC capillary column (60 cm × 0.32 mm i.d.) as an extraction device was connected between the injection needle and injection loop of the autosampler. The capillary column was threaded through a 1/16 inch polyetheretherketone (PEEK) tube with a length of 2.5 cm long and an inner diameter of 330 µm and connected using standard 1/16 inch stainless steel nuts, ferrules, and connectors. CP-Sil 5CB (100% polydimethylsiloxane, film thickness 5 µm),

CP-Sil 19CB (14% cyanopropyl phenyl methylsilicone, film thickness 1.2 µm) (Varian Inc., Lake Forest, CA, USA), Supelco-Wax (polyethylene glycol, film thickness 1.0 µm), Supel-Q PLOT (divinylbenzene polymer, film thickness 17 µm), and Carboxen 1006 PLOT (carbon molecular sieve, film thickness 15 µm) (Supelco, Bellefonte, PA, USA) were used to compare extraction efficiencies. Extraction, desorption, and injection parameters were programmed by the autosampler software (Table S1) [31,32].

4. Conclusions

In this study, we succeeded for the first time in efficiently extracting and concentrating highly polar sulfate conjugates by IT-SPME at acidic pH, and we constructed an automated analysis system coupled online with LC–MS/MS to enable selective and sensitive simultaneous analysis of four sulfated steroid metabolites. The method is easy to apply to the analysis of small volumes of saliva samples without tedious pretreatment except for ultrafiltration. This method may be a useful tool in analyzing the regulation of steroid metabolism and in determining the diagnosis of related diseases.

Supplementary Materials: The following supporting information can be downloaded at: https://www.mdpi.com/article/10.3390/molecules27103225/s1, Figure S1: Schematic diagrams of the automated online IT-SPME/LC–MS/MS system; Figure S2: Structures of sulfated steroid metabolites and their respective stable isotope-labeled compounds as internal standards; Table S1: Program for the IT-SPME process; Table S2: MRM transitions and setting parameters for sulfated steroid metabolites and their stable isotope-labeled compounds.

Author Contributions: Conceptualization, H.K.; methodology, D.N. and H.K.; software, D.N. and H.K.; validation, D.N. and H.K.; formal analysis, D.N. and H.K.; investigation, D.N. and H.K.; resources, D.N. and H.K.; data curation, D.N. and H.K.; writing—original draft preparation, H.K.; writing—review and editing, H.K.; visualization, D.N. and H.K.; supervision, H.K.; project administration, H.K.; funding acquisition, H.K. All authors have read and agreed to the published version of the manuscript.

Funding: This research was funded by a Grant-in-Aid for Basic Scientific Research (C, No. 20K07007) and Smoking Research Foundation (2021).

Institutional Review Board Statement: The study was conducted according to the guidelines of the Declaration of Helsinki and approved by the ethics committee of Shujitsu University (approval code 207; 14 October 2020).

Informed Consent Statement: Informed consent was obtained from all subjects involved in the study.

Data Availability Statement: The data presented in this study are available on request from the corresponding author.

Conflicts of Interest: The authors declare no conflict of interest.

Sample Availability: Samples of the compounds are not available from the authors.

References

1. Schiffer, L.; Barnard, L.; Baranowski, E.S.; Gilligan, L.C.; Taylor, A.E.; Arlt, W.; Shackleton, C.H.L.; Storbecka, K.-H. Human steroid biosynthesis, metabolism and excretion are differentially reflected by serum and urine steroid metabolomes: A comprehensive review. *J. Steroid Biochem. Mol. Biol.* **2019**, *194*, 105439. [CrossRef] [PubMed]
2. Karashima, S.; Osaka, I. Rapidity and precision of steroid hormone measurement. *J. Clin. Med.* **2022**, *11*, 956. [CrossRef] [PubMed]
3. Temerdashev, A.; Dmitrieva, E.; Podolskiy, I. Analytics for steroid hormone profiling in body fluids. *Microchem. J.* **2021**, *168*, 106395. [CrossRef]
4. Mueller, J.W.; Gilligan, L.C.; Idkowiak, J.; Arlt, W.; Foster, P.A. The regulation of steroid action by sulfation and desulfation. *Endocr. Rev.* **2015**, *36*, 526–563. [CrossRef] [PubMed]
5. Geyer, J.; Bakhaus, K.; Bernhardt, R.; Blaschka, C.; Dezhkam, Y.; Fietz, D.; Grosser, G.; Hartmann, K.; Hartmann, M.F.; Neunzig, J.; et al. The role of sulfated steroid hormones in reproductive processes. *J. Steroid Biochem. Mol. Biol.* **2017**, *172*, 207–221. [CrossRef]
6. Sánchez-Guijo, A.; Oji, V.; Hartmann, M.F.; Traupe, H.; Wudy, S.A. Simultaneous quantification of cholesterol sulfate, androgen sulfates, and progestagen sulfates in human serum by LC-MS/MS. *J. Lipid Res.* **2015**, *56*, 1843–1851. [CrossRef]

7. Lee, S.H.; Kim, S.H.; Lee, W.-Y.; Chung, B.C.; Park, M.J.; Choi, M.H. Metabolite profiling of sex developmental steroid conjugates reveals an association between decreased levels of steroid sulfates and adiposity in obese girls. *J. Steroid Biochem. Mol. Biol.* **2016**, *162*, 100–109. [CrossRef]
8. Stárka, L.; Dušková, M.; Hill, M. Dehydroepiandrosterone: A neuroactive steroid. *J. Steroid Biochem. Mol. Biol.* **2015**, *145*, 254–260. [CrossRef]
9. Klinge, C.M.; Clark, B.J.; Prough, R.A. Dehydroepiandrosterone research: Past, current, and future. *Vitam. Horm.* **2018**, *108*, 1–28. [CrossRef]
10. Giacomello, G.; Scholten, A.; Parr, M.K. Current methods for stress marker detection in saliva. *J. Pharm. Biomed. Anal.* **2020**, *191*, 113604. [CrossRef]
11. Pérez-Jiménez, M.M.; Monje-Moreno, J.M.; Brokate-Llanos, A.M.; Venegas-Calerón, M.; Sánchez-García, A.; Sansigre, P.; Valladares, A.; Esteban-García, S.; Suárez-Pereira, I.; Vitorica, J.; et al. Steroid hormones sulfatase inactivation extends lifespan and ameliorates age-related diseases. *Nat. Commun.* **2021**, *12*, 49. [CrossRef] [PubMed]
12. Jiang, X.; Zhong, W.; An, H.; Fu, M.; Chen, Y.; Zhang, Z.; Xiao, Z. Attenuated DHEA and DHEA-S response to acute psychosocial stress in individuals with depressive disorders. *J. Affect. Disord.* **2017**, *215*, 118–124. [CrossRef] [PubMed]
13. Wudy, S.A.; Schuler, G.; Sánchez-Guijo, A.; Hartmann, M.F. The art of measuring steroids: Principles and practice of current hormonal steroid analysis. *J. Steroid Biochem. Mol. Biol.* **2018**, *179*, 88–103. [CrossRef] [PubMed]
14. Gomez, C.; Fabregat, A.; Pozo, Ó.J.; Marcos, J.; Segura, J.; Ventura, R. Analytical strategies based on mass spectrometric techniques for the study of steroid metabolism. *Trends Anal. Chem.* **2014**, *53*, 106–116. [CrossRef]
15. Yan, Y.; Rempel, D.L.; Holy, T.E.; Gross, M.L. Mass spectrometry combinations for structural characterization of sulfated-steroid metabolites. *J. Am. Soc. Mass Spectrom.* **2014**, *25*, 869–879. [CrossRef]
16. Khelil, M.B.; Tegethoff, M.; Meinlschmidt, G.; Jamey, C.; Ludes, B.; Raul, J.-S. Simultaneous measurement of endogenous cortisol, cortisone, dehydroepiandrosterone, and dehydroepiandrosterone sulfate in nails by use of UPLC-MS-MS. *Anal. Bioanal. Chem.* **2011**, *401*, 1153–1162. [CrossRef]
17. Galuska, C.E.; Hartmann, M.F.; Sánchez-Guijo, A.; Bakhaus, K.; Geyer, J.; Schuler, G.; Zimmer, K.-P.; Wudy, S.A. Profiling intact steroid sulfates and unconjugated steroids in biological fluids by liquid chromatography-tandem mass spectrometry (LC-MS-MS). *Analyst* **2013**, *138*, 3792–3801. [CrossRef]
18. Shibata, Y.; Arai, S.; Honma, S. Methodological approach to the intracrine study and estimation of DHEA and DHEA-S using liquid chromatography-tandem mass spectrometry (LC-MS/MS). *J. Steroid Biochem. Mol. Biol.* **2015**, *145*, 193–199. [CrossRef]
19. Lee, S.H.; Lee, N.; Hong, Y.; Chung, B.C.; Choi, M.H. Simultaneous analysis of free and sulfated steroids by liquid chromatography/mass spectrometry with selective mass spectrometric scan modes and polarity switching. *Anal. Chem.* **2016**, *88*, 11624–11630. [CrossRef]
20. Gaudl, A.; Kratzsch, J.; Bae, Y.J.; Kiess, W.; Thiery, J. Liquid chromatography quadrupole linear ion trap mass spectrometry for quantitative steroid hormone analysis in plasma, urine, saliva and hair. *J. Chromatogr. A* **2016**, *1464*, 64–71. [CrossRef]
21. Li, Y.; Wang, D.; Zeng, C.; Liu, Y.; Huang, G.; Mei, Z. Salivary metabolomics profile of patients with recurrent aphthous ulcer as revealed by liquid chromatography-tandem mass spectrometry. *J. Int. Med. Res.* **2018**, *46*, 1052–1062. [CrossRef] [PubMed]
22. Esquivel, A.; Alechaga, É.; Monfort, N.; Ventura, R. Direct quantitation of endogenous steroid sulfates in human urine by liquid chromatography-electrospray tandem mass spectrometry. *Drug Test Anal.* **2018**, *10*, 1734–1743. [CrossRef] [PubMed]
23. Cao, Z.T.; Wemm, S.E.; Liqiao David, H.; Spink, C.; Wulfert, E. Noninvasive determination of human cortisol and dehydroepiandrosterone sulfate using liquid chromatography-tandem mass spectrometry. *Anal. Bioanal. Chem.* **2019**, *411*, 1203–1210. [CrossRef] [PubMed]
24. Olisov, D.; Lee, K.; Jun, S.-H.; Song, S.H.; Kim, J.H.; Lee, Y.A.; Shin, C.H.; Song, J. Measurement of serum steroid profiles by HPLC-tandem mass spectrometry. *J. Chromatogr. B Analyt. Technol. Biomed. Life Sci.* **2019**, *1117*, 1–9. [CrossRef] [PubMed]
25. Rustichelli, C.; Monari, C.; Avallone, R.; Bellei, E.; Bergamini, S.; Tomasi, A.; Ferrari, A. Dehydroepiandrosterone sulfate, dehydroepiandrosterone, 5α-dihydroprogesterone and pregnenolone in women with migraine: Analysis of serum levels and correlation with age, migraine years and frequency. *J. Pharm. Biomed. Anal.* **2021**, *206*, 114388. [CrossRef]
26. Kataoka, H. Automated sample preparation using in-tube solid-phase microextraction and its application—A review. *Anal. Bioanal. Chem.* **2002**, *373*, 31–45. [CrossRef]
27. Kataoka, H.; Ishizaki, A.; Nonaka, Y.; Saito, K. Developments and applications of capillary microextraction techniques: A review. *Anal. Chim. Acta* **2009**, *655*, 8–29. [CrossRef]
28. Kataoka, H. In-tube solid-phase microextraction: Current trends and future perspectives. *J. Chromatogr. A* **2021**, *1636*, 461787. [CrossRef]
29. Kataoka, H.; Matsuura, E.; Mitani, K. Determination of cortisol in human saliva by automated in-tube solid-phase microextraction coupled with liquid chromatography-mass spectrometry. *J. Pharm. Biomed. Anal.* **2007**, *44*, 160–165. [CrossRef]
30. Saito, K.; Yagi, K.; Ishizaki, A.; Kataoka, H. Determination of anabolic steroids in human urine by automated in-tube solid-phase microextraction coupled with liquid chromatography-mass spectrometry. *Pharm. Biomed. Anal.* **2010**, *52*, 727–733. [CrossRef]
31. Yasuhara, R.; Ehara, K.; Saito, K.; Kataoka, H. Automated analysis of salivary stress-related steroid hormones by online in-tube solid-phase microextraction coupled with liquid chromatography–tandem mass spectrometry. *Anal. Methods* **2012**, *4*, 3625–3630. [CrossRef]

Passive smoking not only increases the risk of cancer [9–15] but has also been reported to be associated with the development of cardiovascular disease [16–23], diabetes [6,24,25], metabolic syndrome [26], psychiatric disorders (depression) [27] and cognitive decline [28]. This is a serious problem associated with a variety of adverse health effects, especially for non-smoking children and pregnant women [29,30]. For example, health hazards caused by passive smoking in children are more severe than in adults, bringing increased risk of growth retardation [31,32] and sudden infant death syndrome (SIDS) [33,34] in fetuses born to mothers exposed to tobacco smoke. In addition, passive smoking in infancy has been linked to obesity, diabetes and metabolic syndrome in adulthood [35,36]. Since passive smoking is unintentional and unavoidable in the presence of smokers in the close relatives, there is an urgent need for effective measures to reduce passive smoking worldwide from both public health and clinical aspects [29]. Therefore, it is essential to determine the facts of biological exposure to prevent health hazards caused by passive smoking [1,3,37], however, the current status and effects of passive smoking in various lifestyle behaviors in different environments are not fully understood.

Tobacco-related compounds such as nicotine, its metabolite cotinine and tobacco-specific nitrosamines have been used as biomarkers of exposure to tobacco smoke [6]. However, their half-lives in urine and blood range from a few hours to 3–4 days, and these compounds are rapidly eliminated after exposure to tobacco smoke [6]. In addition, their concentrations in these matrices are much lower in passive smokers than in active smokers, making them unsuitable for assessing the effects of long-term exposure to environmental tobacco smoke in non-smokers. In contrast, hair samples have been frequently used to assess and monitor the bioaccumulation due to long-term exposure to drugs, environmental pollutants, etc. [38–41], as the amounts of these compounds in the hair are less affected by daily exposure and variations in metabolism than those in other biological matrices [6,42]. Furthermore, hair samples can be collected easily and less invasively and can be stored at room temperature for up to five years [6,43]. Recently, we developed a simple and sensitive method for the determination of nicotine and cotinine [44,45] and tobacco-specific nitrosamines [46] in hair by in-tube solid-phase microextraction (SPME) with on-line coupling to liquid chromatography-tandem mass spectrometry (LC-MS/MS) [47–49] and assessed differences in levels of these compounds in active and passive smokers. In the present study, we measured hair nicotine and cotinine levels in healthy non-smokers and asked self-reported lifestyle questions to assess the risk from the actual status of passive smoking in different lifestyle environments.

2. Results

The automated on-line in-tube SPME/LC−MS/MS method [45] is sufficiently selective, sensitive and precise for nicotine and cotinine analysis (Table S1). This method was successfully applied to the analysis of nicotine and cotinine at pg levels per 1 mg of hair samples without any interference peaks. Typical chromatograms obtained from 200 pg mL^{-1} standard solution and a sample corresponding to 0.2 mg of hair were shown in Figure 1. In this study, the analysis method is the same as in the previous paper [45], but the number of non-smokers was increased from 58 to 110, and a questionnaire on lifestyle was administered to analyze the relationship with the exposure level. In a previous study [45], the levels of nicotine and cotinine in the hair of eight smokers were 43.12 ng mg^{-1} and 655 pg mg^{-1}, respectively, and those of 20 non-smokers (not even passive smokers) were 1.86 ng/mg and 8 pg mg^{-1}, respectively. The levels of nicotine and cotinine were 23 and 82 times higher in smokers than in non-smokers, respectively. In the present study, the contents of nicotine and cotinine in the hair of 110 non-smokers differ by more than 100-fold in concentration, and were 1.38 ± 1.36 ng mg^{-1} and 12.8 ± 13.7 pg mg^{-1}, respectively. In particular, the levels of hair nicotine and cotinine in 46 non-smokers (not even passive smokers) were 1.11 ng mg^{-1} and 8 pg mg^{-1}, respectively, similar to the previous results. As shown in Table 1 (Question 1 and 2), however, there was no significant difference in these contents by sex and age of participants.

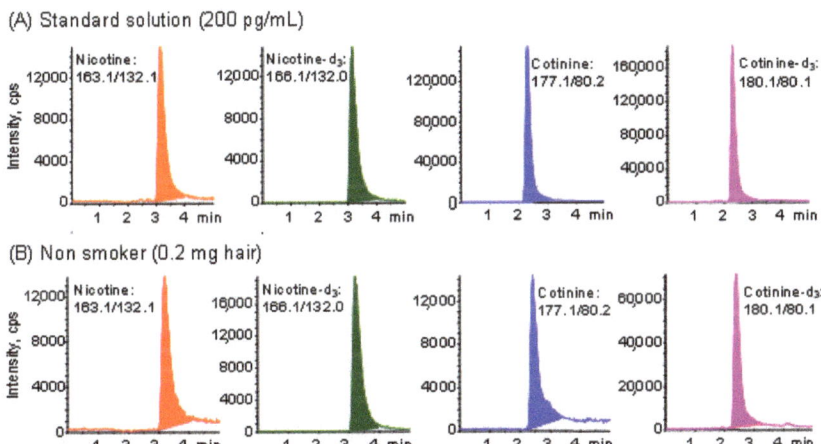

Figure 1. Typical MRM chromatograms obtained from (**A**) standard solution and (**B**) hair sample of non-smoker by in-tube SPME LC−MS/MS. Analytical conditions are described in the Experimental section.

Table 1. Nicotine and cotinine contents in non-smoker's hair based on lifestyle questionnaires.

Question/Answer	n	Content in Hair/Mean (Min.~Max.)	
		Nicotine (ng mg^{-1})	Cotinine (pg mg^{-1})
1. Sex			
Male	30	1.63 (0.20~7.72)	14.4 (4.8~43.9)
Female	80	1.29 (0.07~4.99)	12.2 (2.7~99.1)
P [1] (Male/female)		0.120	0.234
2. Age			
29 years old and under	79	1.39 (0.07~5.40)	12.1 (2.7~99.1)
30 years old and over	31	1.37 (0.17~7.72)	14.6 (4.3~59.3)
P (Younger/older)		0.477	0.198
3. Daily sleeping time			
Less than 6 h	30	1.19 (0.07~3.89)	10.8 (4.1~59.3)
More than 6 h	80	1.45 (0.10~37.72)	13.6 (2.7~99.1)
P (Shorter/longer)		0.188	0.171
4. Stress awareness			
Sometimes [2]	68	1.39 (0.07~7.72)	13.8 (2.7~99.1)
Frequently [2]	42	1.38 (0.17~4.99)	11.2 (4.4~59.3)
P (Sometimes/frequently)		0.483	0.169
5. Passive smoking awareness			
Never	46	1.11 (0.07~4.55)	8.1 (2.7~21.0)
Sometimes	55	1.52 (0.16~7.72)	13.8 (3.8~59.3)
Frequently	9	1.94 (0.72~3.60)	31.0 (6.6~99.1)
P (Never/sometimes)		0.064	0.001
P (Never/frequently)		0.023	0.00001
6. Exposure to other smoke			
Never	55	1.35 (0.12~5.40)	14.0 (2.7~99.1)
Sometimes~Always	55	1.42 (0.07~7.72)	11.6 (3.8~47.0)
P (Never/yes)		0.397	0.177
7. Frequency of tea drinking			
Sometimes/Frequently	46	1.49 (0.17~7.72)	12.4 (4.0~59.3)
Always	64	1.30 (0.07~5.40)	13.1 (2.7~99.1)
P (Less/more)		0.241	0.409

Table 1. *Cont.*

Question/Answer	n	Content in Hair/Mean (Min.~Max.)	
		Nicotine (ng mg^{-1})	Cotinine (pg mg^{-1})
8. Fat and fatty food intake			
Sometimes	8	1.82 (0.37~7.72)	12.7 (5.1~35.2)
Frequently	81	1.27 (0.07~4.99)	11.8 (2.7~99.1)
Always	21	1.64 (0.20~5.40)	16.9 (3.8~47.0)
P (Frequently/always)		0.405	0.227
9. Vegetable intake			
Sometimes	31	1.27 (0.10~4.54)	9.7 (2.7~32.5)
Frequently	64	1.36 (0.07~7.72)	15.1 (4.0~99.1)
Always	15	1.71 (0.20~4.99)	9.4 (3.8~22.0)
P (Sometimes/always)		0.135	0.450
10. Consumption of vegetables			
Raw	55	1.29 (0.07~7.72)	12.0 (2.7~99.1)
Boiled	21	1.30 (0.12~3.60)	13.0 (4.3~59.3)
Pan-fried	34	1.57 (0.10~5.42)	14.0 (3.8~67.0)
P (Raw/pan-fried)		0.232	0.390
P (Boiled/pan-fried)		0.192	0.249
11. Spice use			
Sometimes	36	1.54 (0.12~5.40)	12.7 (4.0~59.3)
Frequently	58	1.41 (0.10~7.72)	13.7 (2.7~99.1)
Always	16	0.93 (0.07~3.48)	9.7 (4.9~31.7)
P (Sometimes/always)		0.055	0.169
12. Meat intake			
Sometimes	14	1.20 (0.28~4.99)	10.5 (5.1~31.7)
Frequently	67	1.47 (0.07~7.72)	13.5 (2.7~99.1)
Always	29	1.28 (0.12~3.60)	12.2 (4.8~67.0)
P (Sometimes/always)		0.414	0.311
13. Consumption of meat			
Boiled	9	0.81 (0.30~2.19)	7.6 (5.1~13.7)
Pan-fried/Deep-fried	28	1.02 (0.10~3.25)	9.6 (2.7~31.7)
Grilled	73	1.59 (0.07~7.72)	19.1 (4.0~99.1)
P (Boiled/grilled)		0.071	0.096
P (Fried/grilled)		0.034	0.053
14. Seafood intake			
Sometimes	48	1.40 (0.12~5.40)	12.0 (2.7~47.0)
Frequently	54	1.37 (0.07~7.72)	14.1 (3.8~99.1)
Always	8	1.24 (0.17~3.44)	8.9 (5.6~1.45)
P (Sometimes/always)		0.478	0.205
15. Consumption of seafood			
Raw	21	1.12 (0.12~4.99)	17.9 (4.4~99.1)
Boiled	33	1.38 (0.29~5.40)	12.6 (4.0~43.9)
Grilled	56	1.48 (0.07~7.72)	11.0 (2.7~67.0)
P (Raw/grilled)		0.155	0.036
P (Boiled/grilled)		0.369	0.228

[1] Probability (significant difference T-test). [2] Sometimes: 3 days and under per week; frequently: 4 days and over per week.

The nicotine and cotinine contents obtained from hair analyses of 110 non-smokers were compared with data of reported lifestyle factors, such as daily sleeping time, stress, passive smoking, food consumption frequency and type. As shown in Table 1, the levels of nicotine or cotinine in hair were not affected by sleeping time or degree of stress. In this table, "Sometimes" means three days and under per week and "Frequently" means four days and over per week. In the questionnaire survey of passive smoking frequency among 110 non-smokers, 46 reported never, 55 sometimes and 9 frequent or always. Their hair nicotine contents were 1.11 ± 1.11, 1.52 ± 1.54 and 1.94 ± 1.18 ng mg^{-1}, and hair cotinine contents were 8.1 ± 4.2, 13.8 ± 11.8 and 31.0 ± 31.9 pg mg^{-1}, respectively. Both compounds were detected in the hair of non-smokers who reported never being exposed to environmental tobacco smoke, but at significantly lower concentrations than in those who reported exposure. Whisker-box plots displaying the medians and interquartile ranges of hair nicotine and cotinine are presented in Figure 2A,B. These results indicate that the greater the awareness of passive smoking, the greater the accumulation of both compounds in the hair. Furthermore, the levels of nicotine and cotinine in hair were not affected by the presence or absence of exposure to non-cigarette smoke, such as cooking or wood burning, indicating that passive smoking could be detected selectively by hair analysis.

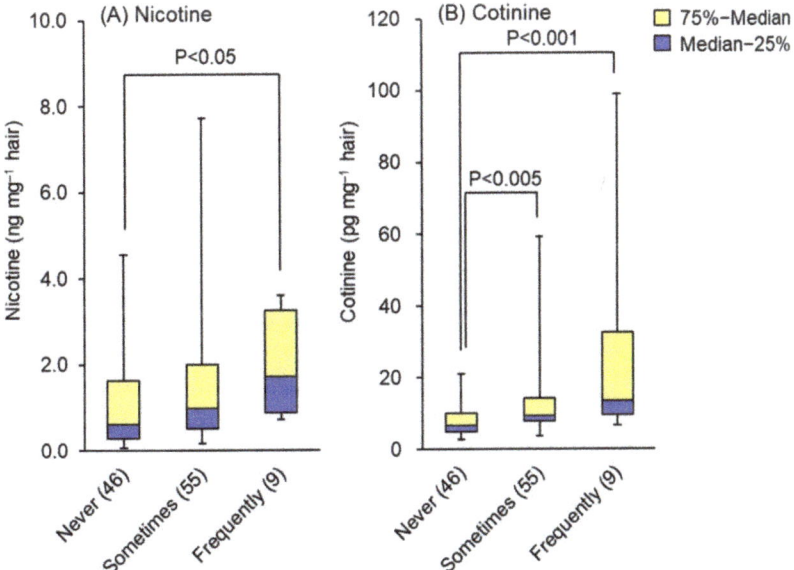

Figure 2. Effects of passive smoking awareness on the levels of (**A**) nicotine and (**B**) cotinine in hair. Data are presented as whisker-box plots displaying medians and interquartile ranges. Not at all: no awareness of passive smoking at all; sometimes: 3 days and under per week; frequently: for 4 days and over per week. The number of subjects is shown in parentheses. Top of each box, 75th percentile; bottom of each box, 25th percentile; solid center line, 50th percentile; error bars, non-outlier range.

There were no significant differences in hair nicotine and cotinine levels between frequencies of food intake, but those who consumed more spices tended to have lower levels. On the other hand, there were no significant differences in hair nicotine and cotinine levels depending on the frequency of meat consumption, but these levels tended to increase in those who ate grilled meat more often. Hair nicotine contents by consumption of boiled (9), fried (28) and grilled (73) meat were 0.81 ± 0.59, 1.02 ± 0.80 and 1.59 ± 1.55 ng mg^{-1}, and hair cotinine contents were 8.6 ± 2.7, 9.6 ± 6.3 and 19.1 ± 20.5 pg mg^{-1}, respectively. Whisker-box plots of medians and interquartile ranges of hair nicotine and cotinine are shown in Figure 3A,B. These results may be affected by the tobacco smoke environment in places where grilled meat is eaten, such as grilled meat restaurants, since there was almost no effect of nicotine and cotinine from smoke other than tobacco smoke, such as smoke generated by cooking.

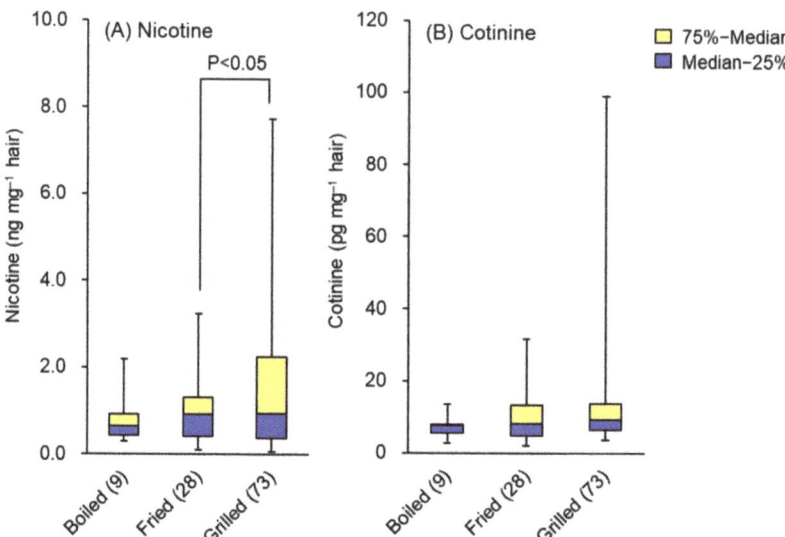

Figure 3. Effects of cooking method of meat on the levels of (**A**) nicotine and (**B**) cotinine in hair. Data are presented as whisker-box plots displaying medians and interquartile ranges. The number of subjects is shown in parentheses. Top of each box, 75th percentile; bottom of each box, 25th percentile; solid center line, 50th percentile; error bars, non-outlier range.

3. Discussion

Using the on-line in-tube SPME/LC-MS/MS method, the effects of tobacco smoke exposure can be assessed selectively and sensitively by analyzing as little as 1 mg of hair. Concentrations of nicotine and cotinine in hair reflect long-term exposure to tobacco smoke and are useful to assess the health effects of environmental tobacco smoke even in people unaware of passive smoking. Although cotinine concentrations in the hair of non-smokers were much lower than nicotine concentrations, there was a marked difference in cotinine concentrations between those who were aware of passive smoking and those who were not. This means that cotinine is more reflective of the reality of passive smoking than nicotine.

The consumption of food affected the accumulation of nicotine and cotinine in the hair. Those who consumed spices more frequently had lower levels of nicotine and cotinine in their hair, suggesting that spices help to remove tobacco compounds from the body. This may be related to the preventive effect of spices such as curcumin on lung cancer [50]. As nicotine and cotinine levels in hair are not affected by cooking smoke, high levels of nicotine and cotinine in the hair of non-smokers who eat grilled meat more frequently suggest that they may be affected by passive smoking in places such as grill restaurants and barbecue. Indeed, in Japan, smoking is relatively common in environments where groups of people drink alcohol and eat grilled meat. People unaware of passive smoking may have been unconsciously exposed to low concentrations of environmental tobacco smoke over long periods of time since nicotine and cotinine were detected in their hair. Therefore, it is important to understand the actual situation of exposure to environmental tobacco smoke because it is impossible to know when and where one will be exposed to passive smoking unless smoking itself is banned, and non-smokers may be forced to smoke passively for long periods of time in a variety of environments. In the future, in order to avoid the risk of passive smoking, it may be possible to prevent health problems caused by passive smoking not only by avoiding exposure to tobacco smoke, but also by improving eating habits and lifestyle habits.

4. Materials and Methods

4.1. Reagents and Standard Solutions

Nicotine and cotinine were purchased from Sigma-Aldrich Japan (Tokyo, Japan). Nicotine-d_3 (isotopic purity 98.4%), and cotinine-d_3 (isotopic purity 99.9%) purchased from Toronto Research Chemicals Inc. (Toronto, Ontario, Canada) were used as an internal standard (IS). Stock solutions of 1 mg mL^{-1} of each compound were prepared by dissolving in methanol. The working solutions were prepared by diluting these stock solutions with distilled water to the required concentration. These prepared solutions were stored at 4 °C in refrigerator until use. Methanol and distilled water were of LC–MS grade, while all other chemicals were of analytical reagent grade.

4.2. On-Line Automated Analysis System and Analytical Conditions

Nicotine and cotinine were measured using an on-line automated analysis system comprising in-tube SPME coupled with LC-MS/MS equipped with a Model 1100 series LC (Agilent Technologies, Böblingen, Germany) and an API 4000 triple quadruple mass spectrometer (Applied Biosystems, Foster City, CA, USA) by the previously reported method [45]. LC−MS/MS data were analyzed using Analyst Software 1.3.1 (Applied Biosystems). An outline of the system is shown in Figure S1 (refer to Supplementary Materials).

A Carboxen 1006 PLOT capillary column (Carboxen molecularsives, film thickness 17 μm, 60 cm × 0.32 mm i.d., Supelco, Bellefonte, PA, USA) was used as an in-tube SPME device and was connected between the injection needle and the injection loop of the LC autosampler. All controls for extraction, desorption and injection were programmed by the autosampler software (Table S2, Supplementary Materials).

A Polar-RP 80A column (50 mm × 2.0 mm i.d., particle size 4 μm, Phenomenex; Torrance, CA, USA) was used for LC separation at a column temperature of 30 °C, with 2.5 mM ammonium formate/methanol (25/75, v/v) as the mobile phase, at a flow rate of 0.2 mL min^{-1} [45]. Electro spray ionization (ESI)–MS/MS of nicotine, cotinine and their stable isotope-labeled compounds was performed in positive ion mode at 4000 V and 600 °C by multiple reaction monitoring (MRM), and quantification and confirmation were performed by MRM of the protonated precursor molecular ions [M+H]$^+$ and the related product ions for each compound by the previously reported method [45]. These MRM transitions and optimal MS/MS parameters are shown in Table S3.

Using this analysis system, calibration curves for nicotine and cotinine were linear in the range of 5–1000 pg mL^{-1} by comparing peak height ratios with each stable isotope-labeled IS, and detection limits (signal to noise ratio of 3) were 0.45 and 0.13 pg mL^{-1}, respectively. Validation data of this method are shown in Table S1.

4.3. Preparation and Analysis of Hair Samples and Lifestyle Questionnaires

Hair samples were provided by 110 healthy Japanese non-smoker volunteers (30 men and 80 women, aged 18–68). All volunteers gave informed consent in writing to the experimental protocol approved by the ethics committee of Shujitsu University (approval code 147; 13 October 2017). The collected hair samples were stored in amber glass desiccator at room temperature until use.

Approximately 10 mg of hair was collected from the back of each subject's head, washed three times with 1 mL dichloromethane by sonication to remove external nicotine and cotinine from the hair surface, and stored in an amber glass desiccator at room temperature until use. About 1–2 mg of dried hair cut into small pieces with scissors was weighed into a 7 mL screw-cap vial, to which 1.0 mL of distilled water and 4 μL of the mixed IS solution containing 200 pg of nicotine-d_3 and cotinine-d_3 were added, and the vial was heated and extracted at 80 °C for 30 min with the cap. An aliquot of the extract was transferred to a 2.0-mL autosampler vial and made up to 1 mL with distilled water, and then used for in-tube SPME/LC−MS/MS analysis. The contents of nicotine and cotinine were calculated as ng mg^{-1} and pg mg^{-1} hair, respectively, using calibration curves of each compound.

For all non-smoker subjects, the frequencies of lifestyle behaviors, such as passive smoking, food intake and cooking method were self-reported by questionnaire. Significant differences in the contents of nicotine and cotinine in the hair samples of non-smokers between the scores of these items were analyzed by Student's t-test.

5. Conclusions

We measured hair nicotine and cotinine levels using an on-line in-tube SPME/LC-MS/MS method in 110 Japanese non-smokers who completed to a self-reported lifestyle questionnaire and assessed the effects of environment and lifestyle on these levels. Hair nicotine and cotinine levels were significantly higher in people who were highly aware of passive smoking, and the risk of passive smoking was found to be influenced by the type of food intake and the dietary environment. Nicotine and cotinine in hair are useful biomarkers for assessing the effects of passive smoking due to long-term exposure to environmental tobacco smoke, and our method can analyze the actual exposure with or without awareness of passive smoking. The results of this study show that improving eating habits and lifestyle in various environments are important to avoid the risk of passive smoking.

Supplementary Materials: The following are available online: Figure S1: Outline of on-line in-tube SPME LC-MS/MS system; Table S1: Linearity, sensitivity and precisions of nicotine and cotinine by in-tube SPME LC–MS/MS; Table S2: Program for the in-tube SPME process; Table S3: MRM transitions and setting parameters for nicotine, cotinine and their stable isotope-labeled compounds.

Author Contributions: Conceptualization, H.K.; methodology, H.K., S.K. and M.M.; software, H.K.; validation, H.K., S.K. and M.M.; formal analysis, H.K., S.K. and M.M.; investigation, H.K., S.K. and M.M.; resources, H.K.; data curation, H.K.; writing—original draft preparation, H.K.; writing—review and editing, H.K.; visualization, H.K.; supervision, H.K.; project administration, H.K.; funding acquisition, H.K. All authors have read and agreed to the published version of the manuscript.

Funding: This research was funded by Smoking Research Foundation.

Institutional Review Board Statement: The study was conducted according to the guidelines of the Declaration of Helsinki, and approved by the ethics committee of Shujitsu University (approval code 147; 13 October 2017).

Informed Consent Statement: Informed consent was obtained from all subjects involved in the study.

Data Availability Statement: The data presented in this study are available on request from the corresponding author.

Conflicts of Interest: The authors declare no conflict of interest.

Sample Availability: Samples of the compounds are not available from the authors.

References

1. Hu, Q.; Hou, H. (Eds.) *Tobacco Smoke Exposure Biomarkers*; CRC Press, Taylor & Francis Group: Boca Raton, FL, USA, 2015.
2. Talhout, R.; Schulz, T.; Florek, E.; van Benthem, J.; Wester, P.; Opperhuizen, A. Hazardous compounds in tobacco smoke. *Int. J. Environ. Res. Public Health* **2011**, *8*, 613–628. [CrossRef] [PubMed]
3. Mattes, W.; Yang, X.; Orr, M.S.; Richter, P.; Mendrick, D.L. Biomarkers of tobacco smoke exposure. *Adv. Clin. Chem.* **2014**, *67*, 1–45. [CrossRef] [PubMed]
4. World Health Organization (WHO). Tobacco Fact Sheet. 2020. Available online: https://www.who.int/news-room/fact-sheets/detail/tobacco (accessed on 1 October 2021).
5. Öberg, M.; Jaakkola, M.S.; Woodward, A.; Peruga, A.; Prüss-Ustün, A. Worldwide burden of disease from exposure to second-hand smoke: A retrospective analysis of data from 192 countries. *Lancet* **2011**, *377*, 139–146. [CrossRef]
6. Torres, S.; Merino, C.; Paton, B.; Correig, X.; Ramírez, N. Biomarkers of exposure to secondhand and thirdhand tobacco smoke: Recent advances and future perspectives. *Int. J. Environ. Res. Public Health* **2018**, *15*, 2693. [CrossRef]
7. Hori, M.; Tanaka, H.; Wakai, K.; Sasazuki, S.; Katanoda, K. Secondhand smoke exposure and risk of lung cancer in Japan: A systematic review and meta-analysis of epidemiologic studies. *Jap. J. Clin. Oncol.* **2016**, *46*, 942–951. [CrossRef]
8. Kurahashi, N.; Inoue, M.; Liu, Y.; Iwasaki, M.; Sasazuki, S.; Sobue, T.; Tsugane, S. Passive smoking and lung cancer in Japanese non-smoking women: A prospective study. *Int. J. Cancer* **2008**, *122*, 653–657. [CrossRef]

9. Ni, X.; Xu, N.; Wang, Q. Meta-analysis and systematic review in environmental tobacco smoke risk of female lung cancer by research type. *Int. J. Environ. Res. Public Health* **2018**, *15*, 1348. [CrossRef]
10. Cao, S.; Yang, C.; Gan, Y.; Lu, Z. The health effects of passive smoking: An overview of systematic reviews based on observational epidemiological evidence. *PLoS ONE* **2015**, *10*, e0139907. [CrossRef]
11. Macacu, A.; Autier, P.; Boniol, M.; Boyle, P. Active and passive smoking and risk of breast cancer: A meta-analysis. *Breast Cancer Res. Treat.* **2015**, *154*, 213–224. [CrossRef]
12. Lee, P.N.; Hamling, J.S. Environmental tobacco smoke exposure and risk of breast cancer in nonsmoking women. An updated review and meta-analysis. *Inhal. Toxicol.* **2016**, *28*, 431–454. [CrossRef]
13. Lee, P.N.; Thornton, A.J.; Hamling, J.S. Epidemiological evidence on environmental tobacco smoke and cancers other than lung or breast. *Regul. Toxicol. Pharmacol.* **2016**, *80*, 134–163. [CrossRef]
14. Sheng, L.; Tu, J.W.; Tian, J.H.; Chen, H.J.; Pan, C.L.; Zhou, R.Z. A meta-analysis of the relationship between environmental tobacco smoke and lung cancer risk of nonsmoker in China. *Medicine* **2018**, *97*, e11389. [CrossRef]
15. De Groot, P.M.; Wu, C.C.; Carter, B.W.; Munden, R.F. The epidemiology of lung cancer. *Transl. Lung Cancer Res.* **2018**, *7*, 220–233. [CrossRef]
16. Law, M.R.; Wald, N.J. Environmental tobacco smoke and ischemic heart disease. *Prog. Cardiovasc. Dis.* **2003**, *46*, 31–38. [CrossRef]
17. Ahijevych, K.; Wewers, M.E. Passive smoking and vascular disease. *Cardiovasc. Nurs.* **2003**, *18*, 69–74. [CrossRef]
18. Dunbar, A.; Gotsis, W.; Frishman, W. Second-hand tobacco smoke and cardiovascular disease risk: An epidemiological review. *Cardiol. Rev.* **2013**, *21*, 94–100. [CrossRef]
19. Messner, B.; Bernhard, D. Smoking and cardiovascular disease: Mechanisms of endothelial dysfunction and early atherogenesis. *Arterioscler Thromb. Vasc. Biol.* **2014**, *34*, 509–515. [CrossRef]
20. DiGiacomo, S.I.; Jazayeri, M.A.; Barua, R.S.; Ambrose, J.A. Environmental tobacco smoke and cardiovascular disease. *Int. J. Environ. Res. Public Health* **2018**, *16*, 96. [CrossRef]
21. Pistilli, M.; Howard, V.J.; Safford, M.M.; Lee, B.K.; Lovasi, G.S.; Cushman, M.; Malek, A.M.; McClure, L.A. Association of secondhand tobacco smoke exposure during childhood on adult cardiovascular disease risk among never-smokers. *Ann. Epidemiol.* **2019**, *32*, 28–34. [CrossRef]
22. Khoramdad, M.; Vahedian-Azimi, A.; Karimi, L.; Rahimi-Bashar, F.; Amini, H.; Sahebkar, A. Association between passive smoking and cardiovascular disease: A systematic review and meta-analysis. *IUBMB Life* **2020**, *72*, 677–686. [CrossRef]
23. Akpa, O.M.; Okekunle, A.P.; Asowata, J.O.; Adedokun, B. Passive smoking exposure and the risk of hypertension among non-smoking adults: The 2015-2016 NHANES data. *Clin. Hypertens.* **2021**, *27*, 1–10. [CrossRef]
24. Hayashino, Y.; Fukuhara, S.; Okamura, T.; Yamato, H.; Tanaka, H.; Tanaka, T.; Kadowaki, T.; Ueshima, H. A prospective study of passive smoking and risk of diabetes in a cohort of workers. *Diabetes Care* **2008**, *31*, 732–734. [CrossRef]
25. Wei, X.; Meng, E.; Yu, S. A meta-analysis of passive smoking and risk of developing type 2 diabetes mellitus. *Diabetes Res. Clin. Prac.* **2015**, *107*, 9–14. [CrossRef]
26. Weitzman, M.; Cook, S.R.; Auinger, P.; Florin, T.A.; Daniels, S.; Nguyen, M.; Winickoff, J.P. Tobacco smoke exposure is associated with the metabolic syndrome in adolescents. *Circulation* **2005**, *112*, 862–869. [CrossRef]
27. Hamer, M.; Stamatakis, E.; Batty, G.D. Objectively assessed secondhand smoke exposure and mental health in adults: Cross-sectional and prospective evidence from the Scottish Health Survey. *Arch. Gen. Psychiatry* **2010**, *67*, 850–855. [CrossRef]
28. Chen, R. Association of environmental tobacco smoke with dementia and Alzheimer's disease among never smokers. *Alzheimers Dement.* **2012**, *8*, 590–595. [CrossRef]
29. Bamoya, J.; Navas-Acien, A. Protecting the world from secondhand tobacco smoke exposure: Where do we stand and where do we go from here? *Nicotine Tob. Res.* **2013**, *15*, 789–804. [CrossRef]
30. Al-Sayed, E.M.; Ibrahim, K.S. Second-hand tobacco smoke and children. *Toxicol. Ind. Health* **2014**, *30*, 635–644. [CrossRef]
31. Leonardi-Bee, J.; Smyth, A.; Britton, J.; Coleman, T. Environmental tobacco smoke and fetal health: Systematic review and meta-analysis. *Arch. Dis. Child Fetal Neonatal. Ed.* **2008**, *93*, F351–F361. [CrossRef]
32. Abraham, M.; Alramadhan, S.; Iniguez, C.; Duijts, L.; Jaddoe, V.W.; den Dekker, H.T.; Crozier, S.; Godfrey, K.M.; Hindmarsh, P.; Vik, T.; et al. A systematic review of maternal smoking during pregnancy and fetal measurements with meta-analysis. *PLoS ONE* **2017**, *12*, e0170946. [CrossRef]
33. Anderson, H.R.; Cook, D.G. Passive smoking and sudden infant death syndrome: Review of the epidemiological evidence. *Thorax* **1997**, *52*, 1003–1009. [CrossRef] [PubMed]
34. Adgent, M.A. Environmental tobacco smoke and sudden infant death syndrome: A review. *Birth Defects Res. B Dev. Reprod. Toxicol.* **2006**, *77*, 69–85. [CrossRef] [PubMed]
35. Gould, G.S.; Havard, A.; Lim, L.L.; Kumar, R. Exposure to tobacco, environmental tobacco smoke and nicotine in pregnancy: A pragmatic overview of reviews of maternal and child outcomes, effectiveness of interventions and barriers and facilitators to quitting. *Int. J. Environ. Res. Public Health* **2020**, *17*, 2034. [CrossRef] [PubMed]
36. Braun, M.; Klingelhöfer, D.M.; Oremek, G.M.; Quarcoo, D.; David, A.; Groneberg, D.A. Influence of second-hand smoke and prenatal tobacco smoke exposure on biomarkers, genetics and physiological processes in children-an overview in research insights of the last few years. *Int. J. Environ. Res. Public Health* **2020**, *17*, 3212. [CrossRef] [PubMed]
37. Avila-Tang, E.; Al-Delaimy, W.K.; Ashley, D.L.; Benowitz, N.; Bernert, J.T.; Kim, S.; Samet, J.M.; Hecht, S.S. Assessing secondhand smoke using biological markers. *Tob. Control.* **2013**, *22*, 164–171. [CrossRef]

38. Boumba, V.A.; Ziavrou, K.S.; Vougiouklakis, T. Hair as a biological indicator of drug use, drug abuse or chronic expo-sure to environmental toxicants. *Int. J. Toxicol.* **2006**, *25*, 143–163. [CrossRef]
39. Schramm, K.W. Hair-biomonitoring of organic pollutants. *Chemosphere* **2008**, *72*, 1103–1111. [CrossRef]
40. Al-Delaimy, W.K. Hair as a biomarker for exposure to tobacco smoke. *Tob. Control.* **2002**, *11*, 176–182. [CrossRef]
41. Vogliardi, S.; Tucci, M.; Stocchero, G.; Ferrara, S.D.; Favretto, D. Sample preparation methods for determination of drugs of abuse in hair samples: A review. *Anal. Chim. Acta* **2015**, *857*, 1–27. [CrossRef]
42. Pérez-Ortuño, R.; Martínez-Sánchez, J.M.; Fu, M.; Fernández, E.; Pascual, J.A. Evaluation of tobacco specific nitrosamines exposure by quantification of 4-(methylnitrosamino)-1-(3-pyridyl)-1-butanone (NNK) in human hair of non-smokers. *Sci. Rep.* **2016**, *6*, 25043. [CrossRef]
43. Benowitz, N.L.; Hukkanen, J.; Jacob, P. Nicotine chemistry, metabolism, kinetics and biomarkers. *Handb. Exp. Pharmacol.* **2009**, *192*, 29–60. [CrossRef]
44. Kataoka, H.; Inoue, R.; Yagi, K.; Saito, K. Determination of nicotine, cotinine, and related alkaloids in human urine and saliva by automated in-tube solid-phase microextraction coupled with liquid chromatography-mass spectrometry. *J. Pharm. Biomed. Anal.* **2009**, *49*, 108–114. [CrossRef]
45. Inukai, T.; Kaji, S.; Kataoka, H. Analysis of nicotine and cotinine in hair by on-line in-tube solid-phase microextraction coupled with liquid chromatography-tandem mass spectrometry as biomarkers of exposure to tobacco smoke. *J. Pharm. Biomed. Anal.* **2018**, *156*, 272–277. [CrossRef]
46. Ishizaki, A.; Kataoka, H. In-tube solid-phase microextraction coupled to liquid chromatography-tandem mass spectrometry for the determination of tobacco-specific nitrosamines in hair samples. *Molecules* **2021**, *26*, 2056. [CrossRef]
47. Kataoka, H. In-tube solid-phase microextraction: Current trends and future perspectives. *J. Chromatogr. A* **2021**, *1636*, 461787. [CrossRef]
48. Yamamoto, Y.; Ishizaki, A.; Kataoka, H. Biomonitoring method for the determination of polycyclic aromatic hydrocarbons in hair by online in-tube solid-phase microextraction coupled with high performance liquid chromatography and fluorescence detection. *J. Chromatogr. B* **2015**, *1000*, 187–191. [CrossRef]
49. Kataoka, H.; Ehara, K.; Yasuhara, R.; Saito, K. Simultaneous determination of testosterone, cortisol and dehydroepiandrosterone in saliva by stable isotope dilution on-line in-tube solid-phase microextraction coupled with liquid chromatography-tandem mass spectrometry. *Anal. Bioanal. Chem.* **2013**, *405*, 331–340. [CrossRef]
50. Balcerek, M.; Matławska, I. Preventive role of curcumin in lung cancer. *Prz. Lek.* **2005**, *62*, 1180–1181.

Article

Online In-Tube Solid-Phase Microextraction Coupled to Liquid Chromatography–Tandem Mass Spectrometry for the Determination of Tobacco-Specific Nitrosamines in Hair Samples

Atsushi Ishizaki and Hiroyuki Kataoka *

School of Pharmacy, Shujitsu University, Nishigawara, Okayama 703-8516, Japan; ishizaki@shujitsu.ac.jp
* Correspondence: hkataoka@shujitsu.ac.jp

Abstract: Active and passive smoking are serious public health concerns Assessment of tobacco smoke exposure using effective biomarkers is needed. In this study, we developed a simultaneous determination method of five tobacco-specific nitrosamines (TSNAs) in hair by online in-tube solid-phase microextraction (SPME) coupled to liquid chromatography-tandem mass spectrometry (LC–MS/MS). TSNAs were extracted and concentrated on Supel-Q PLOT capillary by in-tube SPME and separated and detected within 5 min by LC–MS/MS using Capcell Pak C18 MGIII column and positive ion mode multiple reaction monitoring systems. These operations were fully automated by an online program. The calibration curves of TSNAs showed good linearity in the range of 0.5–1000 pg mL^{-1} using their stable isotope-labeled internal standards. Moreover, the limits of detection (S/N = 3) of TSNAs were in the range of 0.02–1.14 pg mL^{-1}, and intra-day and inter-day precisions were below 7.3% and 9.2% (n = 5), respectively. The developed method is highly sensitive and specific and can easily measure TSNA levels using 5 mg hair samples. This method was used to assess long-term exposure levels to tobacco smoke in smokers and non-smokers.

Keywords: tobacco-specific nitrosamines; hair; exposure biomarkers; in-tube solid-phase microextraction (SPME); liquid chromatography-tandem mass spectrometry (LC–MS/MS)

Citation: Ishizaki, A.; Kataoka, H. Online In-Tube Solid-Phase Microextraction Coupled to Liquid Chromatography–Tandem Mass Spectrometry for the Determination of Tobacco-Specific Nitrosamines in Hair Samples. *Molecules* **2021**, *26*, 2056. https://doi.org/10.3390/molecules26072056

Academic Editor: Roberto Mandrioli

Received: 18 March 2021
Accepted: 1 April 2021
Published: 3 April 2021

Publisher's Note: MDPI stays neutral with regard to jurisdictional claims in published maps and institutional affiliations.

Copyright: © 2021 by the authors. Licensee MDPI, Basel, Switzerland. This article is an open access article distributed under the terms and conditions of the Creative Commons Attribution (CC BY) license (https://creativecommons.org/licenses/by/4.0/).

1. Introduction

Active and passive smoking are serious public health concerns because they increase the risk of various cancers, cardiovascular diseases and respiratory diseases [1,2]. A 2020 study by the World Health Organization (WHO) reported that tobacco kills up to half of the world's 1.3 billion tobacco users, with active and passive smoking estimated to kill about 7 million and 1.2 million people per year, respectively [3,4]. In particular, persons with passive smoking have been reported to have a 1.3-fold higher risk of developing lung cancer than those without passive smoking [5]. To prevent the health hazards caused by active and passive smoking, it is essential to understand the exposure level to tobacco smoke, and the development of a sensitive and specific method for measuring effective exposure biomarkers is an urgent issue [1,2,6–9].

Tobacco smoke, which can be broadly classified into gaseous and particulate components, contains about 5300 chemicals, including more than 500 substances associated with mutagenicity and carcinogenesis, such as tobacco-specific nitrosamines (TSNAs) [1,7–10]. TSNAs are formed by the nitrozation by nitrite and nitric acid of tobacco leaf alkaloids, such as nicotine, nornicotine, anatabine and anabasine, in the process of tobacco production and combustion [10–12]. The main TSNAs detected in tobacco products and smoke includes 4-(methylnitrosoamino)-1-(3-pyridyl)-1-butanone (NNK), N'-nitrosonornicotine (NNN), N'-nitrosoanatabine (NAT), and N'-nitrosoanabasin (NAB). NNK, its major metabolite 4-(methylnitrosoamino)-1-(3-pyridyl)-1-butanol (NNAL), and NNN play important roles as cancer inducers [1,8,10]. NNK and NNN are classified as human carcinogens (group 1) in the WHO International Agency for Research on Cancer (IARC), but NAT and NAB as unclassifiable for human carcinogenicity (group 3) [13].

Although the metabolism of TSNAs is not fully understood, about 31% of NNK absorbed into the body is metabolized to NNAL [7,11]. The half-life of NNK is 2.6 h [7], and it is eliminated rapidly from the body after exposure to tobacco smoke. In contrast, the half-life of NNAL is relatively long, ranging from 10 days to 3 weeks in smokers [1,6,9] and 40–45 days in oral tobacco users [1,7,8]. Urinary concentrations of NNAL have been regarded as a biomarker of tobacco smoking and exposure to tobacco smoke [1,6–9,11]. However, urinary concentrations of these compounds are lower in passive than in active smokers, making them unsuitable for assessing the effects of long-term exposure to environmental tobacco smoke in non-smokers. On the other hand, hair samples have been frequently used to assess and monitor the bioaccumulation due to long-term exposure to exogenous compounds, such as environmental pollutants, drugs and carcinogens [14–19], because these compounds can remain trapped in hair shafts and the periods and amount of exposure can be identified depending on their distribution in hair [18]. Another advantage of using hair is that samples can be collected easily and less invasively and can be stored at room temperature for up to five years [7,20]. In addition, the amounts of compounds in the hair are less affected by daily exposure and variations in metabolism than those in other biological matrices [7,21]. However, the levels of TSNAs in the hair are not only significantly lower than those in urine but are much lower in passive smokers than in active smokers [6,7,21–23].

All of the sensitive analysis methods reported for TSNAs in the hair are based on liquid chromatography–tandem mass spectrometry (LC–MS/MS) [21–23]. These methods are sensitive and specific and useful to identify and quantitate TSNAs in hair, but they require relatively large amounts (20–150 mg) of hair samples. Moreover, sample preparation is both tedious and time-consuming, requiring steps, such as liquid–liquid extraction with dichloromethane and solvent evaporation to dryness, or solid-phase extraction, for separation and preconcentration of TSNAs. We recently developed a simple and sensitive method for the simultaneous determination of four TSNAs, excluding NNAL, in main- and side-stream smoke, involving online in-tube solid-phase microextraction (SPME) coupled to LC–MS/MS [24]. In-tube SPME, using an open tubular capillary column with an inner surface coating as an extraction device, is an efficient sample preparation method that allows automation of the extraction and concentration process and can be easily coupled online to HPLC or LC–MS system using column switching technique [25–27]. It not only reduces the use of and exposure to harmful organic solvents but also reduces analysis times and gives higher sensitivity and good precision. We have reported analytical methods for various trace contaminants in hair samples using this technique [28–30]. The present study describes the development of an online in-tube SPME LC–MS/MS method for the simultaneous determination of five TSNAs, including NNAL, in hair samples and applying this method to the assessment of tobacco smoke exposure in smokers and non-smokers.

2. Results and Discussion

2.1. Optimization of In-Tube Solid-Phase Microextraction and Desorption of TSNAs

We previously described the optimization of in-tube SPME conditions for four TSNAs, excluding NNAL [24]. In this study, several parameters, such as type of capillary coating and number and flow-rate of draw/eject cycles, were optimized for 1 ng mL^{-1} each of five TSNAs, including NNAL. Although the peak amount of NNAL was lower than those of the other TSNAs, all five TSNAs could be efficiently extracted into a Supel-Q PLOT capillary by more than 25 repeated draw/eject cycles of 40 µL sample at a flow rate of 0.2 mL min^{-1} (Figures S1 and S2). The absolute extractable amounts of TSNAs onto the capillary column were calculated by comparing peak area counts with the corresponding amount in standard solution directly injected onto the LC columns. Although the extraction yields of NNK, NNN, NAT, NAB and NNAL onto the Supel-Q PLOT column from 1 mL of a standard solution containing 1.0 ng mL^{-1} of each compound were 21.0%, 13.3%, 22.0%, 21.8% and

5.0%, respectively, their coefficients of variation (CVs) were below 5% due to the use of an autosampler. The TSNAs extracted into the stationary phase of the capillary column were dynamically desorbed and introduced directly into the LC column by online mobile phase flow. Since the capillary column was cleaned and conditioned by methanol and mobile phase flow prior to extraction, no carryover of each analyte or matrix component was observed.

2.2. LC–MS/MS Analysis of TSNAs and Their Stable Isotope-Labeled Compounds

TSNAs and their stable isotope-labeled compounds were efficiently ionized in the ESI-positive ion mode. The MS/MS operation parameters, including curtain gas, nebulizer gas stream ion spray voltage, the corresponding potentials (DP and EP), CE, and CXP, were optimized for each TSNAs. Under optimum MS/MS conditions, protonated ions [M + H]$^+$ (Q1 mass) and prominent fragment ions (Q3 mass) for each compound were detected as precursor and product ions, respectively. The MRM transitions for confirmation and quantification and MS/MS parameters set are shown in Table S1. These data were in good agreement with previously reported data [21,22].

A chromatogram of standard TSNAs by in-tube SPME LC–MS/MS is shown in Figure 1. Five TSNAs and their IS compounds were eluted as well-formed peaks within 4 min on a Capcell Pak C18 MGIII column and detected selectively in MRM mode. The CV% of the retention time for each compound was within 5%. The analysis time per sample was about 28 min, allowing automated analysis of about 50 samples per day by operating overnight.

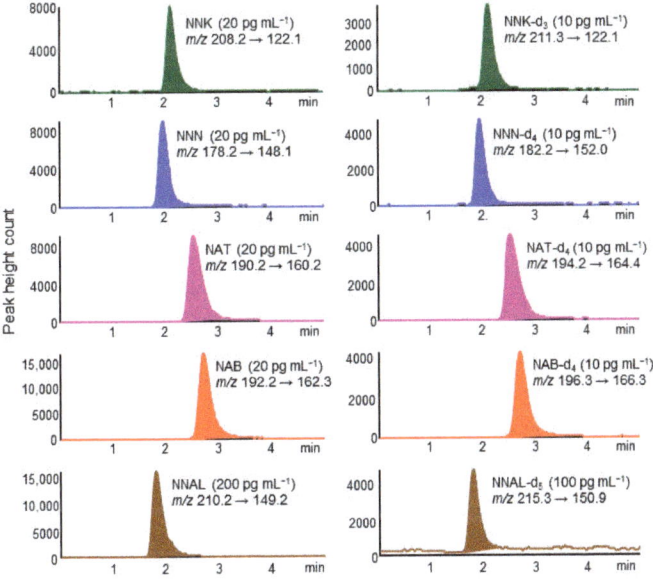

Figure 1. Multiple reaction monitoring (MRM) chromatograms of standard tobacco-specific nitrosamines (TSNAs) and their stable isotope-labeled compounds. In-tube solid-phase microextraction (SPME) LC–MS/MS conditions are described in the Experimental section.

2.3. Analytical Method Validation

Linearity was validated by triplicate analyses each for four TSNAs at eight concentrations (0.5, 1.0, 2.0, 5.0, 10, 20, 50, and 100 pg mL^{-1}) and for NNAL at eight concentrations (5, 10, 20, 50, 100, 200, 500 and 1000 pg mL^{-1}), in the presence of 0.1 ng mL^{-1} each of NNK-d$_3$, NNN-d$_4$, NAT-d$_4$ and NAB-d$_4$, and 1 ng mL^{-1} NNAL-d$_5$. Calibration curves for each compound were linear with correlation coefficients above 0.9998 (Table 1). The CVs of the peak height ratios at each compound's concentration ranged from 0.3 to 17% (n = 3).

Table 1. Linearity and sensitivity of the in-tube SPME LC–MS/MS method for TSNAs.

TSNA	Linearity		LOD [2] (pg mL^{-1})		LOQ [3] (pg mg^{-1})
	Range (pg mL^{-1})	Linearity [1] (R^2)	Direct Injection	In-Tube SPME	In-Tube SPME
NNK	0.5–100	0.9999	1	0.05	0.03
NNN	0.5–100	1	1.8	0.04	0.03
NAT	0.5–100	0.9999	1.1	0.02	0.02
NBT	0.5–100	0.9998	3.5	0.07	0.04
NNAL	5–1000	0.9999	12.9	1.14	0.75

[1] correlation coefficient (n = 24). [2] Limits of detection: pg mL^{-1} sample solution (signal-to-noise ratio of 3). [3] Limits of quantification: pg mg^{-1} hair sample (signal-to-noise ratio of 10).

TSNAs gave superior responses in MRM mode detection, with the LODs (S/N = 3) of TSNAs in standard solutions ranging from 0.02 to 1.14 pg mL^{-1} (Table 1). The in-tube SPME method was about 11 times more sensitive than the direct injection method (5 µL injection). The LOQs (S/N = 10) of TSNAs were 0.02–0.04 pg mg^{-1} hair for all of the TSNAs assayed except NNAL (Table 1). The previously reported LOQs for NNK and NNN were 0.10 and 0.25 pg mg^{-1} hair, respectively [20], indicating that our method's sensitivity was more than 2.5-fold higher. In contrast, the LOQ of NNAL was 0.75 pg mg^{-1} hair, while that of previous methods ranged from 0.063 to 0.24 pg mg^{-1} hair [20,21].

Precision and accuracy were assessed at low and high concentrations of 2 and 20 pg mL^{-1} for the four TSNAs except for NNAL and 20 and 200 pg mL^{-1} for NNAL. The precision, expressed as CV (%), was validated by performing five independent analyses on the same day and on five different days. The intra-day and inter-day precisions of these analyses were found to be 2.1–7.3% and 3.0–9.2%, respectively (Table 2). Accuracy was validated by comparing the measured concentrations of analytes in samples with the known concentrations of the analyte added to the samples ((found/added) × 100%). The intra-day and inter-day accuracies of these analyses were found to be 94–115% and 96–119%, respectively (Table 2).

Table 2. Precision and accuracy of the in-tube SPME LC–MS/MS method for TSNAs.

TSNA	Nominal Concentration (pg mL^{-1})	Precision (CV [1] %) (n = 5)		Accuracy (%) (n = 5)	
		Intra-Day	Inter-Day	Intra-Day	Inter-Day
NNK	2	3.6	6.4	100.5	102.5
	20	2.7	3.7	100.7	100.3
NNN	2	7.3	9.2	104.5	104.5
	20	2.7	4.6	104.7	104.2
NAT	2	4.4	5.2	113.5	112.0
	20	2.1	3.0	103.9	105.2
NBT	2	3.1	8.0	114.5	118.5
	20	3.0	7.7	94.2	99.3
NNAL	20	4.2	7.6	95.0	96.3
	200	4.2	7.0	94.3	99.2

[1] CV, coefficient of variation.

These results obtained based on the generally accepted validation criteria recommended in the ICH guidelines [31] show that the method has good linearity, precision and accuracy.

2.4. Application to the Analysis of Hair Samples

Since TSNAs absorbed into the body accumulate in the hair, they can be effective biomarkers for evaluating long-term exposure to tobacco smoke. However, internal TSNAs accumulated in the hair must be separated from external TSNAs deposited on the outer surface of the hair before analysis. To remove the external contaminants, the hair samples are usually prewashed with dichloromethane, methanol or 0.1% sodium dodecyl sulfate [14,16,19,22,28]. We ensured that the external TSNAs could be completely removed by washing the hair samples one each with 0.1% sodium dodecyl sulfate, water and methanol. The internal TSNAs in hair samples were easily extracted into distilled water by heating at 80 °C for 30 min, and the extract could be directly used for in-tube SPME/LC–MS/MS without any other pretreatment.

Stable isotope-labeled compounds as IS were added to hair samples prior to extraction to minimize the influence of matrix effects on the analysis of TSNAs in the samples. Figure 2 shows typical chromatograms obtained from hair samples (corresponding to 2.5 mg) from smokers and non-smoker. The specificity of this method was verified by analyzing both blank hairs (i.e., TSNA-free) from a non-smoker and the same hair spiked with the five TSNAs to check for co-eluting interferences at the retention times of the compounds of interest. These chromatograms showed no interference with the TSNAs, and their IS compounds from hair samples. In addition, the overall recovery rates of TSNAs spiked into hair samples were over 92% (Table 3).

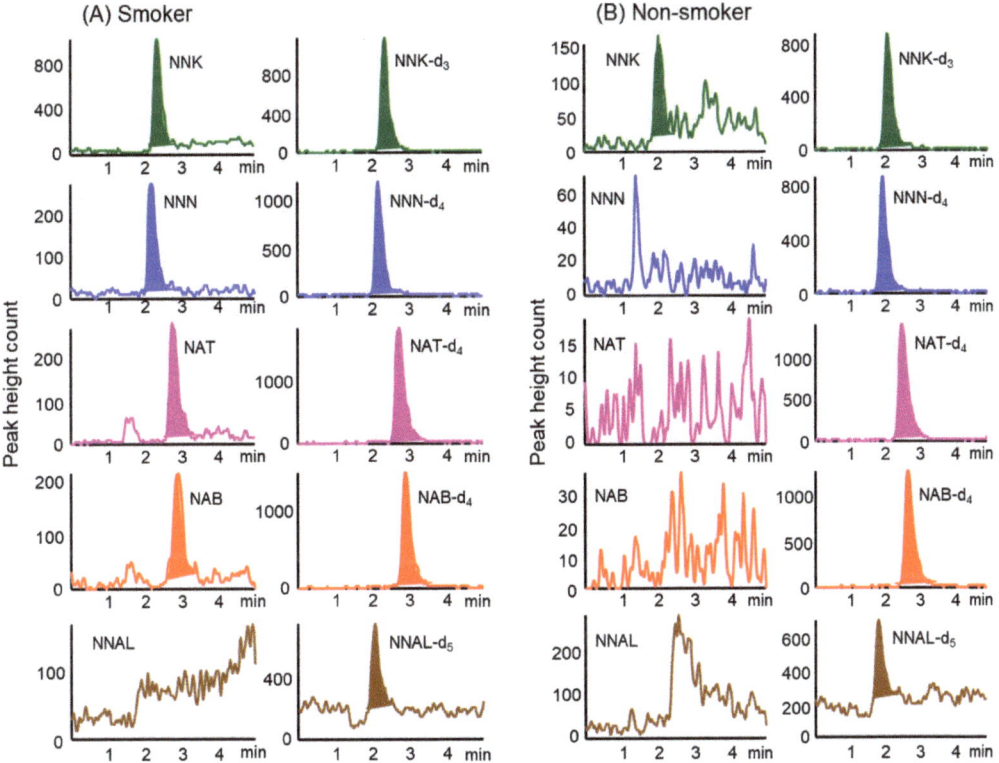

Figure 2. Typical MRM chromatograms were obtained from hair samples of (**A**) smokers and (**B**) non-smoker by in-tube SPME LC–MS/MS. Analytical conditions are described in the Material and Methods section.

Table 3. Recoveries of TSNAs spiked into hair samples of a non-smoker.

TSNA	Concentration (pg mg^{-1} Hair)		Recovery (%)
	Spiked	Mean ± SD (n = 3)	
NNK	0	ND[1]	-
	2	1.85 ± 0.12	92.3
	20	20.0 ± 0.4	100.2
NNN	0	ND	-
	2	1.95 ± 0.09	97.3
	20	19.0 ± 0.8	94.9
NAT	0	ND	-
	2	1.97 ± 0.07	98.4
	20	20.1 ± 0.2	100.4
NAB	0	ND	-
	2	1.92 ± 0.05	96.2
	20	19.7 ± 0.2	98.3
NNAL	0	ND	-
	20	18.7 ± 0.9	93.7
	200	205.4 ± 0.8	102.7

[1] not detectable.

The developed method was used to analyze TSNA concentrations in hair samples from 25 smokers and 29 non-smokers. Of the five TSNAs assayed, NNK and NNN were present at higher concentrations in hair samples from smokers than the other compounds (Table 4). Although NAT and NBT were often detected at low concentrations, NNAL was not detected at all. In contrast, NNK and NNN were present at low concentrations in hair samples from some non-smokers, whereas NAT, NBT and NNAL were not detected at all.

Table 4. Detection frequencies and contents of TSNAs in hair samples of smokers and non-smokers.

TSNA	Smokers (n = 24)					Non-Smokers (n = 29)				
	Detection Frequency (%)	Content (pg mg^{-1} Hair)				Detection Frequency (%)	Content (pg mg^{-1} Hair)			
		Mean ± SD	Min.[1]	Med.[1]	Max.[1]		Mean ± SD	Min.	Med.	Max.
NNK	100	0.95 ± 0.96[2]	0.08	0.68	3.97	34	0.05 ± 0.08[2]	0.00	0.00	0.25
NNN	100	0.43 ± 0.85[3]	0.02	0.22	4.44	14	0.02 ± 0.06[3]	0.00	0.00	0.24
NAT	68	0.09 ± 0.09	0.00	0.08	0.35	0	ND	0.00	0.00	0.00
NBT	88	0.13 ± 0.27	0.00	0.06	1.09	0	ND	0.00	0.00	0.00
NNAL	0	ND[4]	ND	ND	ND	0	ND	0.00	0.00	0.00
Total	100	1.61 ± 1.55[5]	0.11	1.08	6.72	34	0.08 ± 0.11[5]	0.00	0.00	0.31

[1] Min., minimum; Med., median; Max., maximum. [2,3,5] $p < 0.01$, probability (significant difference t-test between smokers and non-smokers). [4] Not detectable.

Although analysis of 150 mg samples of hair from smokers showed the presence of NNAL at concentrations of 0.27–0.67 pg mg^{-1} [22], analysis of 20 mg samples of hair from non-smokers failed to detect any NNAL [21]. Our study, however, found that NNAL was undetectable in hair samples from both smokers and non-smokers. The concentrations of NNK and NNN in the hair samples from smokers were significantly higher than those from non-smokers ($p < 0.01$), and total TSNA concentrations in hair samples were about 20 times higher in smokers than in non-smokers. These results indicate that NNK and NNN in hair are effective biomarkers to assess long-term exposure to tobacco smoke.

3. Materials and Methods

3.1. Reagents and Standard Solutions

Five standard TSNAs, NNK, NNN, NAT, NAB and NNAL, and their stable isotope-labeled compounds, NNK-d_3 (isotopic purity 99.9%), NNN-d_4 (isotopic purity 97.8%), NAT-d_4 (isotopic purity 98.0%), NAB-d_4 (isotopic purity 99.7%) and NNAL-d_5 (isotopic purity 97.8%) as each internal standard (IS) (supplementary Figure S3), were purchased from Toronto Research Chemicals Inc. (North York, ON, Canada). Stock solutions of 1.0 mg mL^{-1} of each compound were prepared by dissolving in LC–MS grade acetonitrile and diluted with distilled water to the required concentration prior to use. The mixed standard solution consisted of 1 ng mL^{-1} each of NNK, NNN, NAT and NAB and 10 ng mL^{-1} NNAL, whereas the mixed IS solution consisted of 0.1 ng mL^{-1} each of NNK-d_3, NNN-d_4, NAT-d_4 and NAB-d_4, and 1 ng mL^{-1} NNAL-d_5. These prepared solutions were stored at 4°C. Methanol and distilled water as mobile phases were of LC–MS grade, while all other chemicals were of analytical reagent grade.

3.2. Preparation of Hair Samples

Hair samples were provided by 54 healthy Japanese volunteers (46 men and 8 women, aged 23–68 tea), including 25 smokers and 29 non-smokers. Approximately 10 mg of hair was collected from the back of each subject's head, washed with 0.1% sodium dodecyl sulfate, water and methanol, and stored in an amber glass desiccator at room temperature until use. About 5 mg of hair cut into small pieces with scissors was weighed into a 7 mL screw-cap vial, to which 0.1 mL of water and 0.1 mL of the mixed IS solution were added, and the vial was heated and extracted at 80 °C for 30 min with the cap. The extract was cooled to room temperature and filtered through a 45 μm hydrophilic PTFE syringe filter (Shimadzu GLC Ltd., Tokyo, Japan). For in-tube SPME LC–MS/MS analysis, 0.1 mL of the filtrate was taken into a 2.0 mL autosampler vial with the septum, and the total volume was made up to 0.5 mL with distilled water. The concentrations of each TSNA in hair were calculated using a calibration curve constructed from the ratios of peak heights of each TSNA to the peak heights of their IS compounds.

3.3. LC–MS/MS Analysis

LC–MS/MS analysis was essentially performed as described in our previous work [24] using an Agilent Technologies Model 1100 series LC system and an Applied Biosystems API 4000 triple, quadruple mass spectrometer. A Capcell Pak C18 MGIII column (100 mm × 2.0 mm, particle size 5 μm; Shiseido, Tokyo, Japan) was used as a separation column. The LC conditions were as follows: column temperature, 40 °C; mobile phase, 5 mM ammonium acetate/methanol containing 0.1% acetic acid (50/50, v/v); flow rate, 0.2 mL min^{-1}. Electrospray ionization (ESI)-MS/MS conditions were as follows: turbo ion spray voltage and temperature, 5000 V and 600 °C; ion source gases (GS1 and GS2) flows, 50 and 80 L min^{-1}; curtain gas (CUR) flow, 40 L mL^{-1}, collision gas (CAD) flow, 4.0 L min^{-1}. Multiple reaction monitoring (MRM) transitions in positive ion mode and other setting parameters, including dwell time, declustering potential (DP), entrance potential (EP), collision energy (CE), and collision cell exit potential (CXP), are shown in Supplementary Table S1. Analyst Software 1.3.1 (Applied Biosystems, Foster City, CA, USA) was used for LC–MS/MS data analysis.

3.4. In-Tube SPME

In tube, SPME was essentially performed as described in our previous works [24,29]. A GC capillary column (60 cm × 0.32 mm i.d.) as an extraction device was connected between the injection needle and injection loop of the autosampler. The capillary column was threaded through a 1/16 inch polyetheretherketone (PEEK) tube with a 2.5 cm long, 330 μm inner diameter and connected using standard 1/16 inch stainless steel nuts, ferrules and connectors. Supel-Q PLOT (Supelco, Bellefonte, PA, USA), Carboxen 1010 PLOT (Supelco), CP-Sil 5CB (Varian Inc., Lake Forest, CA, USA), CP-Sil 19CB (Varian), CP-Wax

52CB (Varian), and Quadrex 007-5 (Quadrex Corporation, Woodbridge, CT) were used to compare extraction efficiencies. The control of extraction, desorption, and injection was programmed by the autosampler software (Table S2) [24,29].

4. Conclusions

The automated online in-tube SPME LC–MS/MS method developed in this study enabled continuous extraction and enrichment of five TSNAs and their sensitive and selective simultaneous analysis. The method is easy to apply to the analysis of a few milligrams of hair samples without tedious pretreatment. Therefore, the proposed method can be a useful tool for biomonitoring smoking levels and for assessing long-term exposure to tobacco smoke over days to months.

Supplementary Materials: The following are available online, Figure S1: Structures of the five TSNAs and their stable isotope-labeled TSNAs as internal standards; Figure S2: Schematic diagrams of the automated online in-tube SPME/LC–MS/MS system; Figure S3: Effects of capillary coatings on the in-tube SPME of TSNAs. TSNAs were extracted by 30 draw/eject cycles of 40 µL of standard solution (1 ng mL^{-1}) at a flow rate of 200 µL min^{-1}; Figure S4: Effects of the number of draw/eject cycles on the in-tube SPME of TSNAs. TSNAs were extracted on a Supel-Q PLOT capillary column by the indicated number of draw/eject cycles of 40 µL of standard solution (1 ng mL^{-1}) at a flow rate of 200 µL min^{-1}; Table S1: MRM transitions and setting parameters for TSNAs and their stable isotope-labeled compounds; Table S2: Program for the in-tube SPME process.

Author Contributions: Conceptualization, A.I. and H.K.; methodology, A.I. and H.K.; software, A.I. and H.K.; validation, A.I. and H.K.; formal analysis, A.I.; investigation, A.I. and H.K.; resources, A.I. and H.K.; data curation, A.I.; writing—original draft preparation, A.I. and H.K.; writing—Review and editing, H.K.; visualization, A.I. and H.K.; supervision, H.K.; project administration, H.K.; funding acquisition, H.K. All authors have read and agreed to the published version of the manuscript.

Funding: This research was funded by Smoking Research Foundation.

Institutional Review Board Statement: The study was conducted according to the guidelines of the Declaration of Helsinki and approved by the ethics committee of Shujitsu University.

Informed Consent Statement: Informed consent was obtained from all subjects involved in the study.

Data Availability Statement: The data presented in this study are available on request from the corresponding author.

Conflicts of Interest: The authors declare no conflict of interest.

Sample Availability: Samples of the compounds are not available from the authors.

References

1. Hu, Q.; Hou, H. (Eds.) *Tobacco Smoke Exposure Biomarkers*; CRC Press, Taylor & Francis Group: Boca Raton, FL, USA, 2015.
2. Mattes, W.; Yang, X.; Orr, M.S.; Richter, P.; Mendrick, D.L. Biomarkers of tobacco smoke exposure. *Adv. Clin. Chem.* **2014**, *67*, 1–45. [PubMed]
3. World Health Organization [WHO]. Tobacco Fact Sheet. 2020. Available online: https://www.who.int/news-room/fact-sheets/detail/tobacco (accessed on 1 March 2021).
4. Omare, M.O.; Kibet, J.K.; Cherutoi, J.K.; Kengara, F.O. A review of tobacco abuse and its epidemiological consequences. *J. Public Health* **2021**. [CrossRef] [PubMed]
5. Hori, M.; Tanaka, H.; Wakai, K.; Sasazuki, S.; Katanoda, K. Secondhand smoke exposure and risk of lung cancer in Japan: A systematic review and meta-analysis of epidemiologic studies. *Jpn. J. Clin. Oncol.* **2016**, *46*, 942–951. [CrossRef] [PubMed]
6. Avila-Tang, E.; Al-Delaimy, W.K.; Ashley, D.L.; Benowitz, N.; Bernert, J.T.; Kim, S.; Samet, J.M.; Hecht, S.S. Assessing secondhand smoke using biological markers. *Tob. Control.* **2013**, *22*, 164–171. [CrossRef]
7. Torres, S.; Merino, C.; Paton, B.; Correig, X.; Ramírez, N. Biomarkers of exposure to secondhand and thirdhand tobacco smoke: Recent advances and future perspectives. *Int. J. Environ. Res. Public Health.* **2018**, *15*, 2693. [CrossRef]
8. Hecht, S.S.; Stepanov, I.; Carmella, S.G. Exposure and metabolic activation biomarkers of carcinogenic tobacco-specific nitrosamines. *Acc. Chem. Res.* **2016**, *49*, 106–114. [CrossRef]
9. Yuan, J.-M.; Butler, L.M.; Stepanov, I.; Hecht, S.S. Urinary tobacco smoke–constituent biomarkers for assessing risk of lung cancer. *Cancer Res.* **2014**, *74*, 401–411. [CrossRef]

10. Konstantinou, E.; Fotopoulou, F.; Drosos, A.; Dimakopoulou, N.; Zagoriti, Z.; Niarchos, A.; Makrynioti, D.; Kouretas, D.; Farsalinos, K.; Lagoumintzis, G.; et al. Tobacco-specific nitrosamines: A literature review. *Food Chem. Toxicol.* **2018**, *118*, 198–203. [CrossRef]
11. Veena, S.; Rashmi, S. A review on mechanism of nitrosamine formation, metabolism and toxicity in vivo. *Int. J. Toxicol. Pharmacol. Res.* **2014**, *6*, 86–96.
12. Gupta, A.K.; Tulsyan, S.; Bharadwaj, M.; Mehrotra, R. Grass roots approach to control levels of carcinogenic nitrosamines, NNN and NNK in smokeless tobacco products. *Food Chem. Toxicol.* **2019**, *124*, 359–366. [CrossRef]
13. International Agency for Research on Cancer [IARC]. *IARC Monographs on the Evaluation of Carcinogenic Risks to Humans*; IARC: Lyon, France, 2012; Volume 100E, Available online: https://monographs.iarc.fr/wp-content/uploads/2018/06/mono100E.pdf (accessed on 1 March 2021).
14. Kintz, P.; Salomone, A.; Vincenti, M. (Eds.) *Hair Analysis in Clinical and Forensic Toxicology*; Academic Press: London, UK, 2015.
15. Boumba, V.A.; Ziavrou, K.S.; Vougiouklakis, T. Hair as a biological indicator of drug use, drug abuse or chronic exposure to environmental toxicants. *Int. J. Toxicol.* **2006**, *25*, 143–163. [CrossRef] [PubMed]
16. Schramm, K.W. Hair-biomonitoring of organic pollutants. *Chemosphere* **2008**, *72*, 1103–1111. [CrossRef] [PubMed]
17. Appenzeller, B.M.; Tsatsakis, A.M. Hair analysis for biomonitoring of environmental and occupational exposure to organic pollutants: State of the art, critical review and future needs. *Toxicol. Lett.* **2012**, *210*, 119–140. [CrossRef]
18. Al-Delaimy, W.K. Hair as a biomarker for exposure to tobacco smoke. *Tob. Control.* **2012**, *11*, 176–182. [CrossRef]
19. Vogliardi, S.; Tucci, M.; Stocchero, G.; Ferrara, S.D.; Donata Favretto, D. Sample preparation methods for determination of drugs of abuse in hair samples: A review. *Anal. Chim. Acta* **2015**, *857*, 1–27. [CrossRef] [PubMed]
20. Benowitz, N.L.; Hukkanen, J.; Jacob, P. Nicotine chemistry, metabolism, kinetics and biomarkers. *Handb. Exp. Pharmacol.* **2009**, *192*, 29–60.
21. Pérez-Ortuño, R.; Martínez-Sánchez, J.M.; Fu, M.; Fernández, E.; Pascual, J.A. Evaluation of tobacco specific nitrosamines exposure by quantification of 4-(methylnitrosamino)-1-(3-pyridyl)-1-butanone (NNK) in human hair of non-smokers. *Sci. Rep.* **2016**, *6*, 25043. [CrossRef]
22. Yao, L.; Yang, J.; Guan, Y.F.; Liu, B.Z.; Zheng, S.J.; Wang, W.M.; Zhu, X.I.; Zhang, Z.D. Development, validation, and application of a liquid chromatography-tandem mass spectrometry method for the determination of 4-(methylnitrosamino)-1-(3-pyridyl)-1-butanol in human hair. *Anal. Bioanal. Chem.* **2012**, *404*, 2259–2266. [CrossRef]
23. Clemens, M.M.; Cardenas, V.M.; Fischbach, L.A.; Cen, R.; Siegel, E.R.; Eswaran, H.; Ekanem, U.S.; Policherla, A.; Moody, H.L.; Magann, E.F.; et al. Use of electronic nicotine delivery systems by pregnant women II: Hair biomarkers for exposures to nicotine and tobacco-specific nitrosamines. *Tob. Induc. Dis.* **2019**, *17*, 50. [CrossRef]
24. Ishizaki, A.; Kataoka, H. A sensitive method for the determination of tobacco-specific nitrosamines in mainstream and sidestream smokes of combustion cigarettes and heated tobacco products by online in in-tube solid-phase microextraction coupled with liquid chromatography-tandem mass spectrometry. *Anal. Chim. Acta* **2019**, *1075*, 98–105.
25. Kataoka, H. Automated sample preparation using in-tube solid-phase microextraction and its application—A review. *Anal. Bioanal. Chem.* **2002**, *373*, 31–45. [CrossRef]
26. Kataoka, H.; Ishizaki, A.; Saito, K. Recent progress in solid-phase microextraction and its pharmaceutical and biomedical applications. *Anal. Methods* **2016**, *8*, 5773–5788. [CrossRef]
27. Kataoka, H. In-tube solid-phase microextraction: Current trends and future perspectives. *J. Chromatogr. A* **2021**, *1636*, 461787. [CrossRef] [PubMed]
28. Kataoka, H.; Inoue, T.; Saito, K.; Kato, H.; Masuda, K. Analysis of heterocyclic amines in hair by on-line in-tube solid-phase microextraction coupled with liquid chromatography–tandem mass spectrometry. *Anal. Chim. Acta* **2013**, *786*, 54–60. [CrossRef] [PubMed]
29. Yamamoto, Y.; Ishizaki, A.; Kataoka, H. Biomonitoring method for the determination of polycyclic aromatic hydrocarbons in hair by online in-tube solid-phase microextraction coupled with high performance liquid chromatography and fluorescence detection. *J. Chromatogr. B Anal. Technol. Biomed. Life Sci.* **2015**, *1000*, 187–191. [CrossRef] [PubMed]
30. Inukai, T.; Kaji, S.; Kataoka, H. Analysis of nicotine and cotinine in hair by on-line in-tube solid-phase microextraction coupled with liquid chromatography-tandem mass spectrometry as biomarkers of exposure to tobacco smoke. *J. Pharm. Biomed. Anal.* **2018**, *156*, 272–277. [CrossRef] [PubMed]
31. ICH Harmonized Tripartite Guideline, ICH Q2 (R1). Validation of Analytical Procedures: Text and Methodology. In Proceedings of the International Conference on Harmonization of Technical Requirements for Registration of Pharmaceuticals for Human Use.

Article

Smart Titanium Wire Used for the Evaluation of Hydrophobic/Hydrophilic Interaction by In-Tube Solid Phase Microextraction

Yuping Zhang [1,2,*], Ning Wang [2], Zhenyu Lu [2], Na Chen [2], Chengxing Cui [2] and Xinxin Chen [2]

[1] College of Chemistry and Materials Engineering, Hunan University of Arts and Science, Changde 415000, China
[2] College of Chemistry and Chemical Engineering, Henan Institute of Science and Technology, Xinxiang 453000, China; wangning911420@163.com (N.W.); luzy_1988@163.com (Z.L.); woniunana22@163.com (N.C.); chengxingcui@hist.edu.cn (C.C.); chenxinxin920@163.com (X.C.)
* Correspondence: zhangyuping@hist.edu.cn or beijing2008zyp@163.com

Abstract: Evaluation of the hydrophobic/hydrophilic interaction individually between the sorbent and target compounds in sample pretreatment is a big challenge. Herein, a smart titanium substrate with switchable surface wettability was fabricated and selected as the sorbent for the solution. The titanium wires and meshes were fabricated by simple hydrothermal etching and chemical modification so as to construct the superhydrophilic and superhydrophobic surfaces. The micro/nano hierarchical structures of the formed TiO_2 nanoparticles in situ on the surface of Ti substrates exhibited the switchable surface wettability. After UV irradiation for about 15.5 h, the superhydrophobic substrates became superhydrophilic. The morphologies and element composition of the wires were observed by SEM, EDS, and XRD, and their surface wettabilities were measured using the Ti mesh by contact angle goniometer. The pristine hydrophilic wire, the resulting superhydrophilic wire, superhydrophobic wire, and the UV-irradiated superhydrophilic wire were filled into a stainless tube as the sorbent instead of the sample loop of a six-port valve for on-line in-tube solid-phase microextraction. When employed in conjunction with HPLC, four kinds of wires were comparatively applied to extract six estrogens in water samples. The optimal conditions for the preconcentration and separation of target compounds were obtained with a sample volume of 60 mL, an injection rate of 2 mL/min, a desorption time of 2 min, and a mobile phase of acetonile/water (47/53, v/v). The results showed that both the superhydrophilic wire and UV-irradiated wire had the highest extraction efficiency for the polar compounds of estrogens with the enrichment factors in the range of 20–177, while the superhydrophobic wire exhibited the highest extraction efficiency for the non-polar compounds of five polycyclic aromatic hydrocarbons (PAHs). They demonstrated that extraction efficiency was mainly dependent on the surface wettability of the sorbent and the polarity of the target compounds, which was in accordance with the molecular theory of like dissolves like.

Keywords: titanium wire; superhydrophobic; superhydrophilic; in-tube solid-phase microextraction; high-performance liquid chromatography

1. Introduction

As an efficient sample preparation technique, solid-phase microextraction (SPME) can integrate sampling, preconcentration, extraction, and sample injection into one step [1,2]. At present, it has been widely applied in the field of environmental analysis, drug monitoring, food testing, and biological analysis in order to remove impurities and enrich the trace target compounds in real samples [3–6]. It uses the adsorption of a sorbent to extract analytes from a sample matrix; these analytes are then desorbed from the sorbent and directed into an analytical instrument, such as gas chromatography (GC), high-performance liquid

chromatography (HPLC), etc. [7–10]. The material used in the extraction coating is an important factor because the extraction process is achieved through a distribution equilibrium between the target compounds in the sample solution and the extraction coating.

At present, most commercial SPME fibers are made of fused silica, which is not only expensive, but also easy to break and swell in organic solvents [11,12]. Therefore, it is urgent to find a kind of fiber with thermal, chemical, and mechanical stability, excellent selectivity, and high sensitivity to overcome these problems. In the past two decades, many studies have focused on high-strength metal substrates, such as aluminum wire [13], silver wire [14], zinc wire [15], platinum wire [16], copper wire [17], and stainless steel wire [18], that were modified with different kinds of organic, inorganic, and hybrid coatings, which exhibited good bending properties and chemical and mechanical stabilities. Titanium dioxide (TiO_2) has been comprehensively studied and used in various fields due to its chemical and thermal stability, biocompatibility, anti-polluting nature, and high corrosion resistance due to its photoelectrochemical activities [19]. Some studies have found that nanostructured TiO_2 is an excellent adsorbent for organic compounds in SPE and SPME [20,21]. Moreover, it is an import smart material for many practical applications, including self-cleaning, solar cells, lithium-ion batteries, pollutant photodegradation, and oil/water separation [22–25]. In situ fabrication of a TiO_2-nanotube coating on the surface of chemically oxidized Ti wire with hydrogen peroxide solution has been used for SPME of dichlorodiphenyltrichloroethane and its degradation products [26]. Liu et al. fabricated TiO_2 nanotube arrays on the surface of Ti wire for use in the SPME of PAHs. Their results showed that TiO_2 nanotube arrays are capable of extracting PAHs, but the very thin nanotube walls make the coating very fragile and easy to destroy when carrying out SPME [27]. The Du group presented a simple and rapid anodic method for the in-situ fabrication of a novel fiber consisting of Ti wire coated with rod-like TiO_2. It was used for the concentration and determination of trace PAHs and phthalates (PAEs) by SPME, coupled to HPLC with UV detection [28]. The Ouyang group prepared a core-shell TiO_2@C fiber for SPME, which was carried out by the simple hydrothermal reaction of a titanium wire, followed by the coating of amorphous carbon. It was successfully used for the determination of PAHs in the Pearl River water with higher GC responses than commercial PDMS and PDMS/DVB fibers [29]. Although different metal wires have been widely used as the sorbent instead of conventional fragile silica fibers by many researchers, various sorbent coatings were always required in conjunction with the metal wire for SPME in previous reports. The Yan group fabricated a stainless steel wire etched with hydrofluoric acid for SPME [30]. Although the pristine wire had almost no extraction capability toward the tested analytes, the etched wire did exhibit a high affinity to the tested PAHs, with a high enhancement factor in the range of 2541–3981, but no extraction ability to hydrophilic phenol, butanol, and aniline was found. It was suggested that a porous and flower-like structure with Fe_2O_3, FeF_3, Cr_2O_3, and CrF_2 on the surface of the stainless steel wire gave a high affinity to the hydrophobic PAHs due to cation-π interaction [31]. In this case, how to evaluate the hydrophobic interaction effectively between the sorbent and target compounds in real samples is still a big challenge.

In this work, the titanium wires and meshes with different surface wettability were fabricated by a simple hydrothermal digestion and chemical immersion method. The superhydrophilic substrates with flower-like TiO_2 nanoparticles were obtained after being etched by HF and changed to superhydrophobic after being modified using a low-surface-energy material. The micro/nano hierarchical structures and photosensitivity of the formed TiO_2 nanoparticles on the surfaces of the Ti substrates exhibited the switchable wettability. After UV irradiation, the superhydrophobic samples became superhydrophilic. The hydrophilic, superhydrophilic, superhydrophobic, and UV-irradiated superhydrophilic wires were initially selected as the sorbents for online in-tube SPME. The morphologies and element composition of the Ti wires were observed by SEM, EDS, and XPS, and their surface wettability was measured using the Ti mesh by a contact angle goniometer. The extraction tube filled with the prepared Ti wire was connected to the injection valve of

HPLC. Six common estrogenic hormones were selected as the target analytes to investigate the extraction efficiency. The online analytical method was established and used for the determination of six estrogens in water samples using the optimal conditions. More importantly, the hydrophobicity interactions between the wire surfaces and the samples were further investigated by the selection of sorbents with different surface wettabilities and the target compounds of hydrophobic PAHs.

2. Results and Discussion

2.1. Characterization

The morphological properties and surface elemental compositions of four kinds of wires were characterized by SEM, EDS, and XRD. As can be seen in Figure 1(a1), the pristine wire had a smooth surface with few grooves. After hydrothermal digestion in strong acids at a high temperature, some small holes appeared, the surface was scattered with flower-like TiO_2 nanostructures in Figure 1(b1), and the increase in roughness led to an increase in adsorption sites in the extraction. Further chemical modification with a low surface energy caused an observed decrease in surface wettability, which formed a superhydrophobic surface in Figure 1(c1). After UV irradiation for Ti wire c, the superhydrophobic surface transformed to superhydrophilic in Figure 1(d1). As seen in the cross-section of the wires in Figure 1(a2), no coating was observed for the smooth pristine wire. For the modified wires, the thickness of the coating was thin due to the formation of multiple layers of nanoflowers on the surface in Figure 1(b2–d2). There were some internal and external gaps in the nanoflower clusters. This coating morphology probably increases its specific surface area, improving the mass transfer and extraction efficiency.

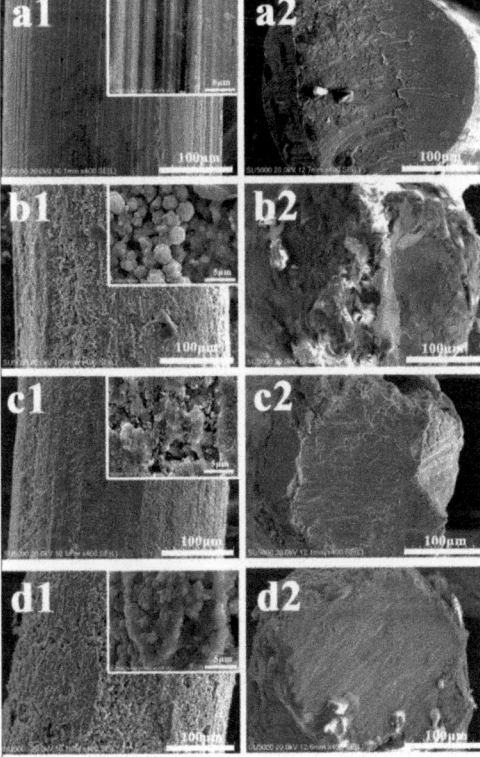

Figure 1. SEM for the pristine wire (**a1,a2**), the superhydrophilic wire (**b1,b2**), the superhydrophobic wire (**c1,c2**), and the UV-irradiated superhydrophilic wire (**d1,d2**).

EDS further proved the elemental changes for four kinds of Ti wires in Figure 2. The pristine Ti wire was mainly composed of Ti and O elements (a). After chemical etching, the oxygen element content increased from 27.7 to 50.2% (b). Carbon, silicon, and fluorine elements were observed after further modification of FOTS, which demonstrated thin organic films surrounding the Ti wire surface (c). Further UV irradiation led to the formation of hydrophilic domains on the TiO_2 surface. When photosensitive TiO_2 on the wire surface is irradiated by UV light, pairs of electron holes appear on its surface, which can react with lattice oxygen to form surface oxygen vacancies. More monolayers and multilayers of water molecules can be kinetically attached to these vacancies by molecular adsorption. It was proven by the increase in oxygen element from 42.8 to 50.0% (d).

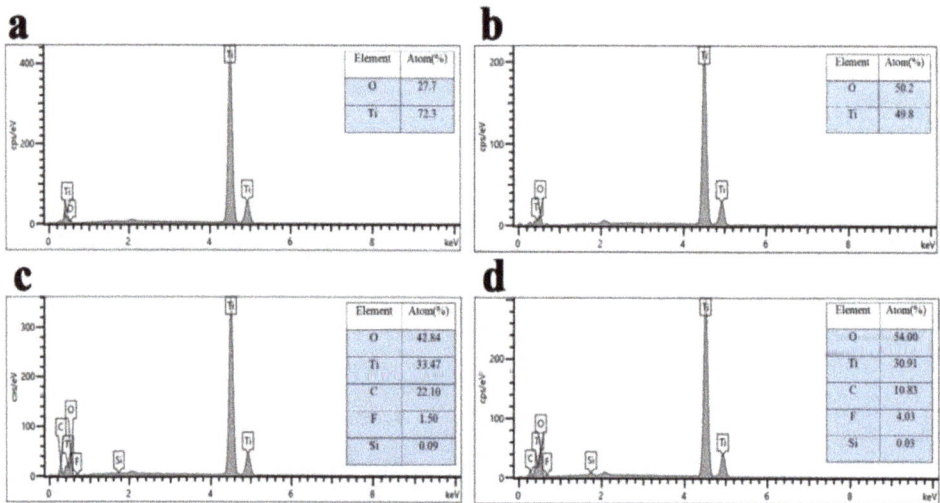

Figure 2. EDX for the pristine wire (**a**), the superhydrophilic wire (**b**), the superhydrophobic wire (**c**) and the UV-irradiated superhydrophilic wire (**d**).

XRD was used to investigate the crystal structure and phase of four kinds of Ti meshes. Figure 3 presents the XRD patterns of Ti substrates with different surface wettabilities. The samples were in the anatase phase, and the diffraction peaks were indexed to that of anatase (JCPDS No. 21-1272) [32]. No characteristic peaks from impurities were detected within experimental error, indicating the formation of anatase nanostructures.

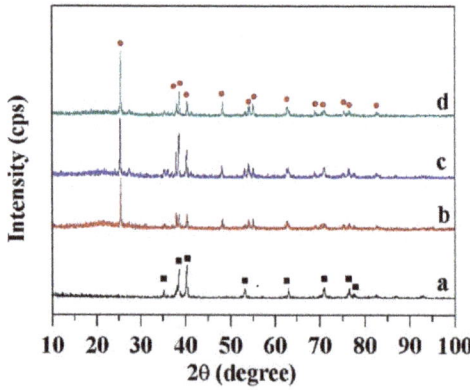

Figure 3. XRD patterns of Ti substrates. Symbol identification: (**a**) the pristine mesh, (**b**) the superhydrophilic mesh, (**c**) the superhydrophobic mesh, and (**d**) the UV-irradiated superhydrophilic mesh.

In the 2θ scan range 10°–100°, all show characteristic peaks of Ti(100), Ti(002), Ti(101), Ti(102), Ti(110), Ti(103), Ti(112), and Ti(201) at 35.1°, 38.4°, 40.2°, 53.0°, 62.9°, 70.6°, 76.2°, and 77.3°, respectively. These data for the pristine (a) are in good agreement with the Ti crystallographic data (Reference code:00-001-1198). In comparison, all show characteristic peaks of $TiO_2(101)$, $TiO_2(103)$, $TiO_2(004)$, $TiO_2(112)$, $TiO_2(200)$, $TiO_2(105)$, $TiO_2(211)$, $TiO_2(204)$, $TiO_2(116)$, $TiO_2(220)$, $TiO_2(107)$, $TiO_2(301)$, and $TiO_2(312)$ at 25.3°, 36.9°, 37.8°, 38.6°, 48.0°, 53.9°, 55.1°, 62.7°, 68.8°, 70.3°, 74.1°, 76.0°, and 83.1°. The 13 peaks are due to the formation of Ti oxides in the crystalline phase, which are in accordance with the TiO_2 crystallographic data (Reference code:03-065-5714). Although the presence of crystal structures is not a differentiated factor for the hydrophobicity or hydrophilicity of the surfaces, a significant difference was obviously found between the pristine Ti surface (a) and other three Ti surfaces with nano TiO_2 particles in Figure 3b–d.

2.2. Evaluation of Surface Wettability for the Prepared Wires

The preparation of superhydrophobic surfaces on hydrophilic metal substrates requires surface micro/nano-structures and low surface energy modification. As shown in Figure 4, there are three typical models to illustrate the surface states, which were developed by Young [33], Wenzel [34], and Cassie and Baxter [35].

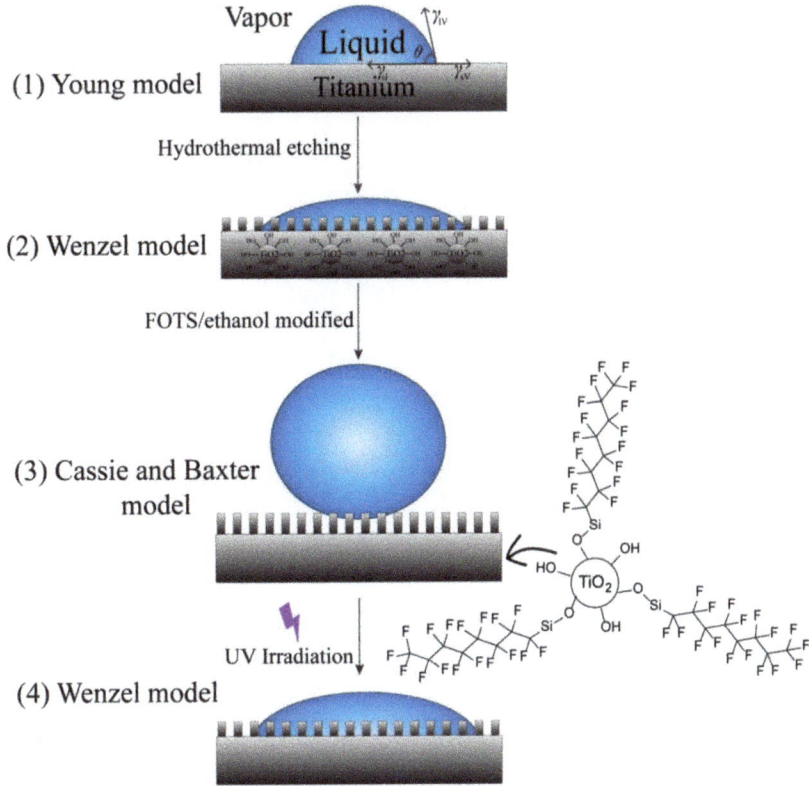

Figure 4. Model representation of the surface wettability for the preparation of Ti wires and meshes. The resulting Ti meshes with 3 typical surface states: (**1**) Young Model, (**2,4**) Wenzel Model, and (**3**) Cassie and Baxter Model.

Before the hydrothermal process, the surface of the Ti substrate is relatively smooth with a hydrophilic property, which is in accordance with Model 1. After hydrothermal treatment by HF about 8 h, the Ti substrate was chemically etched, and the surface became rougher with more porous structures. A few of nanoparticles were also found on the surface with superhydrophilicity, which is close to Model 2. The readily hydrothermal procedure afforded the in-situ synthesis of TiO_2 nanowires on the Ti substates and provided a desirable support for the further modification of a low-surface-energy material. TiO_2 nanoparticles are hydrophilic, which results from the large number of hydroxyl groups on their surface. After the Ti substrate was self-assembled by FOTS, the surface energy was apparently reduced [36]. In this case, the silicon ethoxide groups (Si–Cl) in FOTS were hydrolyzed to silanol groups (Si–OH). FOTS, as a low-surface-energy material then reacted with the –OH groups of the TiO_2 surface. The superhydrophobic coating by the Si–O–Ti bond is in accordance with Mode 3. The superhydrophobic surface was changed to superhydrophilic after UV irradiation, which was in the state of Mode 1.

The Ti meshes that were prepared using an identical method were used for the comparative determination of surface wettability instead of Ti wires. The pristine mesh displayed a static water contact angle of about 72° (Figure 5a). After it was etched by 20 mM HF for 6 h by hydrothermal digestion, the water droplets almost spread on the mesh completely (Figure 5b). When the etched mesh was immersed in FOTS for 1 h, the surface became superhydrophobic, with a WCA of 158° (Figure 5c). After UV irradiation for about 15.5 h, it changed to superhydrophilic, with a CA of nearly zero degrees (Figure 5d). A large group of hydrophilic and oleophilic domains was thus created on the surface of the Ti substrate. Moreover, the surface wettability of the Ti substrate could be reversibly switched between superhydrophilicity and hydrophobicity under the alternation of UV light illumination and long-term dark storage, independent of their photocatalytic activities [37,38].

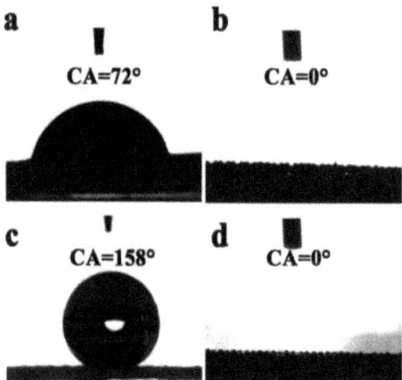

Figure 5. WCAs of the pristine Ti mesh (**a**), the etched Ti mesh (**b**), the FOTS-modified Ti mesh (**c**) and the UV-irradiated superhydrophilic mesh (**d**).

2.3. Extraction Efficiency for the Polar Compounds

Four kinds of Ti wires were inserted into the extraction tube for the comparative investigation of extraction efficiency. Benefiting from the much larger surface area of the subsequent TiO_2 around their surface and different surface energy after hydrothermal etching and chemical modification, the superhydrophilic wire (b) and UV-irradiated superhydrophilic wire (d) exhibited higher extraction efficiency than the pristine and superhydrophobic wires in Figure 6 (left). Two kinds of superhydrophilic Ti wires showed a small enrichment for the first four compounds, including bisphenol A, estradiol, ethynylestradiol, and estrone, and a particularly large enrichment for the last two compounds, including diethylstilbestrol and hexestrol. The comparative peak areas for each target compound are illustrated in Figure 6 (right). Superhydrophilic materials, due to their intrinsic character of polar nature, are expected to be the perfect option for the extraction of organic species with more polarity.

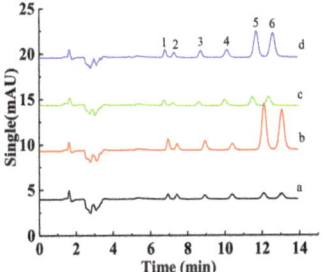 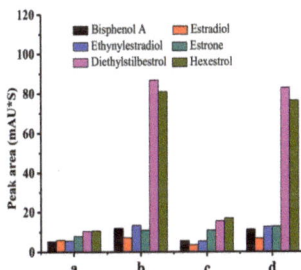

Figure 6. Chromatograms of the target compounds using four kinds of fibers with different surface wettabilities (**Left**) and the comparative peak area for each target compound using 4 kinds of wires for in-tube SPME (**Right**). Symbol identification: (a) pristine wire, (b) superhydrophilic wire, (c) superhydrophobic wire, (d) superhydrophilic wire after UV irradiation. Conditions: sampling volume, 60 mL; sampling rate, 2.00 mL/min; desorption time, 2.0 min. Peak identification: (1) bisphenol A, (2) estradiol, (3) ethynylestradiol, (4) estrone, (5) diethylstilbestrol, and (6) hexestrol. Mobile phase: Water/acetonitrile (53:47, v/v); Detection wavelength, 220 nm, room temperature.

The enrichment factor (EF) values were accurately calculated as follows [3,4,7]: After 20 μL of standard solutions with different concentrations of 0, 0.25, 0.5, 1, 2.5, 5, 10, 25, and 100 mg/L were injected using the sample loop, the calibration curve with a slope of k_1 was then calculated according to the peak area vs. concentration. Similarly, another series of standard solutions with a constant sampling volume of 60 mL and different concentrations of 0, 0.5, 1, 2, 5, and 10 μg/L were prepared. After they were pumped into the extraction tube for preconcentration, each analyte in the standard solution was desorbed and detected through the UV detector. The calibration curve was then achieved with a slope of k_2. The final EF can be calculated according to the following equation: $EF = k_2/k_1$.

2.4. Optimization of Extraction Conditions

To achieve the highest sensitivity, some important factors affecting the extraction efficiency were studied, including the sampling volume, sampling rate, acetonitrile content in the sample, and desorption time [39]. When the peak area tends to be constant with the increase in the sample volume, the extraction equilibrium and the largest extraction efficiency will be obtained. However, an excessive sampling volume consumes more extraction and desorption time. As shown in Figure 7a, the peak areas of the analytes, especially for hexestrol and diethylstilbestrol, increased with the change of the sampling volume from 30 to 80 mL. The increase in the peak areas was unremarkable for other compounds when the sampling volume was more than 60 mL. Compromising the extraction efficiency and time, 60 mL was selected as the optimal sampling volume.

After the same sampling volume of 60 mL was loaded, the sampling rate was investigated in the range of 0.75–2.50 mL/min. It can be seen in Figure 7b that the peak areas of the six analytes changed little when increasing the sampling rate from 0.75 to 2.50 mL/min. While the sampling rate increases beyond 1.00 mL/min, the downward trend occurs for hexestrol and diethylstilbestrol. In order to obtain better extraction efficiency and quick analysis for the six analytes, 2.00 mL/min was selected as the optimal sampling rate for overall consideration.

Organic solvents can promote the solubility of hydrophobic analytes in an aqueous sample and improve analysis repeatability. Herein, acetonitrile was selected as the organic solvent to investigate the effect of organic compounds in the sample on the extraction process. The content of acetonitrile was varied from 0 to 4% (v/v), and the peak areas of the analytes were investigated. As shown in Figure 7c, except for hexestrol and diethylstilbestrol, the extraction efficiency of the other four analytes decreased with the addition of acetonitrile in the sample. The peak area of diethylstilbestrol has a little increasing trend from 1 to 2% and then a decreasing trend from 2 to 4%. In order to obtain satisfactory extraction efficiency, no acetonitrile was added to the samples.

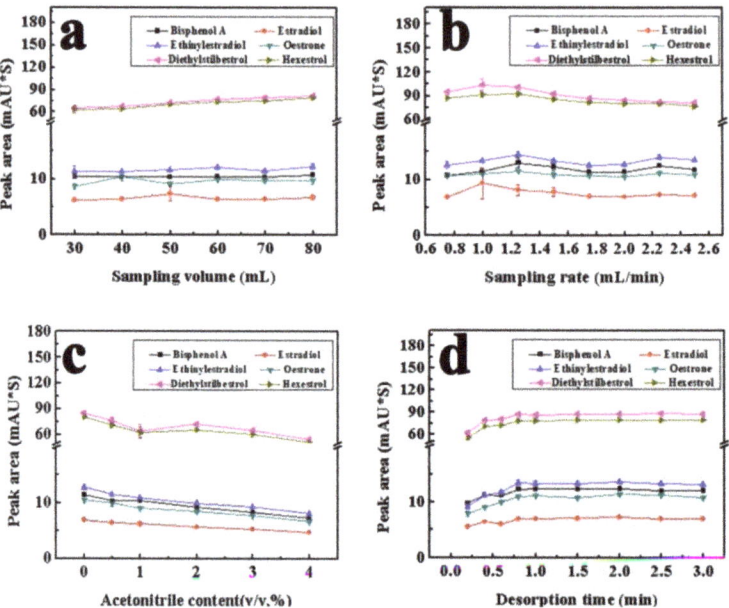

Figure 7. The optimization of extraction conditions using the superhydrophilic fiber as the sorbent for in-tube SPME coupled with HPLC. The concentration of each estrogen is 10 μg/L. Other experimental conditions are the same as Figure 6. Symbols of (**a–d**) stand for the relationship between the change of peak area with the sampling volume, sampling rate, ACN content and desorption time.

A full desorption of all extracted analytes can be obtained after adequate desorption time, which thus reduces the impact of residual analytes on the next extraction. After the extraction was completed, acetonitrile–water (53:47) as the mobile phase flowed through the extraction tube at 1.00 mL/min. The analytes absorbed by the sorbent in the extraction tube were eluted and the desorption process was carried out. The effect of desorption time was varied in the range of 0.20–3.00 min. As shown in Figure 7d, the six estrogen peak areas exhibited an upward trend with the increase of the desorption time. Therefore, 2.00 min was selected as the optimal desorption time.

Based on the above discussion, the optimal conditions of in tube SPME for estrogens are: 60 mL sampling volume, 2 mL sampling rate, 2 min desorption time at a flow rate of 1 mL/min. Under the optimized extraction conditions, the expected extraction performance of superhydrophilic wire (b) for the target analytes was achieved. As shown in Figure 8, the estrogen analytes were not detected for the direct injection of a sample spiked at 10 μg/L. After the concentration of each estrogen compound increased to 250 μg/L, some small peaks of the six estrogens appeared. Compared with the signals after the extraction by the proposed in-tube SPME method, chromatographic peaks become very obvious, especially for diethylstilbestrol and hexestrol. A higher extraction efficiency indicated that there were abundant multiple interactions between the superhydrophilic wire (b) and the target analytes.

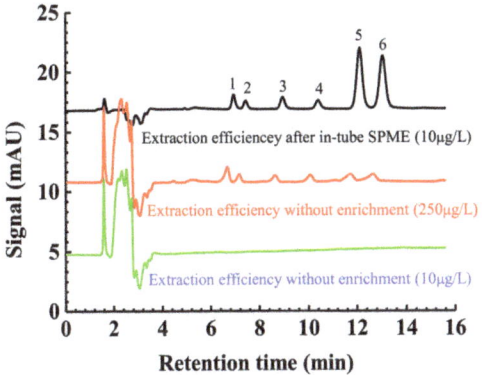

Figure 8. Comparative HPLC chromatograms before and after enrichment using the superhydrophilic wire or in-tube SPME. Peak order and other experimental conditions are the same as Figure 6.

2.5. Method Evaluation and Application to Real Samples

Under the optimal conditions, a series of parameters were investigated to validate the online in-tube-SPME-HPLC method, which included linear range, correlation coefficient (r), inter-day repeatability, intra-day repeatability, limits of detection (LODs), and limits of quantification (LOQs). As shown in Table 1, the analysis method was established with the linear range of 0.5–10.0 µg/L, the correlation coefficient (R^2) of 0.9794–0.9910, and the LOD of 0.08–0.14 µg/L for the six estrogens. The relative standard deviations (n = 3) for the intra-day (≤2.1%) and inter-day (≤4.8%) tests demonstrated that satisfactory repeatability was obtained for the superhydrophilic wire. Each estrogen component on the extraction tube was enriched effectively with their EF values in the range of 20–177.

Table 1. Analytical performances and EF values of six estrogens using in-tube SPME-HPLC method.

Analytes	LODs (µg/L)	LOQs (µg/L)	Linear Ranges (µg/L)	R^2	EF	Extraction Repeatability (n = 3, RSD%) [a]		Preparation Repeatability (n = 3, RSD%) [b]
						Intraday	Interday	
Bisphenol A	0.097	0.323	0.5–10.0	0.9857	20	2.0	4.6	10.4
Estradiol	0.135	0.450	0.5–10.0	0.9886	33	2.1	4.8	8.9
Ethynyl estradiol	0.131	0.437	0.5–10.0	0.9886	28	1.3	3.2	5.2
Estrone	0.079	0.263	0.5–10.0	0.9867	177	0.8	4.5	14.3
Diethylstilbestrol	0.132	0.440	0.5–10.0	0.9794	43	1.7	4.0	2.3
Hexestrol	0.092	0.307	0.5–10.0	0.9910	154	0.9	2.9	9.4

[a] Extraction repeatability was investigated by extracting 10 µg/L of estrogens three times. [b] Preparation repeatability was investigated by extracting 10 µg/L of estrogens with three extraction tubes.

To confirm the applicability of the analytical method, tap water taken from our laboratory were selected for the analysis. As shown in Table 2, none of analytes were detected in the tap water. The spike recoveries of the analytes were determined by the standard addition method. The concentration of addition in the sample was 0, 5, and 10 µg/L. The recoveries of the six analytes were in the range of 73.2–91.1%, indicating that this method could be applied to analyze trace estrogens in real samples.

Table 2. Analysis results and recoveries (%) of six estrogens in tap water.

Analytes	Bisphenol A	Estradiol	Ethynyl Estradiol	Estrone	Diethylstilbestrol	Hexestrol
Recovery (5 µg/L, %)	76.2 5.0	89.4 0.2	86.6 1.6	78.4	91.1 1.5	82.7 2.4
Recovery (10 µg/L, %)	87.7 5.3	76.4 0.6	74.6 0.8	75.3	73.2 4.4	87.4 2.4
Tap water (µg/L)	Not detected	Not detected	Not detected	Not detected	Not detected	Not detected

2.6. Extraction Efficiency for Non-Polar Target Compounds

The correlation of the surface wettability for the sorbent and the hydrophobicity of the target compounds in the sample matrix was further investigated. Superhydrophobic materials were expected to be the ideal sorbent for the enrichment of organic species with the least amount of polarity [40–42]. Herein, the lowest peak areas for five PAHs were obtained using the pristine hydrophilic wire (a) in Figure 9. In comparison, both superhydrophilic wires (b,d) exhibited similar enrichment abilities. It should be noted that the superhydrophobic wire (c) had the highest extraction efficiency to the non-polar compounds of five PAHs. It demonstrated that the surface wettability of the sorbent resulted in different effects for the preconcentration of target compounds. Most artificial superhydrophobic surfaces usually present a rough structure in the micro- or nanoscale and have poor mechanical and chemical stability. Herein, the fabricated superhydrophobic Ti wires can endure a long-time rinse from the mobile phase without a decrease in hydrophobicity and enrichment ability. In order to confirm the truth further, a superhydrophobic Ti mesh was immersed in the mobile phase solution for one week and sonicated discontinuously. The highly hydrophobic property still remained for the dried mesh. It indicated our fabricated superhydrophobic wire with strong mechanic and chemical stability could be used as a robust sorbent for in-tube SPME.

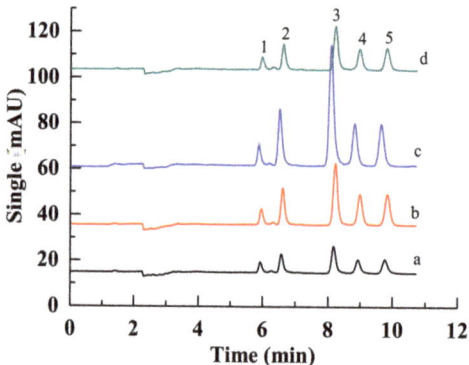

Figure 9. Chromatograms of the target compounds using four kinds of fibers (a, b, c, d) with different surface wettability. The concentration of each compound of PAHs is 10 µg/L. Peak identification: 1. naphthalene, 2. acenaphthylene, 3. acenaphthene, 4. phenanthrene, 5. Anthracene. Water/acetonitrile (25:75, v/v), Detection wavelength, 226 nm. Other conditions are the same as Figure 6.

In general, the enrichment capability is positively related to the absorption or adsorption capability of the coating material [17]. Herein, we first systematically elucidated the relationship between the surface wettability of the sorbent and the polarity of the target analytes by selecting a smart material with switchable surface wettability. It is similar to the theory of "like dissolves like", primarily determining the selectivity of a self-made coating by the hydrophobic/hydrophilic interaction.

3. Materials and Methods

3.1. Materials and Reagents

Ti wire (purity 99.9%) with a diameter of 0.3 mm, Ti mesh (3 cm wide × 6 cm long × 0.8 mm thick) and stainless tube (0.8 mm I.D, 1.6 mm O.D) were purchased from Shenzhen Global Copper and Aluminum Materials Co., Ltd., Shenzhen, China. 1H,1H,2H,2H-Perfluorooctyltrichlorosilane (97%) was purchased from Shanghai Macklin Biochemical Technology Co., Ltd., Shanghai, China. Methanol and Acetonitrile were HPLC grade and were purchased from Tianjin Damao Reagent Company (Tianjin, China). Polar estrogens, including bisphenol A (>99.0%), ethinylestradiol (>98.0%), estradiol (>98%), diethylstilbestrol (>99.0%), hexestrol (>98.0%), and estrone (>98.0%), and non-polar PAHs, including naphthalene, acenaphthylene, acenaphthene, phenanthrene, and anthracene, were purchased from Shanghai Aladdin Biochemical Co., Ltd. (Shanghai, China). Ultrapure water (18.25 MΩ cm, 25 °C) was used for the whole experiment. Water/acetonitrile (53:47, v/v) was used as the mobile phase to elute estrogens at the detection wavelength of 202 nm.

3.2. Apparatus

An EasySep-1020LC pump (Elite instrumental, Dalian, China) was used to transport sample solutions. Analytes were detected by the Agilent 1220 HPLC system equipped with a Hypersil ODS-C18 column (250 × 4.6 mm i.d., 5 μm, Elite instrumental) and a variable wavelength detector (VWD). The prepared wires were characterized by an SEM (quanta 200 environmental scanning electron microscope) and X-ray photoelectron spectra (XPS, Thermo ESCALAB 250XI, Thermo Fisher Scientific, Bohemia, NY, USA). The contact angles were measured by the pendent drop method using a CA goniometer (TST-200, Shen Zhen Testing Equipment Co., Ltd., Shen Zhen, China). The materials were UV-irradiated by an instrument of UV crosslinker (SCIENT03-II, Xin-zhi Biotechnology Co., Ltd., Lingbo, China). Four UV tubes (254 nm) were installed on the top of the UV oven, and the distance between the irradiated samples on the bottom of the UV oven and the upper tube was about 11 cm.

3.3. Standard Solution and Real Samples

The stock solution (100 mg/L) containing five estrogens was prepared with methanol and stored at 4 °C. The working solution was prepared daily by dilution of the stock solution with pure water to 10 μg/L. Tap water taken from our laboratory was selected as the real samples for the evaluation. All water samples were filtered with a 0.45 μm water membrane of cellulose ester before chromatographic analysis.

3.4. Preparation of Ti Wires with Different Surface Wettabilities

The pristine Ti wires and Ti meshes were ultrasonically cleaned by acetone and water consequentially for 5 min and then dried. The cleaned Ti wires (30 cm long, 0.3 mm I.D) and Ti meshes were immersed in 18% HCl solution at 85 °C for 15 min. After that, these samples were placed in a PTFE-lined stainless steel autoclave containing 20 mM hydrofluoric acid. The autoclave was sealed, and the Ti samples were etched under hydrothermal conditions at 160 °C for 8 h [43,44].

In this hydrothermal process, HF not only etches the Ti substrate, providing a Ti source for the formation of TiO_2 nanostructures but also serves as the source of F-dopant. After hydrothermal synthesis, the samples were rinsed thoroughly with deionized water and dried with nitrogen. Then, the superhydrophilic Ti samples were obtained.

The resulting superhydrophilic samples were immersed in 1% ethanol solution of heptadecafluoro-1,1,2,2-tetrahydrodecyl trichlorosilane (FOTS) for 1 h. Superhydrophobic surfaces with flower-like TiO_2 nanostructures were then obtained for the Ti wires and meshes. The superhydrophobic surfaces were changed to superhydrophilic after UV irradiation for about 15.5 h using a UV crosslinker.

Afterwards, as shown in Figure 10(left), the pristine wire (a), the superhydrophilic wire (b), the superhydrophobic wire (c), and the UV-irradiated superhydrophilic wire (d) with a length of 30 cm were filled into stainless-steel tubes (0.75 mm i.d. and 1/16 inch o.d.).

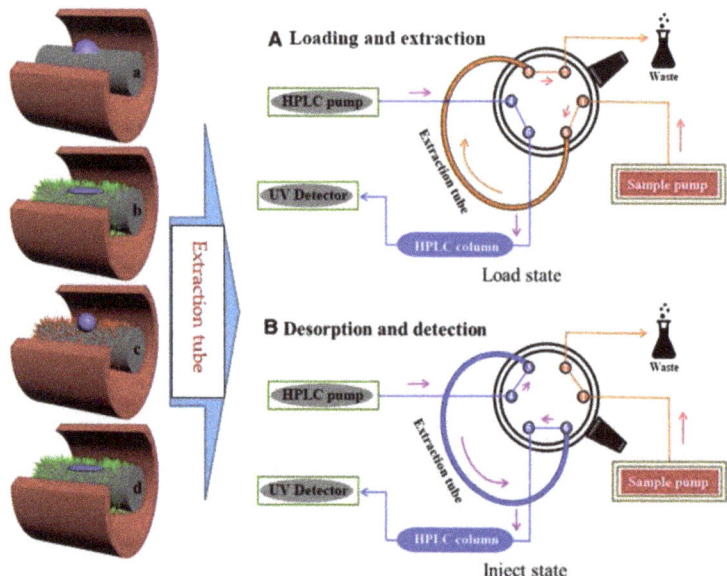

Figure 10. (**Left**) The fabricated wire with different surface wettability was inserted into the microextraction tube. Symbol identification: (a) the pristine wire, (b) the superhydrophilic wrie, (c) the superhydrophobic wire and, (d) the UV-irradiated superhydrophilic wire. (**Right**) Schematic illustration of in-tube SPME–HPLC online system.

3.5. Extraction and Analysis Procedure

As shown in Figure 10(left), the extraction tube was connected to the six-port valve instead of the sample loop. The six-port high pressure valve was switched between the load and inject states to perform the extraction and desorption processes. In the load state, the sample solution was transported through the extraction tube via an EasySep-1020LC pump, and the target analytes were adsorbed by the selected wire. The sample was pumped and flushed from pore 1 to pore 6, then pressured into the extraction tube and entered pore 3. After the preconcentrated target compounds in sample were continually entrapped by the extraction materials, the remaining solution was flushed to the waste bottle from pore 2. After the rotor was rotated about 60°, the system changed to the injection state. In this case, the mobile phase flowed through the extraction tube from pore 4 to pore 3, the absorbed target compounds were desorbed by the mobile phase, and the mobile phase with the desorbed compounds was pumped to the separation column from pore 6 to pore 5 for UV detection, consequently. The extraction and the desorption could be completed by the switch between the load and inject modes in Figure 10(right).

4. Conclusions

In summary, flower-like nano TiO_2 surfaces with switchable wettability were successfully used to investigate the hydrophobic interaction for in-tube SPME. The prepared Ti wires with different surface wettabilities were first compared as a novel extraction support for an online in-tube SPME-HPLC. The results showed that the superhydrophilic wire (b) and the UV-irradiated superhydrophilic wire (b) exhibited the highest extraction efficiency for polar compounds of estrogens compared with the pristine hydrophilic wire (a) and superhydrophobic wire (c). In contrast, the superhydrophobic wire (b) had the higher extraction efficiency for non-polar PAHs compared with the other three wires. We firstly demonstrated that the surface wettability of sorbent might affect the enrichment of target compounds with different polarity effectively. How to understand the theory of like dissolves like from the surface wettability of the sorbent should be further investigated in our future work.

Author Contributions: Conceptualization, Y.Z. and N.W.; methodology, N.C.; software, C.C.; validation, Z.L., X.C. and N.C.; formal analysis, N.W.; investigation, N.W.; resources, Y.Z.; data curation, N.W.; writing—original draft preparation, Y.Z.; writing—review and editing, Y.Z. and C.C.; visualization, N.W.; supervision, Y.Z.; project administration, N.C.; funding acquisition, Y.Z. All authors have read and agreed to the published version of the manuscript.

Funding: This project was supported by the National Nature Science Foundation of China (No. 22074029), the Scientific Innovation Team in Henan Province (No. C20150020) and the Major Project of Science and Technology of Xinxiang City (No. 21ZD005).

Institutional Review Board Statement: Not applicable.

Informed Consent Statement: Not applicable.

Data Availability Statement: Not applicable.

Conflicts of Interest: The authors report no declaration of interest.

Sample Availability: Samples of the compounds are available from the authors.

References

1. Kataoka, H.; Ishizaki, A.; Saito, K. Recent progress in solid-phase microextraction and its pharmaceutical and biomedical applications. *Anal. Methods* **2016**, *8*, 5773–5788. [CrossRef]
2. Kataoka, H. SPME techniques for biomedical analysis. *Bioanalysis* **2015**, *7*, 2135–2144. [CrossRef] [PubMed]
3. Kataoka, H.; Ise, M.; Narimatsu, S. Automated on-line in tube solid-phase microextraction coupled with high performance liquid chromatography for the analysis of bisphenol A, alkylphenols, and phthalate esters in foods contacted with plastics. *J Sep. Sci.* **2002**, *25*, 77–85. [CrossRef]
4. Wen, Y.; Zhou, B.-S.; Xu, Y.; Jin, S.-W.; Feng, Y.-Q. Analysis of estrogens in environmental waters using polymer monolith in-polyether ether ketone tube solid-phase microextraction combined with high-performance liquid chromatography. *J. Chromatogr. A* **2006**, *1133*, 21–28. [CrossRef] [PubMed]
5. Müller, V.; Cestari, M.; Palacio, S.M.; De Campos, S.D.; Muniz, E.C.; De Campos, E. Silk Fibro in nanofibers electrospun on glass fiber as a potential device for solid phase microextraction. *J. Appl. Polym. Sci.* **2015**, *132*, 41717. [CrossRef]
6. Lord, H.L.; Pawliszyn, J. Method optimization for the analysis of amphetamines in urine by solid-phase microextraction. *Anal. Chem.* **1997**, *69*, 3899–3906. [CrossRef]
7. Fernandez-Amado, M.; Prieto-Blanco, M.C.; Lopez-Mahía, P.; Muniategui-Lorenzo, S.; Prada-Rodríguez, D. Strengths and weaknesses of in-tube solid-phase microextraction: A scoping review. *Anal. Chim. Acta* **2016**, *906*, 41–57. [CrossRef] [PubMed]
8. Kataoka, H. Recent developments and applications of microextraction techniques in drug analysis. *Anal. Bioanal. Chem.* **2010**, *396*, 339–364. [CrossRef] [PubMed]
9. Yamamoto, Y.; Ishizaki, A.; Kataoka, H.; Chromatogr, J. Biomonitoring method for the determination of polycyclic aromatic hydrocarbons in hair by online in-tube solid-phase microextraction coupled with high performance liquid chromatography and fluorescence detection. *J. Chromatogr. B* **2015**, *1000*, 187–191. [CrossRef] [PubMed]
10. Augusto, F.; Carasek, E.; Silva, R.G.C.; Rivellino, S.R.; Batista, A.D.; Martendal, E. New sorbents for extraction and microextraction techniques. *J. Chromatogr. A* **2010**, *1217*, 2533–2542. [CrossRef] [PubMed]
11. Arthur, C.L.; Pawliszyn, J. Solid phase microextraction with thermal desorption using fused silica optical fibers. *Anal. Chem.* **1990**, *62*, 2145–2148. [CrossRef]
12. Guo, M.; Song, W.-L.; Wang, T.-E.; Li, Y.; Wang, X.-M.; Du, X.-Z. Phenyl-functionalization of titanium dioxide-nanosheets coating fabricated on a titanium wire for selective solid-phase microextraction of polycyclic aromatic hydrocarbons from environment water samples. *Talanta* **2015**, *144*, 998–1006. [CrossRef] [PubMed]
13. Djozan, D.; Assadi, Y.; Haddadi, S.H. Anodized aluminum wire as a solid-phase microextraction fiber. *Anal. Chem.* **2001**, *73*, 4054–4058. [CrossRef] [PubMed]
14. Sungkaew, S.; Thammakhet, C.; Thavarungkul, P.; Kanatharana, P. A new polyethylene glycol fiber prepared by coating porous zinc electrodeposited onto silver for solid-phase microextraction of styrene. *Anal. Chim. Acta* **2010**, *664*, 49–55. [CrossRef] [PubMed]
15. Djozan, D.; Abdollahi, L. Anodized Zinc Wire as a solid-phase microextraction fiber. *Chromatographia* **2003**, *57*, 799–804. [CrossRef]
16. Bagheri, H.; Mir, A.; Babanezhad, E. An electropolymerized aniline-based fiber coating for solid phase microextraction of phenols from water. *Anal. Chim. Acta* **2005**, *532*, 89–95. [CrossRef]
17. Hashemi, P.; Shamizadeh, M.; Badiei, A.; Poor, P.Z.; Ghiasvand, A.R.; Yarahmadi, A. Amino ethyl-functionalized nanoporous silica as a novel fiber coating for solid-phase microextraction. *Anal. Chim. Acta* **2009**, *646*, 1–5. [CrossRef]

18. Zheng, J.; Liang, Y.; Liu, S.; Ding, Y.; Shen, Y.; Luan, T.; Zhu, F.; Jiang, R.; Wu, D.; Ouyang, G. Ordered mesoporous polymers in situ coated on a stainless steel wire for a highly sensitive solid phase microextraction fiber. *Nanoscale* **2015**, *7*, 11720–11726. [CrossRef]
19. Alotaibi, A.M.; Williamson, B.A.D.; Sathasivam, D.; Kafizas, A. Enhanced photocatalytic and antibacterial ability of Cu-doped anatase TiO_2 thin films: Theory and experiment. *ACS Appl. Mater.* **2020**, *12*, 15348–15361. [CrossRef]
20. Wang, S.-T.; Huang, W.; Lu, W.; Yuan, B.-F.; Feng, Y.-Q. TiO_2-based solid phase extraction strategy for highly effective elimination of normal ribonucleosides before detection of 2′-deoxynucleosides/low-abundance 2′-O-modified ribonucleosides. *Anal. Chem.* **2013**, *85*, 10512–10518. [CrossRef]
21. Zhao, S.; Wang, S.-Y.; Yan, Y.; Wang, L.; Guo, G.-S.; Wang, X.-Y. GO-META-TiO_2 composite monolithic columns for in-tube solid-phase microextraction of phosphopeptides. *Talanta* **2019**, *192*, 360–367. [CrossRef] [PubMed]
22. Fan, K.; Zhang, W.; Peng, T.-Y.; Chen, J.-N.; Yang, F. Application of TiO_2 fusiform nanorods for dye-sensitized solar cells with significantly improved efficiency. *J. Phys. Chem. C* **2011**, *115*, 17213–17219. [CrossRef]
23. Hussain, M.; Ceccarelli, R.; Marchisio, D.L.; Fino, D.; Russo, N.; Geobaldo, F. Synthesis, characterization, and photocatalytic application of novel TiO_2 nanoparticles. *Chem. Eng. J.* **2010**, *157*, 45–51. [CrossRef]
24. Mu, Q.-H.; Li, Y.-G.; Zhang, Q.-H.; Wang, H.-Z. Template-free formation of vertically oriented TiO_2 nanorods with uniform distribution for organics-sensing application. *J. Hazard. Mater.* **2011**, *188*, 363–368. [CrossRef] [PubMed]
25. Yang, X.-Y.; Peng, H.-L.; Zou, Z.-M.; Zhang, P.; Zhai, X.-F.; Zhang, Y.-M.; Liu, C.-W.; Liu, D.; Gui, J.-Z. Diethylenediamine-assisted template-free synthesis of a hierarchical TiO_2 sphere-in-sphere with enhanced photocatalytic performance. *Dalton Trans.* **2018**, *47*, 16502–16508. [CrossRef] [PubMed]
26. Cao, D.-D.; Lu, J.-X.; Liu, J.-F.; Jiang, G.-B. In situ fabrication of nanostructured titania coating on the surface of titanium wire: A new approach for preparation of solid-phase microextraction fiber. *Anal. Chim. Acta* **2008**, *611*, 56–61. [CrossRef]
27. Liu, H.-M.; Wang, D.-A.; Ji, L.; Li, J.-B.; Liu, S.-J.; Liu, X.; Jiang, S.-X. A novel TiO_2 nanotube array/Ti wire incorporated solid-phase microextraction fiber with high strength, efficiency and selectivity. *J. Chromatogr. A* **2010**, *1217*, 1898–1903. [CrossRef]
28. Li, Y.; Ma, M.-G.; Zhang, M.; Yang, Y.-X.; Wang, X.-M.; Du, X.-Z. In Situ anodic growth of rod-like TiO_2 coating on a Ti wire as a selective solid-phase microextraction fiber. *RSC Adv.* **2014**, *4*, 53820–53827. [CrossRef]
29. Wang, F.-X.; Zheng, J.; Qiu, J.-L.; Liu, S.-Q.; Chen, G.-S.; Tong, Y.-X.; Zhu, F.; Ouyang, G.-F. In situ hydrothermally grown TiO_2@C core–shell nanowire coating for highly sensitive solid phase microextraction of polycyclic aromatic hydrocarbons. *ACS Appl. Mater.* **2017**, *9*, 1840–1846. [CrossRef] [PubMed]
30. Xu, H.-L.; Li, Y.; Jiang, D.-Q.; Yan, X.-P. Hydrofluoric acid etched stainless steel wire for solid-phase microextraction. *Anal. Chem.* **2009**, *81*, 4971–4977. [CrossRef] [PubMed]
31. Ma, J.C.; Dougherty, D.A. The cation–π interaction. *Chem. Rev.* **1997**, *97*, 1303–1324. [CrossRef] [PubMed]
32. Wu, G.-S.; Wang, J.-P.; Thomas, D.F.; Chen, A.-C. Synthesis of F-doped flower-like TiO_2 nanostructures with high photoelectrochemical activity. *Langmuir* **2008**, *24*, 3503–3509. [CrossRef] [PubMed]
33. Young, T. An essay on the cohesion of fluids philosophical transactions. *R. Soc. London* **1805**, *95*, 65–87.
34. Wenzel, R.N. Resistance of solid surfaces to wetting by water. *Ind. Eng. Chem.* **1936**, *28*, 988–994. [CrossRef]
35. Cassie, A.B.D.; Baxter, S. Wettability of porous surface. *Trans. Faraday Soc.* **1944**, *40*, 546–551. [CrossRef]
36. Xu, W.; Song, J.; Sun, J.; Lu, Y.; Yu, Z. Rapid fabrication of large-area, corrosion-resistant superhydrophobic Mg alloy surfaces. *ACS Appl. Mater.* **2011**, *3*, 4404–4414. [CrossRef] [PubMed]
37. Lu, Y.; Song, J.-L.; Liu, X.; Xu, W.-J.; Xing, Y.-J. Preparation of superoleophobic and superhydrophobic titanium surfaces via an environmentally friendly electrochemical etching method. *ACS Sustain. Chem. Eng.* **2013**, *1*, 102–109. [CrossRef]
38. Qin, L.; Jie, Z.; Lei, S.; Pan, Q. A smart "strider" can float on both water and oils. *ACS Appl. Mater.* **2014**, *6*, 21355–21362. [CrossRef] [PubMed]
39. Li, C.-Y.; Sun, M.; Ji, X.-P.; Han, S.; Wang, X.-Q.; Tian, Y.; Feng, J.-J. Carbonized cotton fibers via a facile method for highly sensitive solid-phase microextraction of polycyclic aromatic hydrocarbon. *J. Sep. Sci.* **2019**, *42*, 2155–2162. [CrossRef]
40. Baktash, M.Y.; Bagheri, H. A Superhydrophobic Silica aerogel with high surface area for needle trap microextraction of chlorobenzenes. *Microchim. Acta* **2017**, *184*, 2151–2156. [CrossRef]
41. Li, Q.; Sun, X.; Li, Y.-K.; Xu, L. Hydrophobic melamine foam as the solvent holder for liquid–liquid microextraction. *Talanta* **2019**, *191*, 469–478. [CrossRef] [PubMed]
42. Wei, S.-B.; Kou, X.-X.; Liu, Y.; Zhu, F.; Xu, J.-Q.; Ouyang, G.-F. Facile construction of superhydrophobic hybrids of metal-organic framework grown on nanosheet for high-performance extraction of benzene homologues. *Talanta* **2020**, *211*, 120706. [CrossRef] [PubMed]
43. Movafaghi, S.; Wang, W.; Metzger, A.; Williams, D.D.; Williams, J.D.; Kota, A.K. Tunable superomniphobic surfaces for sorting droplets by surface tension. *Lab Chip* **2016**, *16*, 3204–3209. [CrossRef] [PubMed]
44. Lai, Y.-K.; Huang, J.-Y.; Cui, Z.-Q.; Ge, M.-Z.; Zhang, K.-Q. Recent advances in TiO_2-based nanostructured surfaces with controllable wettability and adhesion. *Small* **2016**, *12*, 2203–2224. [CrossRef] [PubMed]

MDPI
St. Alban-Anlage 66
4052 Basel
Switzerland
Tel. +41 61 683 77 34
Fax +41 61 302 89 18
www.mdpi.com

Molecules Editorial Office
E-mail: molecules@mdpi.com
www.mdpi.com/journal/molecules

www.ingramcontent.com/pod-product-compliance
Lightning Source LLC
LaVergne TN
LVHW070154120526
838202LV00013BA/1139